Wealth from Waste

Wealth from Waste

VOLUME–I

Agricultural, Food and Chemical Processing Waste

S.C. Bhatia
B.E. (Chemical), MBA

ATLANTIC
PUBLISHERS & DISTRIBUTORS (P) LTD

Published by
ATLANTIC
PUBLISHERS & DISTRIBUTORS (P) LTD
B-2, Vishal Enclave, Opp. Rajouri Garden,
New Delhi-110027
Phones : 25413460, 25429987, 25466842

Sales Office
7/22, Ansari Road, Darya Ganj,
New Delhi-110002
Phones : 23273880, 23275880, 23280451
Fax : 91-11-23285873
web : www.atlanticbooks.com
e-mail : info@atlanticbooks.com

Printed in India
at Nice Printing Press, Delhi

Preface

Man has lost the capacity to foresee and forestall. He will end by destroying the earth. Man can hardly even recognise the devils of his own creation. Planet earth is the only known unit in the highly complex universe that has the capability to support and sustain human life. But man has been continuously assaulting his own environment, little realising that he is digging his own grave.

Waste has always been associated with human activity and is a necessary evil in any developmental process. Today, however, the sheer quantity and diversity of wastes generated by industries and municipalities pose serious risks to both human health and the environment. The situation is particularly bad in developing countries such as India. Inefficient technologies, ineffectual policies and insensitivity on the part of the industrial sector, have led to large-scale degradation of the ecosystem by domestic and industrial effluents. It is imperative, therefore, that relevant technologies are evolved to treat and recycle wastes.

This volume dealing with Agricultural waste, Food processing waste, Waste from chemical processing and Allied industries is the first volume of *Wealth from Waste*.

This reference book is an effort to put together the various options available to meet the twin goals of environmental conservation and sustainable development. It comprises a collection of carefully selected articles, from both specialists and entrepreneurs, on what is possible today in the field of waste management.

Proper environmental management and waste minimisation are big problems. These have to be tackled on a global level. People who are ignorant and indifferent to their environment must be educated.

As urbanisation grows, so also the extent and dimension of the waste problem. The quantum and type of waste generated in any urban centre is a function of the size and character of the urban centre. For example, industrial towns generate large quantities of industrial waste, which could be in solid, liquid or gaseous form. As a broad categorisation, the different forms of waste generated in any urban centre would be: (i) household/commercial refuse; (ii) street wastes; (iii) construction and demolition debris; (iv) hospital waste; and (v) industrial waste.

Even within this categorisation, there are most likely to be intra-city differences in the composition and quantum of waste: slum settlements are likely to have a higher degree of organic matter in household waste and street wastes than higher or even middle income areas. Refuse generation is a function of consumption, production and growth, which therefore affect not only the quantum but nature of refuse generated.

The domestic and commercial waste is primarily composed of organic matter, recyclables, toxic substances, soiled waste and dust/fine earth. The organic matter comprises food leftover, fruit and vegetable peels, spoiled foods etc. Recyclables comprise paper, plastic, rubber, metal and glass. Toxic substances comprise paints, aerosols, used batteries, medicines etc., and solid waste includes soiled cotton, syringes, sanitary napkins, baby nappies etc.

Waste management is a noble approach to convert waste into wealth. The wastes and effluents are spoiling the fertility of land and polluting the environment. There has been a lot of thought and exercise to come out with some innovative ideas to control such activity and also to save nature from an undesirable situation and keep the environment clean. Waste management is a new concept that has emerged to cope with the rapid industrialisation in the country to maintain the ecological balance and also to keep the environment clean and pollution free. Waste management aims at controlling or reusing the waste disposed of at various levels from industries, usually to agricultural land and other open areas. This concept is slowly gaining popularity among entrepreneurs, professionals and others attached to various industrial sectors. But progress in this regard has not been significant, and there is a dire need to create more awareness among entrepreneurs, manufacturers, etc., with regard to information regarding different aspects of controlling waste and reusing the same through appropriate technology.

Both waste minimisation and utilisation enable entrepreneurs to scale down the cost of production to make it competitive. It also helps to produce cleaner products. There are four principles of waste minimisation and cleaner production: (i) reduction at the source; (ii) recycling; (iii) product reformulation; and (iv) modification and packaging.

Industrial wastes can be broken down into a number of categories, including raw materials, process wastes, packing wastes and products that have not been manufactured to the required standard. In general, the amount and quality of the refuse generated in any processing operation will depend upon the mechanics of that particular manufacturing operation. No exact statement can be made about the type, quantity and quality of the wastes produced by any particular industry. The wastes produced by steel unit will definitely differ from those produced in the chemical or food processing industry. But the waste materials generated from paper and plastic industries are similar to the packing of paper and plastic materials to be found in household rubbish. Against this, the metal processing factory will obviously generate metallic wastes, but in addition, will produce large quantities of slag, processing chemicals and other residues. Hazardous wastes produced by chemical industries and other specialised industries will depend upon the particular product of the manufacturing process.

The method and capacity of storage, the correct type of collection vehicle, the optimum size of crew and frequency of collection depend mainly on volume and density. However, knowledge of some other physical and chemical characteristics is required for assessing the performance of the existing disposal methods and to evaluate the suitability of various alternative disposal systems. The characteristics of the solid wastes in a city also vary seasonally, as well as with the socio-economic groups and dwellings in different areas. Therefore, large number of samples are to be collected for necessary analysis which should be truly representative—and finally weighted average values are estimated.

Waste management has emerged to cope with the rapid industrialisation in the country to maintain the ecological balance and also to keep the environment clean and pollution free. Waste minimisation means cleaner production. This can be achieved by efficient use of energy, water, raw materials and housekeeping, reduction at source, recycling, product reformulation and modification and packaging.

This reference book highlights various aspects of waste minimisation, disposal of waste and resource recovery that takes place from waste. Recovery of recyclable items, composing and energy recovery are some of these. Thus the main aim of this book is to convert wealth from waste.

This reference book is intended for practising engineers—chemical, mechanical, civil, agriculture, food technologist, plastic and rubber technologist, mining and metallurgical and persons working in thermal power stations with an interest in waste and effluent treatments. This book should also be read not only by those working in the field of environment but also by the health workers, doctors, social scientists, policy makers and even general readers. This volume is divided into three sections.

Section I deals with agricultural wastes. Chapter 1 is devoted to vermicompost and organic manure. Vermicomposting is the process of decomposing organic wastes into manure by the churning action of earthworms. Chapter 2 deals with wood and bark wastes as feedstuffs and fertilisers. Chapter 3 describes the utilisation of coconut shell powder into industrustrial products. Among different sources of renewable energy, biomass residues hold special promise due to their inherent capability to store solar energy and amenability to subsequent conversion to solid, liquid and gaseous fuels. Keeping this in mind Chapter 4 is devoted to utilisation of biomass. Chapter 5 focuses on production of enzyme pectinase and rice husk from agricultural waste. Chapter 6 is devoted to paper, paper board and furfural from agricultural waste.

Section II deals with food processing waste. Chapter 7 is devoted to fruits, vegetables, plantation and forest products. In the preparation of fruits and vegetables for processing purposes large amounts of wastes are generated in factories. Various methods of processing and utilisation of these wastes are discussed in detail. Chapter 8 deals explicitly with animal and poultry products, various methods of waste reduction, treatment and disposal of effluents and utilisation of by-products. Chapter 9 studies the coffee and tea wastes while Chapter 10 is confined to cereals, pulses and oilseeds. Chapter 11 concentrates on wastes from the fermentation industries and suggests their use as a feedstuffs and fertiliser. Likewise, various methods of recovery of by-products from dairy wastes and treatment techniques for dairy wastes are discussed in detail in Chapter 12.

Section III describes the waste from chemical processing and allied industries. In any country, economic development and environment protection should go hand in hand. The environment impact of an existing process can be reduced by using clean technologies. The strategies in developing clean technologies should take into account the life cycle of materials, likely changes envisaged in customer practices, environmental regulation/legislation etc. Against this background Chapter 13 deals with petroleum refinery and petrochemicals and attempts to lay down various methods of recycling and treatment of liquid and solid wastes. Following this, the effluent from sugar and distillery industry that contains large amount of dissolved and undissolved organic matter is precisely analysed in Chapter 14. Various methods of bagasse utilisation, process mud, recovery of potash and vitamin B_{12} are discussed in detail. Chapter 15 concentrates on leather and tannery and approaches to waste minimisation methods, disposal and recovery/reuse of various chemicals in an easily comprehensible manner.

Chapter 16 focuses on pesticides while Chapter 17 deals with paper industry. Chapter 18 concentrates on paint and dye industry along with waste minimisation and recycling methods. Chapter 19 concerns with fertiliser industry. Various methods of recovery, recycle and reuse of wastes constitute the chapter. Pharmaceutical plants discharge highly alkaline and toxic liquid wastes. Removal and recovery of aluminium salts from pharmaceutical effluents and waste minimisation options are the core of Chapter 20. Chapter 21 and 22 elucidates the re-refining of used lubricating oil and reconditioning and reclaimation of transformer (insulating) oil. Various reclaimation and re-refining techniques have been suggested. Control of gaseous wastes, organic solvent vapour, fumes and objectionable odour is one of the most complex problems of waste disposal. Chapter 23 is an in-depth study of control of gaseous wastes and odours through methods such as reduction at source, absorption, chemical oxidation and incineration. Chapter 24 deals with recycling and purification of solvents for laboratory. Chapter 25 focuses on liquid chemical and chlorinated waste treatment. Various methods of liquid chemical waste destruction by incinerator have been recommended and finally Chapter 26 concentrates on waste minimisation in textile industry.

The text in this book is throughout supplemented with diagrams and tables. The treatment of all topics is in a cogent, lucid style aimed at enabling the reader to grasp the information quickly and easily. The references and index which are given in the end will prove useful study-aids to readers.

While painstaking care has gone into producing this useful and exhaustive reference book, the author would welcome any constructive suggestions from students, teachers and professionals in the industry.

S.C. Bhatia

Contents at a Glance

Contents

SECTION I

Agricultural Waste

1

Vermicompost: An Organic Manure

INTRODUCTION

The utility of earthworms, in farming has been known for centuries. These lowly creatures are vital for health of the soil and the plants growing on it. Scientists have established that this beneficial effect is because earthworm eats up vegetable matter, soil etc., and excrete small pellets of finally ground soil (called casts), very rich in nitrogen, phosphorus and potassium (N, P, K) which are the essential plant nutrients. In this process they also upturn the soil, thus providing the much needed air to micro-organisms and the roots of the plants. Earthworms are also known to help in maintaining water balance and other chemical levels in the soil. Due to these properties, earthworm can be used to produce rich manure.

With an eye to all these benefits, increasing attention is being focused on breeding of earthworms (vermiculture) and their subsequent use in preparation of manure called vermicompost.

The production of vermicompost involves breeding of earthworms in a mixture of cowdung, soil and agriculture residue till the whole mass is converted into cast. This cast is then collected to give the vermicompost.

Thus vermicomposting is the process of decomposing organic wastes into manure by the churning action of earthworms. The material obtained after decomposing is called vermicompost. Vermicomposts have many desirable properties conducive for improving the soil fertility and crop yield. The concept of organic farming, which refers cultivation without using chemical fertilisers and pesticides, has given lot of impetus for use of vermicompost. Already handfuls of companies in India and overseas, have made forays into organic farming.

The continuous use of chemical fertilisers to boost the agricultural yield resulted in change in soil structure, loss of soil fertility, loss of micro-nutrients and, overall damage of eco-systems. Also, the products produced by using chemical fertilisers cause lot of human health hazards.

The recent understanding of the ill effects of using chemical fertilisers, especially in the western countries, has given lot of impetus for organic farming. The essence of organic farming is working in harmony with nature and making use of various locally available resources for crop production. This is where vermiculture play a vital role. By virtue of their behavioural properties, the earthworms contribute a lot to improve soil fertility, promote soil aggregation, encourage favourable soil reaction and enrich the soil nutrients.

It is now well known that heavy use of chemicals ruins the soil and its fertility. The time has come to look into the benefits of organic manure. Several methods have been developed by which agricultural residue or waste can be converted into organic manure to replace chemical fertilisers. Of these, vermicomposting is most suited in terms of a balance between cost and effectivity.

Earthworms churn the soil and make it porous, thereby facilitating the proper ratio of air, water and solids in the soil. Earthworms also improve the water filtration rate, bring minerals and make plant nutrients more available. Earthworms also have the capability to neutralise the soil. It has been proved that earthworms maintain the pH at neutral level regardless of whether the existing soil pH is higher or lower. This is made possible by the action of earthworm's calciferous glands and buffering action of carbonic acid.

The burrowing earthworm is sensitive to light. As it burrows, it improves soil porosity, aerates the soil, and excretes ingested soil and organic substances that have undergone complex biochemical changes in the earthworm's intestines. The excretions are called vermicastings and this rich manure has good fertilising properties. It improves the soil's water retention properties, destroys pathogens, encourages growth of useful micro-organisms, increases immunity against pests, reduces soil salinity and maintains an ideal acidic-alkaline balance. Besides all this, it retains cocoons and ensures future generations of earthworms.

TECHNICAL ASPECTS

About 8 MT of vermicompost will be sufficient for 3 acres of farm land and to obtain 8 MT of vermicompost about 12 MT of bio-mass will be required i.e., 1.5 MT feed mixture produces 1 MT of vermicompost in about 3 months time. Best results are obtained from exotic or foreign variety of earth worms namely *Eisenia foetida* and *Eudrilus eugeniae.*

These rapidly convert bio-mass into casts and their breeding is very fast. Endemic or indigenous species like *Perionyx excavatus* and *Perionyx sansibaricus* are also suitable but their conversion rate and breeding is bit slower.

Vermicompost unit require a moisture level of 40–50 per cent and temperature of 20–30°C and these are to be maintained.

Chemical analysis of a typical vermicompost shows that N, P and K content is higher than that of organic manure but as compared to chemical fertilisers their consumption is more. For instance if a crop require 40 kg of nitrogen, this could be provided with 100 kg of urea or about 1.5 MT of vermicompost. In addition to nitrogen, vermicompost will provide additional essential nutrients like phosphorus and potassium with some micro-nutrients which are not available in urea. Moreover, it leads to regeneration of soil fertility thus promoting long term sustainability. Vermicompost can be used mixed with chemical fertilisers. Studies shows better results and good yields this way.

Mineral elements and range of nutrients are given in Tables 1.1 and 1.2.

Table 1.1. Mineral elements.

Mineral elements (mg/100g dry wt.)

Details	*Casts*	*Soil*
C	8550	3925
N	536	350
C: N ratio	16	11

(Contd...)

Details	*Casts*	*Soil*
P (aqua regia)	102	68
P (lactate—soluble)	13.7	2.2
K (soda extract)	1097	799
K (lactate—soluble)	44.6	7.0
pH (H_2O)	5.8	5.0
pH (KCT)	5.4	4.0

Table 1.2. Range of nutrients.

Description	*Quantity*
Organic carbon (%)	9.15 to 19.98
Total nitrogen (%)	0.5 to 1.5
Available phosphorus (%)	0.1 to 0.3
Available potassium (%)	0.15 to 0.56
Available sodium (%)	0.06 to 0.3
Calcium and magnesium (meg/100 grams)	22.67 to 46.70
Copper (ppm)	2.0 to 9.3
Iron (ppm)	5.7 to 9.3
Zinc (ppm)	5.7 to 11.5
Available sulphur (ppm)	128 to 548

MANUFACTURING PROCESS

Following steps are essential for the production of vermicompost.

Pits

These are constructed either above or below the surface using the brick masonry or stone slab or even plastic. A pit of size 1.6 meter long, 1 meter wide and 0.75 meter deep is sufficient for 1 cu. meter volume. In this size of pit about 6,000–7,000 worms can be used. These pits should be covered with thatched roof to maintain the desired temperature and moisture content. Protection from predators is essential and for that a ditch of standing water is made around the pit to prevent ants from entering into the pit and attack worms. Wiremesh will protect from cats, dogs and birds.

Feed Mixture

Any organic waste such as farmyard waste, green waste, sugarcane thrash, coir or pitch, kitchen waste etc., is used as feed mixture. These are mixed with cow dung in 2 : 1 ratio and put into the pit. This feed mix is allowed to decompose for 2–3 weeks. In this period no earthworm is added as due to heat they will die. Vermiculture ecosystem is shown in Fig. 1.1.

Breeding

1 kg earthworm represent 600–1000 numbers of worms. These can convert 45 kg wet biomass in a weeks time yielding about 25 kg of vermicompost. The partial decomposed biomass in the pit is inoculated

with earthworms of chosen species. Regular watering is done to maintain the optimum moisture level of 40–50 per cent. 1.5 MT of feed mix (2/3 agriculture residue + 1/3 cow dung) produces 1 MT of vermicompost in 3 months.

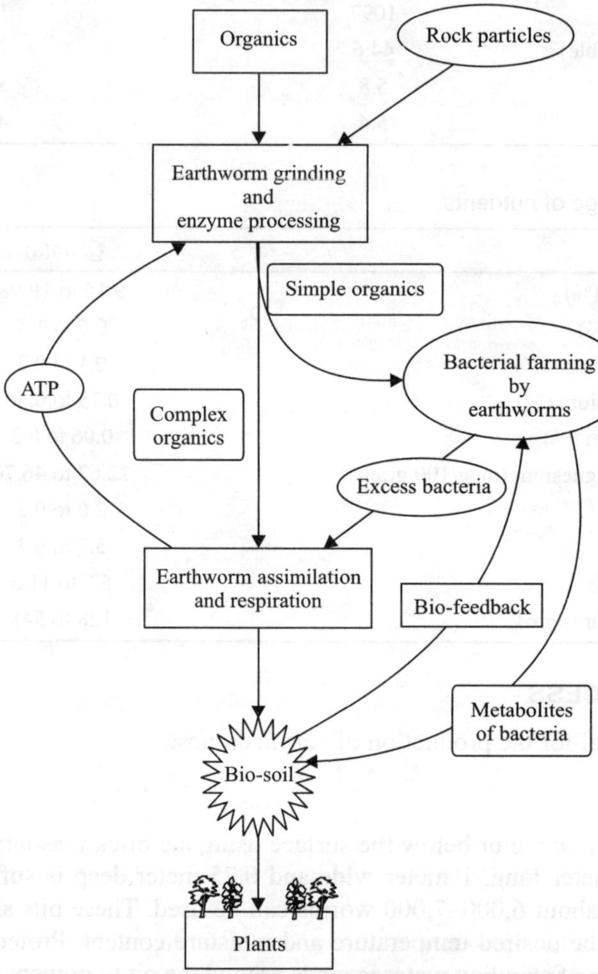

Fig. 1.1. Vermiculture ecosystem.

Collection

Once the feed mixture is seen to largely contain casts, it is dumped into a conical heap and left for few hours. Worms collect at the base and can be easily removed to re-use. The rest dried material is screened through a 3 mm sieve to collect as vermicompost. The quality is further improved if *Neem* cake is used. It enhances the microbial activity. *Karanj* cake can also be added to keep ants away. Addition of bone meal enhances the quality further. Decomposing cultures such as *actinomycetes*, *acetobacter* and *phosphate solubilising bacteria* are added at the beginning to speed up the process.

The succulent waste (agro-waste, weeds, municipal waste, vegetable waste, leaf litter, hedge cutting, etc.) is mixed with cowdung, if available, but the fibrous wastes like rice husk, saw dust, coir pith and hard stubble of sugarcane trash have to be mixed with dung. Apart from these, waste paper, hardboard, old rags, sewage sludge and forestry based industrial wastes can also be used.

The waste is plastered with 1 inch of soil slurry and 2–3 cm diameter holes 30 cm apart is made on the soil slurry for proper aeration for decomposition. This has to be left as such for 2 to 3 weeks initially to provide appropriate environment for the earthworms. Moisture level at the range of 40 to 50 per cent is maintained.

Vemicastings at the rate of 7.5 kg per tonne of waste is applied on the pit. The waste is dumped on the vermicastings and is mulched with organic litter.

Watering is done regularly in order to ensure moisture in the pit.

Final Product (Vermicompost)

The reduction of the waste is slowly continued until the waste is converted to vermicompost. The worms have a muscular gizzard, which grinds their food to particles 2 to 4 microns in size. Normally, it takes 1–2 months to convert all the waste into vermicompost.

The vermicompost thus obtained is highly stable and do not disintegrate. It is packed in 50 kg gunny bags and sold in the market.

VERMICOMPOSTING USING EXOTIC SPECIES OF EARTHWORMS

Exotic species of earthworms have also been used in India for vermicomposting. Internationally three species of earthworms have received acclaim for vermicomposting, they being *Eisenia fetida* and *Eudrilus eugeniae*, which are exotic, and *Perionyx excavatus*, which is endemic. Earthworms, farm bacteria and plants are shown in Fig. 1.2.

Since 1982 *E. eugeniae* has been promoted for waste degradation. *Eudrilus eugeniae* is reported to have surpassed both in feeding and reproductive rates compared to other species of earthworms. *Eisenia fetida,* however, is used in certain areas in India for domestic waste composting. After several experimentations the recommended method is as follows.

Fill the waste material into tanks or as heaps followed by sufficient water to moisten the waste. The tanks can also have holes of 15 mm diameter all-round (walls) at 300 mm apart to provide cross-ventilation. The well-aerated material does not emit any foul smell. After two weeks the earthworms are released on the surface. They enter into the organic matter. The set up is left without disturbance for six weeks. Water is sprinkled occasionally on the surface during the composting process. The compost along with the earthworms is collected. Old jute (gunny) bags can also be used to cover the containers, but the bags themselves will get composted after some time. A crate (600 mm × 45 mm × 600 mm) can hold a population of 1500 adult *Eudrilus engeniae*, and 3000 to 5000 *Eisenia fetida* and *Perionyx excavatus.* Frequent harvesting of *Eudrilus eugeniae* is necessary to reduce population pressure and enables continuous growth of earthworm population.

GENERAL COMPOSTING

A compost heap can also be made layer by layer. In the bottom layer, place twigs, stalks, hay, wood chips or other coarse material, which allows air to flow at the bottom. Follow with a layer of high nitrogen material like manure or grass clippings, leaves, manure, wood shavings, chopped weeds, vegetable and fruit scrapes, nut and egg-shells. Sprinkle the material with water as you build the pile and repeat the sequence. The pile should be at least three feet high to trigger and sustain the required biological reactions. Do not let the pile get over five feet high, in which case the mass may pack down, squeeze out air and slow down decomposition. You can use a black polythene sheet to cover or alternatively use a jute sack as cover over the heap to retain the moisture. This can also be used as a pre-digestion exercise. Once cool (say after about thirty days), the earthworms can be introduced into the heap to proceed with vermicomposting.

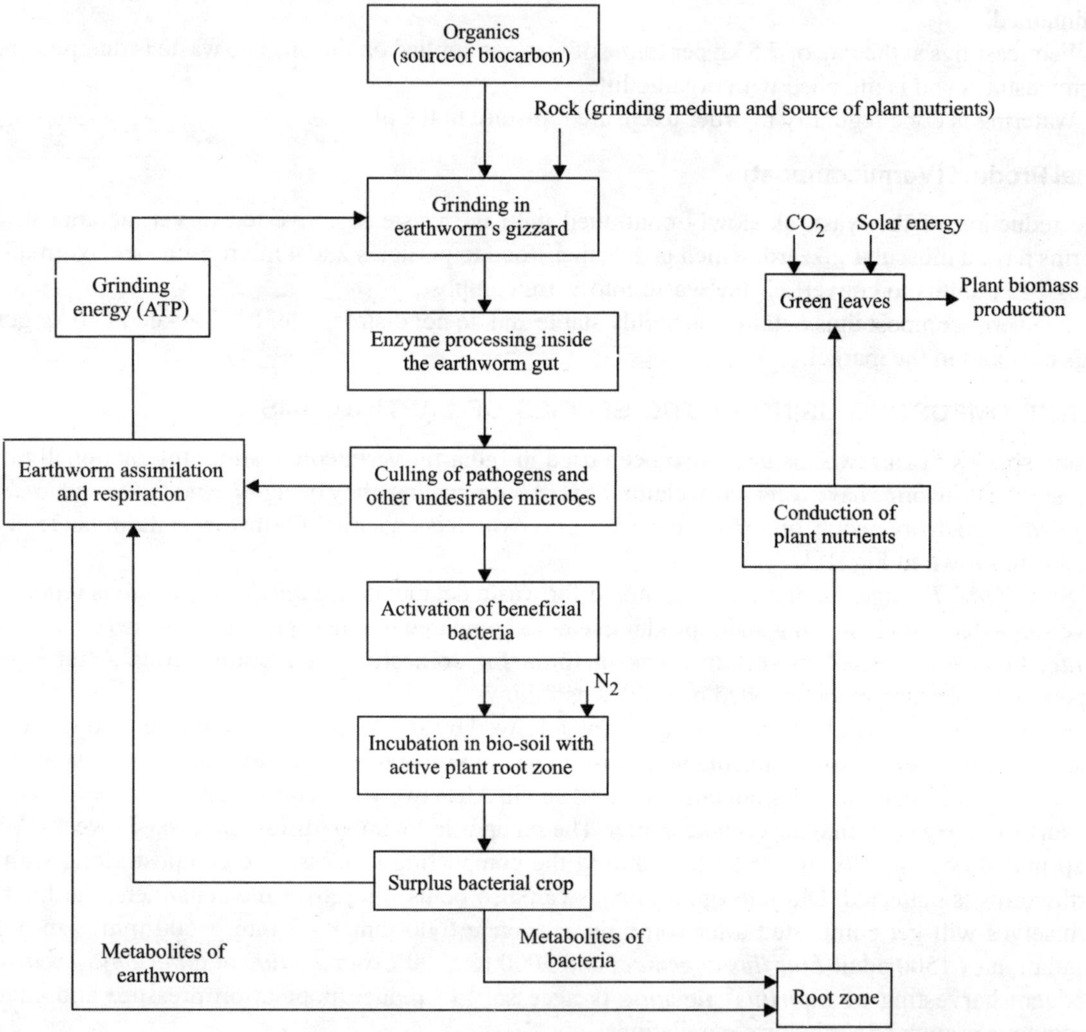

Fig. 1.2. Earthworms, farm bacteria and plants.

WHEN IS THE COMPOST READY?

The compost is ready when the material is moderately loose and crumbly and the colour of the compost is dark brown. It will be black, granular, light weight and humus-rich. To facilitate separating the worms from the compost, stop watering two to three days before emptying the beds. This will force about 80 per cent of the worms to the bottom of the bed. The rest of the worms can then be removed by hand.

The vermicompost is then ready for application. The smell is earth-like. Any bad odour is a sign that fermentation has not reached its final goal and that the bacterial processes are still going on.

You can use the compost in your garden, around shrubs bushes and trees and in potted plants. Compost can be spread on the top of the soil and hoed in lightly to encourage the surface soil life. Roots from tender plants are then able to reach the compost and draw out its nutrients. Vermicomposting is easy to practise, is ecologically safe and economically sound.

BENEFITS OF VERMICULTURE

Benefits of vermiculture are given below:

1. Vermiculture technology is unique because it can utilise organic residues that cannot be utilised by any bio-conversion technology. Moreover, utilisation is complete.
2. Vermiculture offers a high solids retention time, even upto 6–12 months. This is achieved without any increase in hydraulic retention time and bio-reactor size.
3. Vermiculture harnesses a mixed culture of beneficial soil bacteria as the most diverse and productive bio-processing agents. Vermiculture is self regulated.
4. In vermiculture, instantaneous values of F/M of 20 per day is possible.
5. Vermiculture has a productivity of 1 $kg/m^2/day$ for a 0.33 m thick bio-reactor. Volumetric productivity, thus, is 3 $kg/m^3/day$. This is achieved without wastage of substrate and there is no further residue that needs to be tackled.
6. Bacteria are the prime bio-conversion agents in vermiculture. These are selected by earthworms depending upon the changing substrate and demand on end product. Earthworms also carry out continuous upgradation of bacterial culture while they are doing their job.
7. Vermicastings enable farmers to harness all the known and unknown beneficial micro-organisms in the field. Vermicasting is a sustainable bio-fertiliser.

CASE STUDY—NIRULA RESTAURANTS (DELHI)

Entrepreneur and owner of the popular Nirula's chain of restaurants *Deepak Nirula* had a mini wormiculture plant at his sprawling Anand Niketan home and wormiculture plants in Okhla and Noida where he sends the waste from this chain.

The procedure is fairly simple. You pile up the garbage in containers 18 feet high and 2 feet wide and then introduce worms. You add fresh dung and then the worms proliferate by themselves, disintegrating the waste in the process. The compost formed can then be used as manure for your plants'.

Despite his pioneering efforts in wormiculture Nirula has so far never tied up with either MCD or NDMC, the two main civic bodies in the Capital, for mass use of wormiculture. 'Because he has never felt the need to tie up with civic bodies. He believe that any effort that one makes as an individual can work. In the colony where he lives for instance and in the neighbouring areas of Shanti Niketan, central park and west end there is a lot of leaf litter. The municipality never picks up the litter and the malis used to burn the leaves. We now use them in wormiculture,' says *Nirula*.

But how does he reconcile his roles as a corporate and an environmentalist? 'There is no reason why one can't do so. If the corporates don't take responsibility for the city then who will,' says Nirula.

TIPS ON THE PIT

You could make a vermiculture pit in your garden or terrace, or buy ready-made containers for your balcony or kitchen window. If you live in an apartment building, you could get the entire society involved.

Make a vermiculture bin or pit in an empty car park (along the boundary wall) or in any corner of the garden, and train the sweeper or garden to operate the bin.

Your could either build the bin yourself or contact any organisation that specialises in the technology to build it for you. Once the vermiculture pit is functional, ensure that rats, crows and dogs don't get at it. You could use a hinged wire mesh to protect it.

Dig the vermiculture pit at a higher level in your garden (to prevent water-logging). Though the depth and width of the pit can vary according to your needs, it must be at least one metre deep and two metres wide. Line the bottom with a thick layer of soil and spread at least two to five kg of vermicastings over it. A layer of cow dung or 'neem' cake, if available, should be added to this. This should be topped with fresh vegetable waste as soon as the cow dung layer turns black. Finish off with a layer of dried leaves, which keeps any foul odour away.

It takes about six months for your vermiculture compost to be ready for use. Don't overdo the waste. If you detect a foul odour (the pit is basically odourless), reduce the amount of garbage. Even soiled sanitary napkins and pampers (after the plastic linings are removed) can be converted into fertilisers in this bin.

BIO-CONVERSION OF COIR PITH INTO ORGANIC MANURE

Coir pith is a waste generated during decorticating of coconut husk to extract fibre. Since the pith does not decompose naturally, the disposal is very difficult. In India nearly 5 lakh MT of pith is generated every year. Based on the researches conducted at Central Coir Technological Research Institute (CCTRI), Alleppey and other institutions, it has been proved that the pith can be converted into organic manure by using certain micro-organisms. The manure so obtained fares very well in terms of water holding capacity, availability of NPK, etc. It is used as growing medium for ornamental plants, tissue culture, etc. It has good export prospects too, mainly in the western countries.

Extraction of fibre from coir is a major industrial activity in coconut growing areas of South India. After the separation of the fibre nearly 70 per cent of the husk remains as waste-pith. It is a light, fluffy material, which occupies lot of space. Extraction of one kg of coir fibre generates two kg of pith. Since the pith does not decompose naturally, it poses lot of problems for disposal.

In the pith, lignin is complexed with cellulose and has a high C:N ratio of 112:1 which creates a negative effect when applied to the crops directly.

Research work done in Agricultural Universities and Research laboratories has proved that with proper treatment of cocount pith by using micro-organisms, the lignin and cellulose complex can be degraded and decomposed. The decomposed pith contains both major and minor nutrients required for plant growth. This compost is found to be an excellent organic manure and soil conditioner and, can be used for all the crops. The NPK contents of coir pith manure is much higher than other form of organic manure as given in Table 1.3.

Table 1.3. NPK contents of coir pith manure.

Name of manure	*N %*	*P %*	*K %*
Coir pith manure	1.4	2.5	2.0
Sheep dung	0.6	0.5	0.9
Cattle dung	0.4	0.3	0.2
Poultry manure	1.50	1.8	0.8

Product and Its Properties

As already mentioned coir pith manure contains many desirable nutrients essential for plant growth and, to improve soil fertility. The properties of coir pith manure is given in Table 1.4.

Table 1.4. Properties of pith manure.

Description	*Value*
Total pore space (%)	19.19
Water holding capacity (%)	648.31
Moisture (%)	30–40
pH	6.2–6.9
Electrical conductivity (milli-mhos/cm)	0.1–0.25
Organic carbon (%)	20.9
Nitrogen (%)	1.26
Calcium (%)	0.5
Magnesium (%)	0.48
C/N ratio	24:1
Micro-nutrients (ppm)	
Iron	0.09
Manganese	25.00
Zinc	15.80
Copper	6.20

Manufacturing Process

Manufacture of coconut pith manure is very simple and cheap process. It is an outdoor activity wherein the coconut pith is systematically heaped along with some additives. While heaping, some quantity of a specially prepared microbial culture is inoculated. The heap is maintained in a shaded place. Regular watering is done to maintain the moisture. The pith would decompose in a period of 60–90 days and is ready for use.

2

Wood and Bark Wastes as Feedstuffs and Fertilisers

INTRODUCTION

Single cell protein (SCP), or protein biomass produced by micro-organisms, is widely being considered as an important means of alleviating world protein shortage. Feed can represent upto 75 per cent of total beef production costs. Many animal feeding systems already make use of crop residues and waste products as a means of reducing the energy and costs factors. However, it is essential that SCP be shown to be a safe, reliable, and good quality feed for animals and perhaps even as a food for direct consumption by man. The success of SCP will be judged by society on economic considerations and on the acceptability of the final product.

Much of the early interest in SCP was concerned with the use of *n*-alkane components of petroleum, either as gas oil or purified *n*-paraffin fractions (Fig. 2.1). However, the massive increase in crude oil or prices and the problems of contamination have been a major concern. Methanol derived from natural gas is now considered an important feedstock for SCP production, largely on economic grounds and because it is free from contamination with potentially carcinogenic polycyclic aromatics.

Alternative feedstocks are now being sought from more traditional waste organic materials, and this chapter considers, in particular, the use of ligno-cellulosic wastes, namely wood and bark, when considering the suitability of an organic waste for SCP production, the following criteria have been set out by *Maclennan* and are well worth examining before embarking on a waste utilisation programme:

1. It must be cheap.
2. It must be readily degradable micro-biologically.
3. On economic grounds, it is only possible to produce animal feed grade SCP in large-scale plants, typically of 100,000 to 300,000 tonnes/yr/plant.
4. The feedstock should be available on a year round basis and be capable of storage safely and economically.
5. Transport to site of plant must be at low cost.
6. It must be available on a reliable basis and be of constant and predictable quality.

There can be little doubt that cellulose, both from agricultural and forestry sources and from waste material, must constitute a future major source for the production of feed, energy, and chemicals. Cellulose is the most abundant and renewable natural resource available to man. It has been realistically calculated that c 3.3×10^{14} kg/yr of carbon dioxide are fixed on the surfaces of the earth and that approximately 6 per cent of this (i.e., 22 billion tonnes/yr) will be cellulose. Four billion tonnes/yr are readily available for processing, with 2.6 billion tonnes existing as wood and pulp. On a worldwide basis, land plants produce 24 tonnes cellulose/person/yr. The three main naturally occurring sources of cellulose from agriculture and forestry are wood, straw from grains, and sugarcane after removal of sugar.

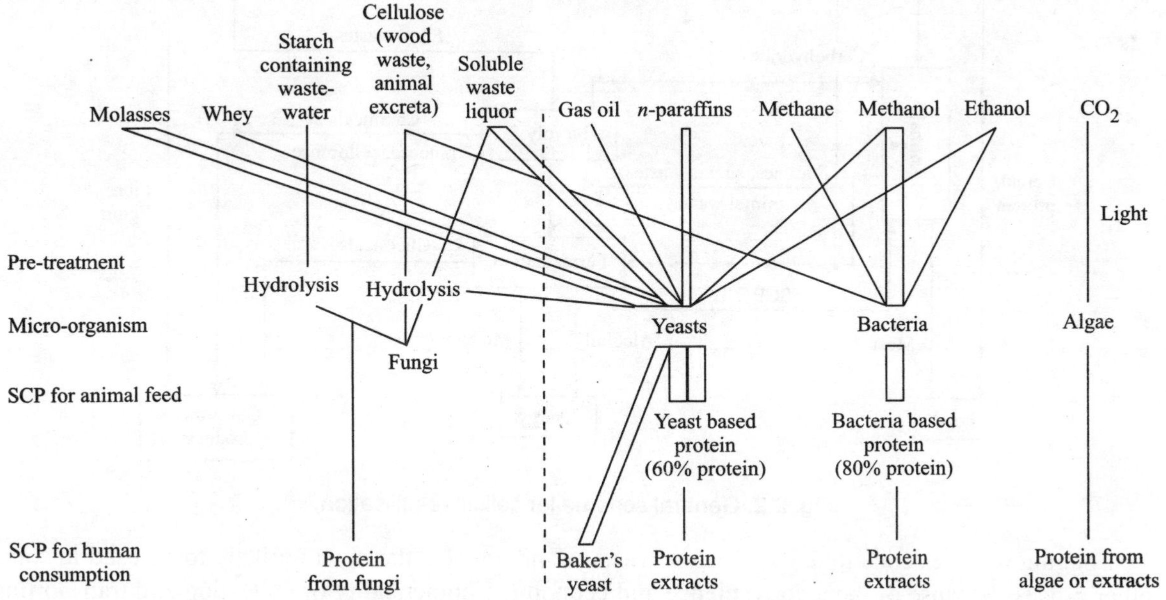

Fig. 2.1. Micro-organisms and substrates for SCP production.

Although cellulose can be readily attacked microbiologically, it invariably occurs in combination with lignin and in a complicated interwoven state. Dry wood has an approximate composition of 40 to 45 per cent cellulose, 20 to 30 per cent hemicellulose, and the rest mostly lignin. Since lignin is only sparingly attacked by micro-organisms, it has caused a serious impediment to the wide acceptance of such materials for SCP production. However, new technologies are being developed which will allow this abundant and renewable source of organic carbon to be used for fermentation purposes (Fig. 2.2). The following sections consider the nature of the availability of lignocellulose substrate derived from timber, the enzymology of cellulose and lignin degradation, and some of the worthwhile fermentation processes that are using lignocellulose substrate for proteinaceous biomass and fertiliser production.

SOURCES OF LIGNOCELLULOSE

Associated with the timber industry worldwide, large tonnages of logging and milling wood residues are annually being generated. Limited markets exists for most of these cellulose-rich by-products. Approximately one-third of the timber tonnage that is harvested will eventually become a residue or potential waste product.

There are three main categories of residue: logging residues (80 per cent—tree tops, branches, etc.), primary manufacturing residues (17 per cent—saw mills, veneer mills, and pulp mills), and secondary manufacturing residues (3 per cent).

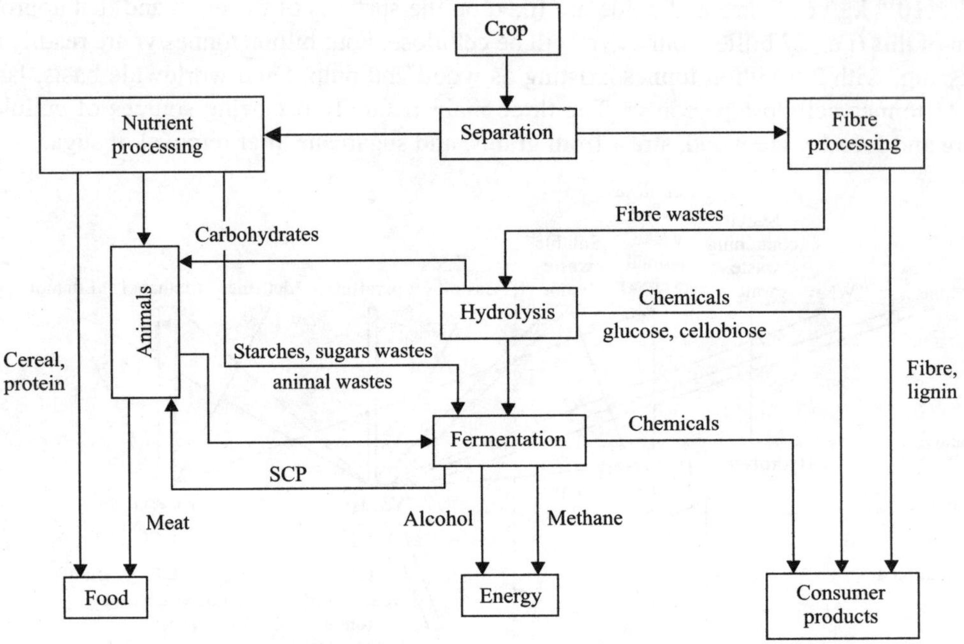

Fig. 2.2. General scheme for cellulose utilisation.

Logging residues constitute the greatest wastage and are also the most unlikely to be used in some other process because of the inconvenience and economic impracticality of collecting and transporting to a central area. Primary manufacturing residues have much more potential for worth-while use because they occur at centralised positions. For examples, in America, in 2001, 150 million tonnes of wood residues and by-products (such as saw dust, sides, edgings, and chips) and 50 million tonnes of bark were available at mills throughout the country. Although much of the milling residues are being profitably used in the pulp and paper industries, and particle board manufacture is expanding each year, it has been calculated that, in America, there are still 30 million tonnes of wood residue and 14 million tonnes of bark that are not being used. Such materials are mainly discarded into water courses or burnt, either way contributing to environmental pollution.

ENZYMOLOGY OF LIGNOCELLULOSE DEGRADATION

A vast literature has been amassed concerning the nature of cellulose degradation by micro-organisms. The studies involving two organisms, *Trichoderma viride* and *Sporotrichum pulverulentum*, have dominated the research efforts.

Studies with *T. viride* have been primarily concerned with relatively pure sources of cellulose and have led to the development of sophisticated fermentation technologies involving enzyme hydrolysis of cellulose (e.g., waste paper) to dextrose sugar. However, current work in America indicates that the problems of recovering crystalline dextrose from enzyme hydrolysis of cellulose materials are at present insurmountable. Thus, it is more probable that the dextrose in solution should be considered as a raw material for the production of ethyl alcohol or as a carbon source for growing filamentous fungi or yeasts. The technological implications of this work have been examined by *Brown* and *Fitzpatrick*.

Eriksson's group in Sweden has realistically considered lignocellulosic materials as potential substrates for fungi; particularly the white rot fungi, which can degrade lignin as well as the other major components of wood. Such fungi (e.g., *Sporotrichum pulverulentum*), will normally attack cellulose and lignin simultaneously when degrading wood. It seems a general factor that when lignin is being microbiologically degraded there must be a parallel degradation of one of the other polysaccharides. Lignin is a three-dimensional polymer compound of one or more phenylpropane monomers. The simplest monomers are hydroxycinnamyl alcohol, coniferyl alcohol, and sinapyl alcohol. Coniferly alcohol is the major component of most conifer lignins and is a significant proportion of the lignins of other species.

Clearly, a knowledge of the mechanisms of enzyme degradation of both cellulose and lignin will be of considerable value for the future development of applied research to use these complex substrates for food purposes. Such an understanding can then be coupled to fermentation technology.

The celluloses of wood are linear polymer of D-anhydrogluco-pyranose units linked by β-1, 4-glucosidic bonds, and are variously folded, bundled, and stabilised by hydrogen bonding. The key enzyme in all cellulolytic systems is *endo*-1,4-β-glucanase, which attacks at random the 1,4-β linkage along which cellulose chain. Different amounts of *endo*-glucanase have been found in different enzyme preparations. *Almin* and *Eriksson* and *Pettersson* distinguished five *endo*-glucanases in *Sporotrichum pulverulentum*, while only two *endo*-glucanases could be demonstrated in culture liquids from *Trichoderma viride*. These enzymes are normally associated with an *exo*-1,4-β-glucanase, which splits off cellobiose or glucose units from the non-reducing end of the cellulose, and one or several 1,4-β-glucosidases which hydrolyse cellobiose to glucose and cellobionic acid to glucose and glucolactone. In *S. pulverulentum*, an additional oxidative enzyme, cellobiose oxidase, is necessary for *in vitro* cellulose degradation. However, no such enzyme has yet been found to be involved in the extracellular enzyme action of *T. viride*. The present understanding of the enzymology of cellulose and lignin breakdown is given in Fig. 2.3.

The enzymology of lignin degradation is less well understood but may involve a laccase and an oxidoreductase, cellobiose: quinone oxidoreductase. The cellobiose: quinone oxidoreductase is important in both the degradation of cellulose and lignin. Studies with cellulase-less mutants have shown that lignin degradation is more favoured by the presence of cellulose and it has been suggested that this is due to the action of the cellobiose: quinone oxidoreductase.

Cellobiose: quinone oxidoreductase reduces quinone or phenoxy radicals in the presence of cellobiose, which is then oxidised to cellobiono-δ-lactone.

The importance of the phenol oxidase, laccase, for lignin degradation is now well accepted. A laccase-less mutant of *S. pulverulentum* could not degrade lignin. The studies on the basis of enzymology of cellulose and lignin degradation will lay the foundation on which fermentation bioconversion processes can be developed.

PRE-TREATMENT OF LIGNOCELLULOSE MATERIALS

Chemical and Physical Treatments

In nature, cellulose invariably is present in the crystalline form and combined with various proportions of lignin. This subtle combination of two of the most common naturally occurring organic materials has been both a great biological blessing and an effective barrier to useful conversion techniques. Lignification restricts enzymological and microbiological access to the cellulose and other polysaccharides, while the crystallinity of the cellulose restricts the rate of all modes of enzymological attack on the cellulose.

Without this unique natural barrier to microbiological attack, the durability and longevity of nature's trees would be greatly reduced.

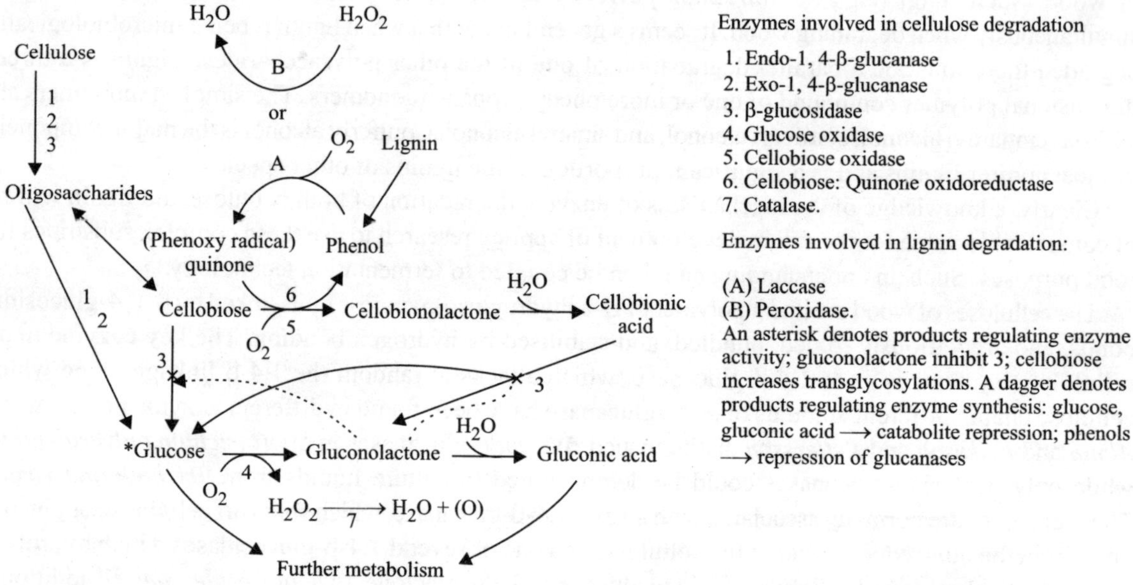

Fig. 2.3. Enzyme mechanisms for cellulose degradation and their extracellular regulation in *S. pulverulentum*.

Therefore, with a view of harnessing the cellulose part of lignocellulosic materials, much effort has been given to pre-treatment techniques which will effectively and economically reduce both the lignification and the crystallinity of these natural polymers. The physical and chemical pre-treatments of lignocellulose materials have been reviewed by *Millett*.

Acid hydrolysis has been increasingly proposed as a major approach to saccharification of cellulose. In this way, cellulose is completely broken down to glucose units which may then form the feedstock of further chemical synthesis or for SCP production. Such techniques are being widely considered for straw and waste paper, but so far not for wooden materials. Dilute alkali pre-treatment (5 to 10 per cent NaOH) may well have a role to play in upgrading the feeding value of forage and certain hardwood wastes. The saponification of intermolecular bonds can cause increased enzymological and microbiological access to the carbohydrate components of cell walls and enhance conversion to more utilisable end-products. Electron irradiation and vibratory ball milling have greatly altered the crystallinity of cellulose.

Delignified wood pulp of all species provides an excellent source of dietary energy for ruminant animals and is an excellent substrate for further microbiological conversion. Removal of only one-third of the lignin of hardwood and two-thirds of softwoods yields a product equivalent to hay in ruminant digestibility. Wood fines, a by-product remaining after hardwood is pulped to produce tissue paper, have been shown to be suitable partial replacement for forage and grains in beef and sheep ratios as a source of dietary energy.

Enzymatic Delignification of Lignocellulose

A novel approach to the delignification of wood has been the subject of a recent patent which makes use of the ability of certain filamentous fungi to form lignin decomposing enzymes that can at least partially delignify wood products. The end-product is not unlike semi-chemical pulp and can then be subjected to mechanical de-fibration or further chemical digestion including a possible utilisation for biomass production.

The fungi involved in this process are white rot basidiomycete fungi capable of completely digesting all the wood components. This destructive ability is well known in forest product management and protective steps are routinely taken to prevent their growth. Such fungi, therefore, possess a complement of enzymes capable of completely degrading the complex molecular structure of lignocellulose. *Eriksson* have demonstrated that it is possible to regulate the enzymic ability of selected fungi in such a way that the fungi specifically attack the lignin components of the wood with little effect on the cellulose or other carbohydrate materials. This control can be engineered by genetic changes, by using special additives, or by the use of cellulase inhibitors.

As a first step, fungi were selected which had a natural ability to form lignin decomposing enzymes and a lesser potential to produce cellulose decomposing enzymes. Such fungi occur generally in the environment and include *Peniophora sanguinea*, *Phellinus isabellinus*, *Polyporus resinous*, *Trametes pini*, and, in particular, *Peniophora cremea* PB1. The latter fungus could digest at least 50 per cent of the lignin content of birch, while only decomposing 2 per cent of the cellulose.

Mutants were also developed for their ability to produce decomposing enzymes but with a greatly reduced ability to produce lignin decomposing enzymes. Such mutants were obtained by irradiating the spores of several white rot fungi—*Polyporus adustus* and *Sporotrichum pulverulentum* in particular, but also *Pheniophora gigantea*, *Trametes cinnabarina*, *Polyporus hirsutus*, *Pycnoporus sunguineus*, *Polyporus versicolor*, and *Polyporus zonatus*. The selective abilities of several cellulase-less mutants of *Polyporus adustus* are shown in Table 2.1 which shows that these mutants, while retaining their ability to decompose lignin, have a greatly reduced ability to decompose cellulose and other carbohydrate components.

Table 2.1. Decomposition of cellulose, glucomannan, xylan, laminarin, pectin and lignin by natural *Polyporus adustus* and by cellulase-less mutants.

			Substrate (%)			
Strain	*Cellulose*	*Xylan*	*Glucomannan*	*Laminarin*	*Pectin*	*Lignin*
Natural	12	24	18	28	21	38
Cel-3	0	0	0	28	21	35
Cel-5	0	0	0	26	21	43
Cel-6	0	0	0	28	21	52
Cel-7	0	12	0	30	23	43
Cel-21	0	0	0	28	23	36
Cel-22	0	0	0	28	27	38
Cel-25	0	0	0	27	23	42
Cel-27	0	5	0	33	23	30
Cel-28	0	14	0	29	21	43
Cel-30	0	12	0	29	23	42
Cel-101	0	0	0	28	22	38
Cel-102	0	0	0	28	21	6
Cel-103	0	0	0	24	21	
Cel-108	0	0	0	28	22	30

Yet another embodiment in this patent makes use of fungi with both lignin and cellulose decomposing enzymes; by the addition of selective inhibitors, the ability of the cellulose enzymes can be reduced without affecting the lignin decomposing enzymes. Compounds that have been used include glucose, arabinose, cellobiose, zylose, mannose, saccharose, ammonium salts, nitrate, asparagine, and casein hydrolysate. Many fungi have been used, including *Pheniophora sunguinea*, *Phellinus isabellinus*, *Trametes pini*, *Polyporus abietinus*, *P. hirsutus*, and *Stereum hirsutus*.

These methods of delignifying wooden materials offer a new and appealing approach to utilising waste lignocellulosic materials. The cellulose product that results from these biological treatments could then readily be further upgraded to serve as animal feed.

UTILISATION OF WASTE LIQUORS OF SULPHITE PULPING INDUSTRY

The residual monosaccharides in the waste liquors of the sulphate pulping industry have long been utilised for the growth of yeast cells, *Candida utilis*. The yeast cells can utilise hexoses and pentoses and produce a nutritious but bitter proteinaceous biomass. There are number of factories throughout the world, mostly concentrated in Eastern Europe, which produce this product.

A new process, the Pekilo process, in which a filamentous fungus *Paecilomyces varioti* is cultivated continuously on the waste liquors of the acid sodium calcium-magnesium or ammonium bisulphite cooking process is now in full operation at the Jamsankoski pulp mill in central Finland. This is the only industrial scale plant for continuous cultivation of fungal mycelium which produces a high-grade protein from a natural renewable carbohydrate source. At present, two 360 m^3 fermenters are in operation, and although they are not yet working to full capacity because of lack of raw material, they have each been able to produce 15 to 16.5 tonnes/24 hr. of dry biomass corresponding to a growth rate of 2.7 to 2.8/m^3/hr. The dried biomass has a crude protein level of 52 to 57 per cent and is sold to feed compounding mills. Feeding and toxicological trials have been successfully carried out in several countries and the Pekilo protein is now recognised officially in Finland as a food ingredient. A simple flow sheet of the Pekilo process is shown in Fig. 2.4 and the chemical composition is given in Table 2.2.

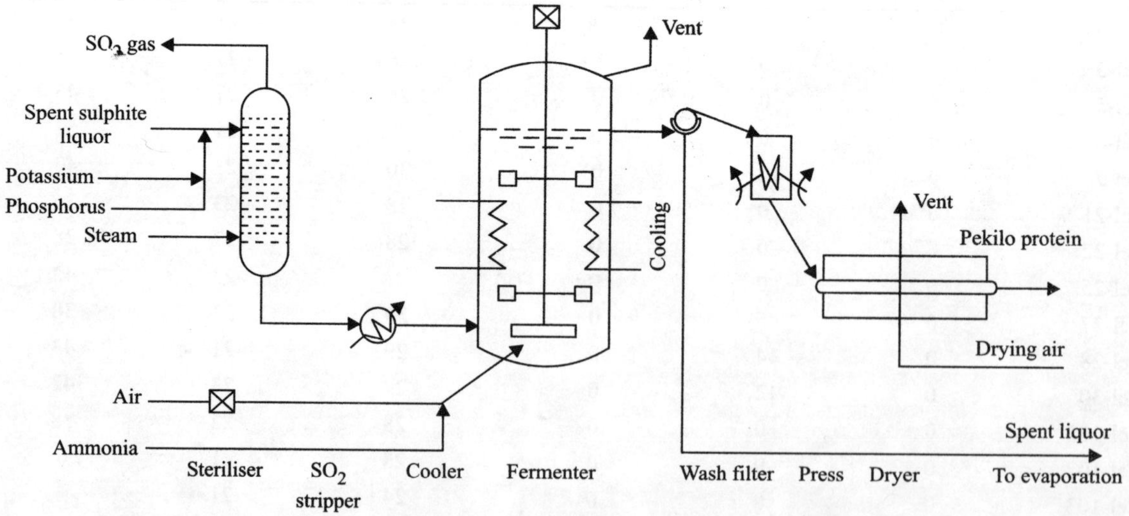

Fig. 2.4. A simplified flow sheet of the Pekilo process.

Table 2.2. Average chemical composition of Pekilo product.

Products	*Composition (g/16 g N)*
Threonine	4.8
Valine	5.0
Methionine	1.6
Isoleucine	4.6
Leucine	7.1
Tyrosine	4.0
Phenylalanine	4.2
Lysine	6.5
	(%)
Moisture, maximum	6.0
Crude fat	1.3
Crude protein	55
Crude fibre	7
N-free extracts	25
Ash	6
Nucleic acids	10
	(ppm)
Thiamine	7
Riboflavin	70
Pyridoxin	20
Niacin	450
Pantothenic acid	60
Biotin	2
Folic acid	15

The construction and operation of the Pekilo fermentation system has presented many technical difficulties, in particular the problem of oxygen limitation due to the special viscosity characteristics associated with the cultivation of fungal mycelia. A major advantage of the Pekilo process is the very considerable reduction in pollution resulting from the pulp mill. In addition to the monosaccharides, the fungus also utilises the acetic acid in the spent liquor. The BOD of the entire pulp mill has been reduced by at least 50 per cent. It is difficult to compare this process with conventional yeast and bacterial fermentations because, for commercial reasons. However, the Finnish workers consider that the Pekilo protein can be economically produced in small units producing 5,000 to 10,000 tonnes annually, whereas *n*-alkane and methanol based processes are generally considered to require production of 50,000 to 100,000 tonnes annually to be viable. Neither *Candida utilis* or *Paecilomyces varioti* can use water soluble oligomers and insoluble polysaccharides, which are present originally in spent sulphite liquor. For this reason, hydrolysis must be used before fermentation. However, *Eriksson* has developed a symbiotic process whereby *Sporotrichum pulverulentum* degrades the complex polymers to monomers which are then utilised for growth by *Candida utilis*.

TREE BARK AS FERTILISER AND ANIMAL FEED

Bark is that part of the tree that is external to the vascular cambium and is composed of different types of cells, mostly rich in organic molecules. On the average, bark comprise about 10 per cent of the fresh weight off most coniferous tress, and in most countries it has been considered a waste material to be disposed of by dumping or incineration. Disposal by destruction is wasteful and must not be accepted without careful scientific consideration. Methods should be explored and developed to turn this otherwise useless material into something of benefit to man and the ecosystem.

Historically, bark has been used by the Romans for floats to assist swimmers, by the Chinese to prepare paper, and in more recent times, on the Continent as bottle casks, insulation boards, and linoleum. The use of bark for heating purposes has had limited value. Some advances have been made in the use of bark as one of the ingredients in hardboard and insulating board manufacture. Chopped or pulverised bark can be used as a deep litter for animal bedding, having good water and ammonia absorbing properties. Chemical extraction of bark has been considered because of the wide range of potentially useful compounds to be found in various species.

Chemically, bark is normally more complex than wood. As with wood, the structural or cell wall components are composed primarily of polysaccharides such as cellulose and hemicellulose and the phenolic compounds lignin. However, a unique group of compounds (the phenolic acids) are also present in bark and these may be separated from lignin by dilute alkali. Present also in bark are waxes, fats, and numerous organic molecules which are readily soluble in neutral solvents.

From a horticultural point of view, various grades of chipped, ground, or pulverised bark have had some use as a mulching cover to suppress weeds and to reduce water loss by evaporation. There are, however, many report which imply a toxic effect of bark to plants. Indeed, this is not surprising when it is shown how many truly toxic molecules are present in bark. Much of the longevity of trees residues in the potency of the bark deposits to microbial and insect life, and many of the compounds are powerful uncouplers of oxidative phosphorylation. The possibility of detoxifying bark by broad spectrum microbial composing has at last been realised and numerous commercial products are now on the market which claim to be the end-product of a composing process. The success of these processes is due in part to the more efficient pulverisation of the bark, reducing it to relatively small sizes and thus allowing more efficient microbial penetration and utilisation. A fuller appreciation of composting processes can be found in most applied microbiological texts. In modern fermentation terminology, the term composting is synonymous with solid state fermentation. Thus, given the correct experimental parameters, it now seems that bark may find an expanding horticultural market as a soil conditioner and fertiliser (Table 2.3). The possibility of using bark as a means of supporting the growth of microbes for biomass production has been considered by several workers. Such studies have, in particular, used solvent extracted tree bark and tree bark extracts. The early studies of *Kuhlman* using *Pinus taeda* and *Lenzites saepierie* were not very successful. However, *Updegraff* and *Grant*, with *Pinus radiata*, and *Nordstrom*, with *Picea abies*, using simple flash cultures, produced respectively 12 mg/g and 37.5 mg/g mycelial protein/g barks. Recently, *Daugulis* and *Bone* have shown that alkaline extracted bark can be degraded in the submerged culture of *Polyporus anceps* and *Phamerochaete chrysoprium*, giving yields of 116 and 136 mg/g bark, representing a substantial improvement over all the previous workers. A non-linear relationship between bark concentration and protein yield was found. This process is now being stepped up to pilot fermentation stage. Thus, at present, bark is finding an increasing horticultural market throughout the world, and in the future, bark may well enter the more lucrative market of animal feed. Time and man's inventive genius will tell. Undoubtedly, lignocellulose complexes of wood and bark are difficult feedstocks for SCP, and for a successful outcome of this problem, there is a strong need for co-operation between biochemists, microbiologists, geneticists and engineers if the ultimate goal—the development of an economically feasible large-scale bioconverstion process based on lignocellulosic waste materials shall be realised.

Table 2.3. Horicultural uses of bark.

Species	*Mulching and landscaping*	*Plunge beds*	*Potting composts*	*Bulb forcing*	*Mushroom cultivation*	*Orchid culture*	*Pollution control*	*Tannin extraction*
Scots pin (*Pinus silvestris*)	Well suited (ornamental)	Suitable	Well suited	All species suitable if lime is added	Well suited if lime is added	Well suited	Suitable	Not recommended
Corsican pine (*Pinus nigra*)	Suitable	Suitable	Well suited	–	Well suited if lime is added	Well suited	Suitable	Not recommended
European larch (*Larix decidua*)	Well suited (ornamental)	Suitable		–		Suitable	Not tested	Suitable
Japanese larch (*Larix leptolepsis*)	Well suited (ornamental)	Suitable		–		Suitable	Not tested	Suitable
Norway spruce (*Picea abies*)	Suitable	Suitable		–		Not recommended	Suitable	Suitable
Sitka spruce (*Picea stitchensis*)	Suitable	Suitable		–		Not recommended	Suitable	Well suited
Douglas fir (*Pseudotsuga menziesii*)	Suitable	Suitable		–		Not recommended	Not tested	Suitable

VANILLIN FROM SAW DUST

Vanillin is an important pleasant smelling aromatic aldehyde which occurs in vanilla beans and gum benzoin in trace amounts. Both vanillin and ethyl vanillin are used most extensively for flavouring of confectionery, baked goods, candies and chocolates. It is also widely used in perfumes, fancy scents, laundry detergents and tobacco dressings. Ethyl vanillin does not occur in nature but is a synthetic product. It is three time more powerful than vanillin as flavouring and perfumery agent. More recently pulverised vanillin has come in the market for flavouring and perfumery industry.

Vanillin can be prepared from eugenol, safrole and related compounds which are abundantly present in clove oil, cinnamon oil and safros oil. Production of vanillin from eugenol consists of alkaline isomerisation to form isoeugenol which on oxidation gives vanillin. Safrole also undergoes alkaline isomerisation followed by rupture of methylenedioxy ring on boiling with alcoholic alkali. The resulting product is methylated to give isoeugenol and isocahvibetol. The isoeugenol on alkaline oxidation gives vanillin and replacement of methylation step by ethylation gives ethyl vanillin.

Vanillin can also be prepared from guaiacol by various standard methods of introducing aldehyde group in aromatic nucleus.

Vanillin is obtained from lignin of gymnosperm species by oxidation under suitable conditions of temperature and pressure, whereas hard woods on such oxidation produced both vanillin and syringaldehyde, thus only soft wood lignin is suitable for producing vanillin.

Lignosulphonates are obtained as by-product during the chemical pulping of wood by sulphate method adopted for the manufacture of pulp and paper. Bulk of the vanillin is being manufactured from lignin available as lignosulphonate separated as calcium or sodium salts or after fermentation to remove sugars. Marathon paper company of USA is making vanillin from lignosulphonates by one of US patent. Regarding the suitability of lignosulphonate for vanillin, it has been reported that higher sulphur content yield more vanillin (Table 2.4) on oxidation indicating that more percentage of *alpha* carbon positions of the aromatic ring in the lignin molecule have been substituted with sulphonic acid group. This probably gives vanillin on oxidative cleavage.

Table 2.4. Percentage vanillin on lignin basis as related to sulphur content.

Sulphur (%)	*Vanillin on lignin basis (%)*
5.74	2.2
5.12	5.1
6.82	4.2
7.59	4.8
7.97	5.3
9.62	6.8

In India sulphite liquor of soft wood is not available. Indian paper mills are mainly using bamboo and other hard wood species. The chemical pulping is done by alkaline sulphate method. Even the alkaline black liquor containing the lignin is not available as the paper mills are burning it to recover the inorganic chemicals. Only a couple of Indian pulp mills are using sulphite process for pulping using mainly bamboo and some hard woods. The lignosulphonate of these materials is not suitable for vanillin from technical and economic reasons.

Further the technical process is a well guarded secret of the manufacturing companies. From the literature survey it appears that the main objective of most of the investigators has been to elucidate the lignin structure rather than the process development. Commercial processes adopt the use of air for oxidation. All the vanillin affluent from the oxidation of lignosulphonate is returned to the manufacturers of lignin products. Yield of vanillin obtained from different materials is in the range of 5–27 per cent as reported in Table 2.5. Very little work regarding the process development on the manufacture of vanillin has been done in India. An Indian patent for manufacture of vanillin from groundnut shells describes nitrobenzene oxidation under alkaline medium claiming 3.0 per cent vanillin on groundnut shell basis. The highest yield of vanillin from bamboo black liquor has been reported to be 0.545 per cent on lignin basis using the nitrobenzene method.

Table 2.5. Yield of vanillin from various lignin materials.

Lignin source	*Oxidant*	*Yield (lignin basis)*
Spruce saw dust	CuO	18–20
Pine saw dust	CuO	14–16
Spruce wood	Ph NO_2	25.8
Spruce wood	Ph NO_2	27.5
Soft wood saw dust	Ph NO_2	19.1
Soft wood thiolignin	Ph NO_2	5–6
Sulphate black liquor	CuO	12.5
Red pine thiolignin	Ph NO_2	6.6
Red pine alkali lignin	Ph NO_2	10.5
Soft wood alkali lignin	Air	6.9
Soft wood alkali lignin	Ph NO_2	6.02
Spruce saw dust	Air/$CuSO_4$	6.2–8.2
Romanian coniferous saw dust	Ph NO_2	5.12
Pine saw dust	Ph NO_2	14.0
Spruce wood saw dust	CuO	20.0

Deodar (*Cedrus deodara*) saw dust available from saw mills is commercially used for the distillation of essential oil. After the essential oil recovery the saw dust is disposed of as fuel. The saw dust left after distillation of essential oil was taken from one of the local commercial units and used as a source of lignin for investigation for the preparation of vanillin. Large number of experiments were carried out on different scales, viz., 1, 10 and 50 L. The oxidation of saw dust was carried out in an alkaline medium using nitrobenzene, cupric oxide, copper sulphate and air as oxidants. The saw dust was heated with reagents under suitable reaction conditions in an autoclave. The alkaline reaction mixture after removal of oxidising agent was acidified and filtered. The filtrate was then extracted with benzene or chloroform. The extract was concentrated. Vanillin was separated from the concentrated extract by usual methods and subsequently purified by re-crystallisation. In some of the experiments the extract was treated with hot water. The hot water extract was subjected to extraction with ligroin which on concentration gave light yellow crystalline needles of vanillin.

Varying yields of vanillin (Table 2.6) were obtained under different isolation as well as reaction conditions. The vanillin obtained had melting point 81–82°C and chemically pure as shown by TLC and

GLC techniques. Further the sample of the vanillin made out of deodar wood lignin was sent to commercial biscuit factory as well as to a firm dealing with flavouring chemicals. Both the organisations reported that the vanillin is acceptable for flavouring and is of the required purity. It can be seen from the Table 2.5, that even by using different species of saw dust after pre-extraction with alcohol, benzene 1:1 and after using different reaction conditions the yield of vanillin ranged between 2.55–3.5 per cent. Most of the work done on vanillin from lignin is patented. However, the yield of vanillin obtained by air oxidation method using essential oil distilled saw dust is comparable to the yield reported on commercial scale which is a big breakthrough as air oxidation eliminates the formation of undesirable by-products. Thus there is a possibility of obtaining vanillin from deodar saw dust be aerial oxidation process. Commercial manufacturer are using lime instead of caustic soda with air as oxidant and are getting 6–8 per cent vanillin on lignin basis (from Ca lignosulphonate). Work for the modification of process in this direction is under progress.

Table 2.6. Yield of vanillin from alkaline oxidation studies of saw dust.

Material	*Oxidant*	*Yield (Klason lignin basis) (%)*
Deodar saw dust (oil extracted)	$PhNO_2$	6.5
Deodar saw dust (oil extracted)	CuO	3.5
Deodar saw dust (oil extracted)	$CuSO_4$ and air	6.1–7.2
Deodar saw dust (oil extracted)	Air	6.6–7.0
Spruce wood saw dust	$PhNO_2$	6.5
Spruce wood soda lignin	Air	3.3
Spruce wood saw dust	Air	4.95
Fir saw dust	CuO	1.75–2.04
Pine saw dust (alcohol/benzene extracted)	CuO	2.1

3

Utilisation of Coconut Shell Powder into Industrial Products

INTRODUCTION

Coconut shell is a material obtained in the copra processing. Large quantities of shells are consumed in traditional chulas and country ovens. The shell can be converted into an industrial product by pulverising in to fine powder. The shell powder is used in many industrial products, such as mosquito coil, agarbathi etc. Compression moulded products are the few products wherein coconut shell powder is extensively used. The shell powder project can be set up in various capacities based on the availability of shell, end use segment, etc.

Coconut is a tropical palm, which grows abundantly in the coastal regions of India. It has now been plated as an exclusive crop, other than the normal practice of planting in farm borders and bunds. This coupled with the introduction of short term and better yielding varieties have increased the production of coconut to a great extent.

The flesh inside the seed of the coconut is used for oil extraction or edible purposes. The shell, which covers the flesh, is hard and rigid which has all along been used as fuel. The shell is also carbonised for the manufacture of activated carbon. However, manufacture of activated carbon from coconut shell is a costly process and requires large quantity of shell to be viable.

The economical way to utilise coconut shell for value added products, is to convert it into powder form by pulverising. The powder is used as filled material in thermosetting plastic components, mosquito coil, agarbathies, etc. It is also used in manufacture of synthetic resins and adhesives. The filler is added with the basic raw material in order to increase the volume without affecting its properties. It is added to save the raw materials cost.

TECHNICAL ASPECTS

Characteristics of Coconut Shell

Coconut shells are similar in composition to hard woods, though the cellulose content is lower than lignin. The composition of various constituents of the shell is given in Table 3.1. The shell occupies 15 per cent of the coconut. The shells occupy a lot of volume, more than proportional to their weight; in other words, the bulk density of the shell is high. This makes it transport intensive.

Table 3.1. Composition of various constituents of shell.

Constituent	*Composition (%)*
Lignin	29.40
Pentosans	27.70
Cellulose	26.60
Moisture	6.00
Solvent extractives	4.20
Uronic anhydrides	3.50
Ash	0.60

Characteristics of finished products

The coconut shell powder is light brown in colour and clear in appearance. It should be free from pith and other dust particles. Moisture should be less than 10 per cent. The powder should be free flowing. Generally, the fineness of the powder would range 75–325 mesh size. While the bigger sized particles are used as fillers, the smaller ones are used for making synthetic resins. The size and quality should be maintained to consumer standards.

MANUFACTURING PROCESS

The manufacturing of shell powder involve the following processes.

Manual Breaking

The coconut shells are broken into smaller pieces using wooden mallets. This ensures better production in pulveriser. The size should be 1 inch across. This can be fed into a beater type disintegrator for further crushing.

Micro pulverising

This processes a cyclone separator. This machine pulverises the coconut into smaller powder. The cyclone separator separates the bigger pieces. These are fed into the micro pulveriser again.

Sieving

The pulverised powder is sieved to obtain different size of the powder as specified by the customers. Sieve shaker is used to separate particles into various required sizes.

Quality control and standards

Appearance	It should be light brown and clear
Moisture	Less than 10 per cent. Higher moisture content will inhibit free flowing nature
Size	Pass through 80–100, sieve for bakelite fillers
	Pass through 200–300 sieve for resin glues

Uses

The final product of size 80–100 mesh is used in compression mouldings and thermoset mouldings of polymers like phenol formaldehyde and bakelite. The coconut shell powder of size 200–300 mesh is used in the manufacture of synthetic resin glues.

Details of Manufacturing Process

In the processing of coconut into copra, roughly equal weights of copra, coconut shells, and husks are produced. The coconut husks are usually further processed into coir fibre (mattress and bristle fibres) if a coir factory is located in the vicinity of the coconut plantations. Roughly again, half of the husk goes into fibre and half ends up as waste coir dust during the defibring process.

The principal sources of wastes from coconut therefore are coconut shells and coconut husks. Present methods of utilising these wastes are rather primitive. In most parts of the world especially in philippines the coconut husk is used as fuel in native copra drying kilns locally known as 'tapahan'. The 'tapahan' is simply a big hole dug in the ground inside on which the husks are burned. Over the hole is a slatted platform on which the coconuts split into halves, are laid to dry in the hot gases and smoke that emanate from the rather inefficient combustions of coconut husks. After 36 to 38 hours the coconut is dried into copra in lots of 1000 to 2000 nuts.

The coconut shells recovered from copra-making are commonly burned in primitive carbonisation retorts to produce charcoal. The gas by-products of carbonisation are discarded, a practice which is wasteful, considering that 66 per cent of the coconut shell is volatile matter and only slightly over 20 per cent is recovered as charcoal.

In certain regions of the world where coconut plantations supply the raw material for desiccated coconut factories, the coconuts are dehusked before being sold to the factory. The husks are in turn sold to coir defibring plants. The defibring process utilise the retting method whereby the husks are soaked in water for some 12 hours before being fed to the defibring machines. The resulting fibres therefore are wet and require drying by natural means (under the sun) and/or by mechanical means. At present, mechanical dryers for coir utilise diesel fuel.

Production output in coir factories is limited by three factors: (i) the supply of coconut husks; (ii) the rate of disposal of the waste coir dust; and very recently; and (iii) the availability of diesel fuel for mechanical drying.

The problem of waste coir dust disposal is particularly acute in that this voluminous waste now occupies large portions of factory sites. In many instances, the availability of other dumping sites dictates whether a factory can continue production once its current dumping site is filled up.

Many of the problems that beset the coconut industry may be solved by a more efficient utilisation of its waste products.

Producer Gas from Coconut Shells

Experiments were carried out various parts of the world to gasify coconut shells with air and steam. An experimental, updraft gas producer with a one-square-foot grate area was used. The effects on performance of steam content in the blast, the fuel bed depth, and the rate of gasification were studied. The parameters used to evaluate gasification performance were cold gas thermal efficiency and percentage decomposition of combined water.

Coconut shell contains about 50 per cent combined water by weight, as can be calculated from the ultimate analysis of coconut shell shown in Table 3.2.

Because of the considerably amount of combined water, the combustion zone in the reactor during gasification did not become too hot as to cause clinkering of the ash, and therefore addition of steam in the blast was not necessary for clinker control. However, addition of steam increased the hydrogen content of the gas, hence improved its calorific value and the efficiency of gasification. From stoichiometric calculations it was found that there was more water coming out with the gas products in the form of vapours than water entering the reactor in the form of steam. This was necessarily so because of the incomplete decomposition of the combined water to coconut shell. In effect, steam in the blast went through the reactor undecomposed and after reaction, was increased by an amount equal to the combined water that remained undecomposed.

Table 3.2. Ultimate analysis and heating value of coconut shell and coir dust compared with wood.

Ultimate analysis, % (dry, ash-free basis)	*Coconut shell*	*Coir dust*	*Wood*
C	49	46	50
H	6	7	6
O	44	46	43
N	1	1	1
Gross calorific value Btu/lb.			
From Dulong's equation	8020	7200	–
From bomb calorimeter test	8090	7260	8010
Net calorific value, Btu/lb.	7400	6580	7430

Figure 3.1 shows the experimental results of test runs to determine the effects on gasification of steam content in the blast. It can be seen that the percentage decomposition of combined water increased with percentage steam in the blast upto a certain point, from 45 per cent decomposition when there was no steam to an optimum 80 per cent decomposition at 9 per cent steam in the blast.

Similarly, the cold gas thermal efficiency increased with blast steam content from 67 per cent at no steam to an optimum 82 per cent at 9 per cent steam. Cold gas thermal efficiency was taken to be the ratio of the net calorific value in Btu/cu. ft. of the producer gas saturated at 60°F of the heating value of coconut shell plus the heat content of steam in the blast to produce one cubic foot of gas.

Figure 3.2 shows the influence of gasification rate on thermal efficiency and percentage decomposition of combined water. The gasification rate was directly measured by weighing the fuel fed and consumed in the gas producer during a given period of time. The actual rate varied from 20 to 120 lb/hr/sq ft. The optimum condition was at a gasification rate of 40 lb/hr/sq ft. at 9 per cent steam in the blast.

The cold gas efficiency was 82 per cent and the percentage decomposition of the combined water was 76 per cent. The effect of fuel bed depth was also studied by obtaining data in test runs at different bed depths, varying the depth from 8 inches to 48 inches in increments of 4 inches. It was found that no significant correlation existed between fuel bed depth and gasification performance. In 11 test runs at a gasification rate of 80 lb/hr/sq ft. and 9 per cent steam in the blast, the mean value of thermal efficiency was 72 per cent with a standard deviation of 2.5, the average combined water decomposition was 62 per cent with a standard deviation of 10, and the average net calorific value of the gas was 117 Btu/cu. ft. with a standard deviation of 6.

Producer Gas from Coir Dust

The same experimental gas producer as used for coconut shell was also used to gasify coir dust. The initial moisture content of the waste coir dust was 85 per cent, wet basis, and it was necessary to expel some of the water by mechanically pressing the coir dust to form compressed briquettes of about 1 × 2 × 2.5 inches on the edges with moisture content reduced to about 50 per cent.

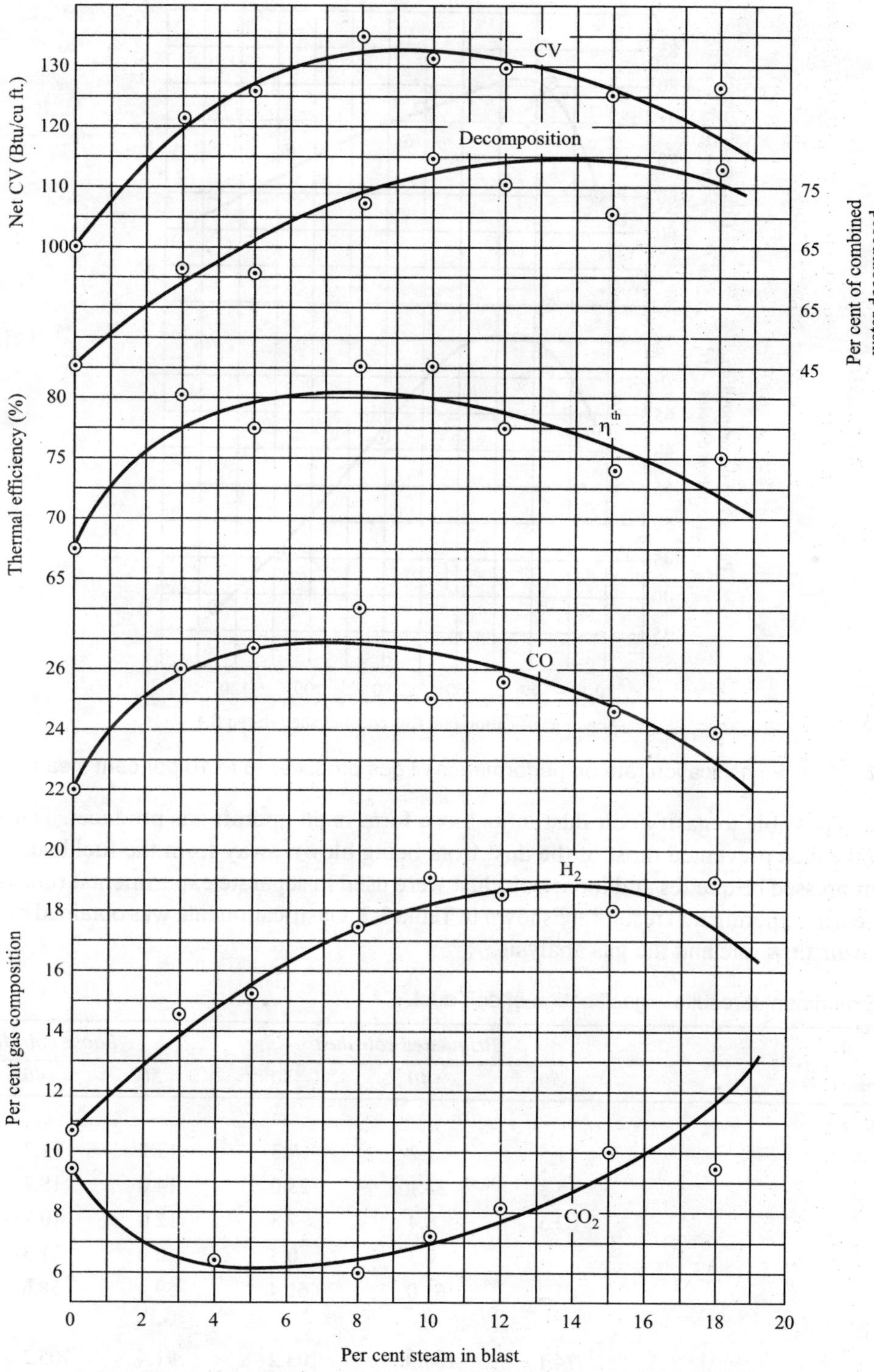

Fig. 3.1. Effects of steam content in blast (gasification rate = 40 lbs/ hr/ sq. ft.)

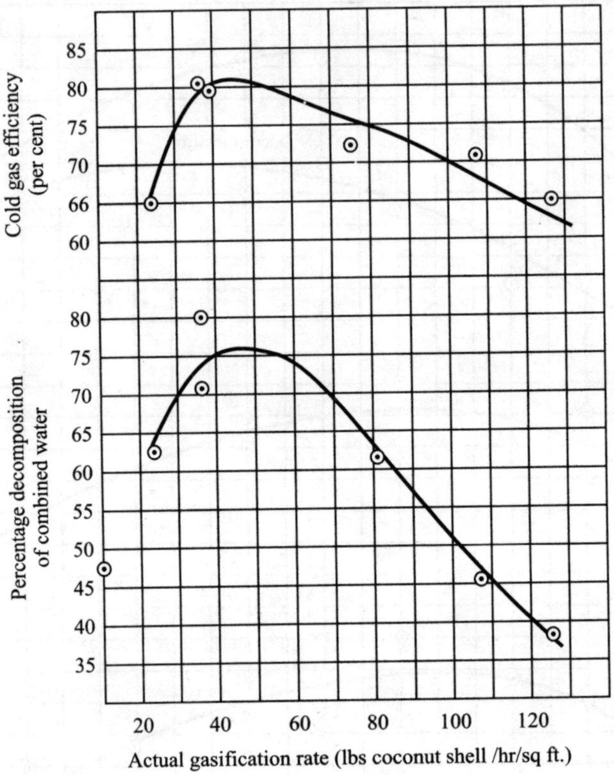

Fig. 3.2. Effect of gasification rate on performance of gas producer (8 to 10 per cent steam in blast).

It was also possible to gasify coir dust in its loose form in an updraft gas producer. The short fibres mixed with the dust prevented most of the dust from being blown away from the fuel bed.

Both compressed briquettes and loose coir dust were used in separate experimental runs lasting from 4 to 6 hours each. Experimental results are shown in Table 3.3. Gasification rate was obtained by calculation knowing the air flow rate and the gas analysis.

Table 3.3. Experimental results—gasification of coir dust.

	Briquetted coir dust			*Loose coir dust*		
Moisture content, %	*50*	*40*	*30*	*50*	*40*	*30*
Gas analysis, %						
CO_2	11.5	7.2	7.3	13.7	11.2	11.4
CO	15.8	23.1	23.0	14.0	18.2	18.7
H_2	7.5	8.4	7.8	12.0	10.7	15.2
CH_4	–	0.3	0.5	0.9	1.3	0.6
N_2	65.2	61.0	61.4	59.4	58.6	54.1
CV, Btu/cu. ft.						
Gross CV	74.1	103.0	103.2	91.4	105.2	113.6

(Contd...)

	Briquetted coir dust			Loose coir dust		
Moisture content, %	*50*	*40*	*30*	*50*	*40*	*30*
Net CV	70.3	98.6	98.7	84.5	98.6	105.5
Thermal efficiency, %	55.5	69.5	69.0	62.2	68.8	74.0
Combined water decomposition, %	18.7	20.5	19.7	43.6	37.8	50.0
Gasification rate, lb/hr/sq ft. (as fired)	72.9	67.9	52.4	70.2	62.2	56.7

It was observed that the thermal efficiency, the calorific value, and the percentage decomposition of combined water were higher when gasifying loose coir dust than when gasifying briquetted coir dust.

It was observed further that when the fuel contained moisture above 50 per cent (wet basis), gasification performance was unreliable. Every time the producer was charged with fresh fuel there was a tendency for the gas to become incombustible, and about 10 minutes' wait after charging was necessary for the gas to burn again.

At moisture contents of 40 per cent or less, the gas produced burned continuously even right after charging. Frequent poking of the fuel bed, however, was necessary to bring down the fuel to the grate, as the coir dust bed had a tendency to stick to the side of the reactor walls even after the bottom part of the bed had been consumed.

Producer Gas as Fuel for Copra Drying

A mechanical dryer for copra (Fig. 3.3) was designed to utilise raw producer gas from coconut shells as fuel, dimensions of the drying chamber being 16 ft. × 4 ft. × 5 ft. Direct heating was employed, i.e., producer gas together with the tar was burned in a furnace producing a smokeless flame, and the combustion products came into direct contact with the material being dried. To predict its performance, a mathematical model of the dryer system was developed.

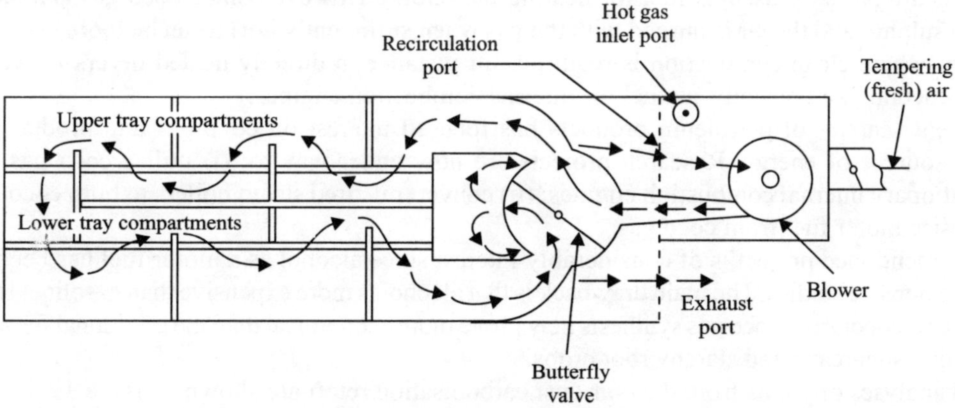

Fig. 3.3. Copra dryer showing direction of flow of air. (Direction of flow is reversed when the butterfly valve is rotated through an angle of 90 degrees from position shown.

The results of theoretical calculations predicting the total drying time and efficiency at various loads of nuts and blower rates, to remove 48 per cent moisture from an initial 51 per cent, wet basis are shown in Table 3.4.

Table 3.4. Theoretical results of dryer performance.

Dryer capacity No. of nuts	*Blower rate W_3 lb/min*	*Inlet temperature T_3, °F*	*Recirculation (RC, %)*	*Exhaust air relative humidity (RH, %)*	*Drying time (hours)*	*Efficiency (%)*
1000	90	180	64	40	12	66
1000	90	180	92	60	16	79
2000	100	180	63	50	18	68
2000	150	180	85	48	16	73
2000	150	200	87	54	12	79

Coco Gas and Its Uses

Coco gas is the term used for the gas by-products of carbonisation of coconut shells and coir dust when heated in closed retorts in the absence of air. As previously mentioned, the present method of charcoal production is quite wasteful since only 20 to 25 per cent of the coconut shell is recovered as charcoal and gas by-products are discarded. Research projects on improved designs of carbonisation retorts and the proper utilisation of the gas by-products are presently being undertaken. For instance, in a study on an integrated system of copra drying and charcoal production, a carbonisation retort is being designed to burn dirty or waste lubrication oil. (The waste oil burners have been designed and successfully tested to burn waste oil at rates of 2 to 20 gallons per hour). The charge of coconut shells are to be heated indirectly by hot combustion gases that pass through flues that vertically traverse the inside of the vertical retort. The coco gas by-products will be used as fuel for copra dryers.

Studies on a pilot plant scale of the performance of a horizontal retorts for carbonising coir dust have been in progress for several years now, the coir dust char output being either in the form of briquetted charcoal or in powder form which can readily be processed into activated carbon. The coco gas by-products are partially used as fuel for heating the retorts. However, since coco gas is a clean fuel (there is no sulphur and the tar is burned with the gas when sufficiently hot) it can be more useful as fuel in processes where clean combustion is required, for instance in directly heated dryers for copra and desiccated coconut, and recently, as fuel for internal combustion engines.

The recent scarcity of petroleum products has focused interest on coco gas and producer gas as alternative sources of energy. Research projects are now under way to: (i) utilise coco gas in large (60 hp) stationary internal combustion engines; (ii) convert oil-fired steam boilers to burn coco gas; and (iii) synthesise motor fuel from coco gas.

The last mentioned project is of considerably interest since alcohol as a motor fuel has been proved feasible by extensive studies. The main drawback is that alcohol is more expensive than gasoline. However, production of alcohol from coco gas synthesis may prove more economical than the traditional fermentation methods using sugarcane and starchy root crops.

Typical analyses of yields from the coir dust carbonisation retort are shown in Table 3.5.

Table 3.5. Yields of products from carbonisation of coir dust and wood.

Product	*Coir dust %*	*Wood %*
Charcoal	40.9	41.3
Pyroligneous liquor	28.0	31.7

(Contd...)

Product	*Coir dust %*	*Wood %*
Tar	3.5	10.0
Non-condensable gas	27.6	17.0
Composition of non-condensable gas, %		
CO_2	18.7	54.0
CO	22.5	28.0
H_2	47.6	1.0
CH_4	10.4	15.0
C_nH_m	–	–
Air	0.8	–

ADSORPTION OF PHENOL BY ACTIVATED CARBON DERIVED FROM COCONUT SHELL

Phenol removal from synthetic waste-water by adsorption on activated carbon produced from coconut shell having impregnation ratio 0.5 was used to determine the effects of different dosage, pH and effect of time on adsorption. The dosage data obtained was used in the construction of Freundlich and Langmuir isotherms. The maximum removal is obtained at pH 2.0.

Phenol comes under the volatile organic compounds. The three most significant products from phenol are phenolic resins, bisphenol A and caprolactum. The largest single use of phenol is phenolic resins, which are produced by the condensation of phenol with an aldehyde i.e., formaldehyde. The resins are used widely in major industries particularly in the construction, automotive and appliance industries. The phenolic resins are used for the manufacture of plywood adhesives. In addition to this it is used as a chemical intermediate for the manufacture of herbicides, insecticides and dyestuff.

Phenol occur as a major contaminant in the effluent of industrial process such as refining, coke and coal processing, resin manufacture, foundries and paint stripping operations.

Phenol is highly toxic, protein degenerating compound. Phenol is absorbed by skin and causes severe blistering. Contact with large quantities can cause paralysis of central nervous system, and ultimately death. Skin tumours have also been reported.

To treat phenolic waste many methods are available, these include recovery, incineration, adsorption, biological treatment and chemical oxidation. In this study suitability of activated carbon from coconut shells for synthetic phenol adsorption and its characteristics are discussed.

Experimentation

Preparation of synthetic waste-water

The synthetic waste-water for the present study was prepared. Concentration of phenol solution was 500 ppm.

Spectrophotometer

Systronics UV-vis spectrophotometer was used for all measurements.

Preparation of coconut shell active carbon

The coconut shell activated carbon used in this study was prepared by pulverising the coconut shell into powder, washed and dried in the oven. It was then kept in the furnace upto 600°C for two hours and then stored, and finally activated by chemial method. In chemical activation, degree of impregnation is an important factor. The known quantity of stored coconut shell powder was mixed with barium chloride in the impregnation ratio of 0.5.

Required quantity of water was added to this mixture and kept for boiling till water evaporates, then the slurry type mixture is kept in oven at 110°C for 24 hours to remove moisture. The activated carbon was washed with hydrochloric acid followed by hot distilled water for about 6 times to remove the activating agent and HCl respectively. Washed carbon was dried at 110°C and then sieved for required size 300 μ and packed in polythene bag.

Characteristics of coconut shell carbon

Before experimentation on coconut shell carbon, it is required to know the characteristics such as moisture content, ash content, pH, decolourising power and surface area of the carbon derived from coconut shell (Table 3.6).

Table 3.6. Characteristics of prepared carbon.

Type of carbon	*IR*	*pH*	*Moisture content (%)*	*Decolouring power (mg/g)*	*Ash content (%)*	*Surface area (sq. m/gm)*
$BaCl_2$ activated	0.5	6.0	6.50	24.25	17.68	630.80

Batch studies at optimum dosage

To determine the optimum dosage of activated carbon of coconut shell, the carbon was added to conical flasks in various dosages containing known concentration of phenol solution (500 mg in 1 litre) and adjusted to pH 2.0 ± 0.20 using 0.1 N HCl. The solution in the flask was subjected to stirring for optimum contact time, then filtered and analysed for residual phenol concentration using spectrophotometer is chosen as optimum dosage.

Batch studies at different pH

The effect of pH on phenol was studied by performing equilibrium adsorption at different pH. The pH of the solution was adjusted by using 0.1 N NaOH. The activated carbon derived from coconut shell were mixed and stirred to optimum contact time, then filtered and was analysed for residual phenol concentration.

Effect of time

The effect of time on phenol removal was studied for different time and was analysed for residual phenol concentration.

Results and Discussion

Effect of adsorbent dosage

Effect of adsorbent dosage is studied and graph of phenol removal vs. dosage is plotted as shown in Fig. 3.4. It is observed that as the dosage of activated carbon increases amount of residual phenol decreases sharply and attains minimum. The point, where maximum phenol removal is obtained is taken as optimum dosage. Using this data Freundlich and Langmuir isotherm are constructed in Figs. 3.5 and 3.6.

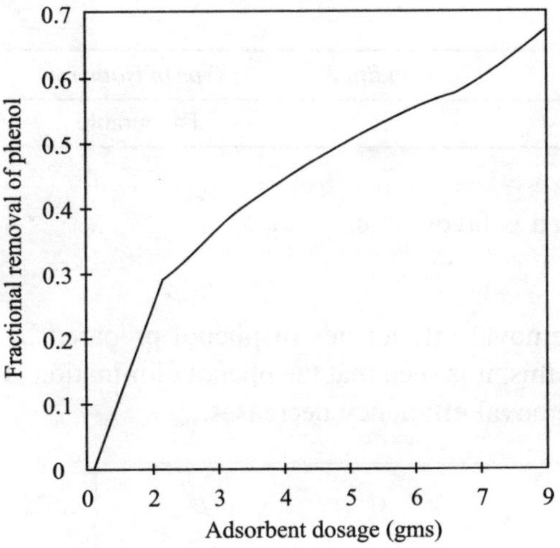

Fig. 3.4. Effect of absorbent dosage.

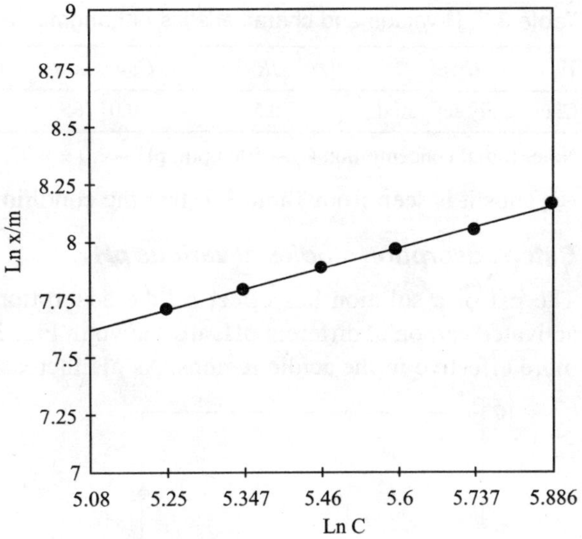

Fig. 3.5. Freundlich isotherm.

Freundlich isotherm: It is given by

$$x/m = KC^{1/n} \quad \text{... (3.1)}$$

where

x = amount of solute absorbed

m = weight of adsorbent

C = concentration of solute in solution

K and n are constants

Equation (3.1) can be written as,

$$\text{Ln}\,(x/m) = K/n.\ \text{Ln}\,C \quad \text{.. (3.2)}$$

Langmuir isotherm: It is given as

$$x/m = abc/1 + bc \quad \text{... (3.3)}$$

Equations 3.1 and 3.2 are constants. Equation 3.3 can be written as,

$$C/(x/m) = 1/ab + C/a \quad \text{... (3.4)}$$

The important characteristics of Langmuir isotherm can be explained in terms of dimensionless constant separation factor R, which is defined by

$$R = \frac{1}{(1+a)\,C_o}$$

where,

a = Langmuir constant

C_o = initial solute concentration

This parameter indicates the shape of the isotherm and if $0 < R < 1$ then, the isotherm is favourable. Table 3.7 shows the R-values and characteristics of Langmuir isotherm.

Table 3.7. R-values and characteristics of Langmuir isotherm.

Type of carbon	*IR*	*Constant α*	*R*	*Condition*	*Type of isotherm*
Chemically activated	0.5	0.01765	0.002	0 < R < 1	Favourable

Note: Initial concentration C_o = 500 ppm, pH = 2.0 ± 0.02.

Thus it is seen from Table 3.7, that the condition obtained is favourable.

Batch adsorption studies at various pH

The pH of a solution has cffect on the adsorption. The removal efficiencies of phenol prepared by activated carbon at different pH, are shown in Fig. 3.7 from this, it is seen that the phenol elimination is more effective in the acidic regions. As pH increases the removal efficiency decreases.

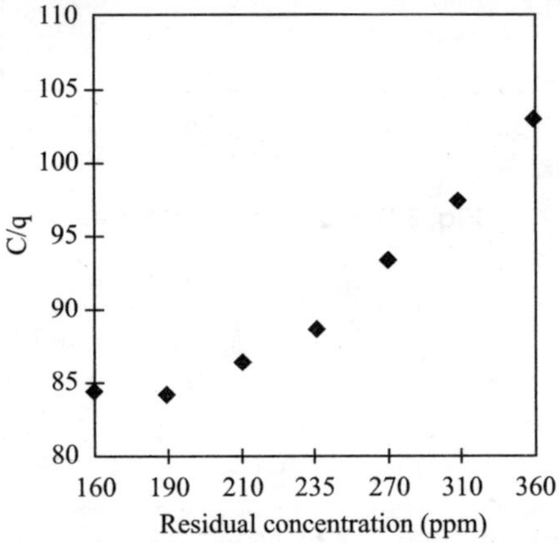

Fig. 3.6. Langmuir isotherm.

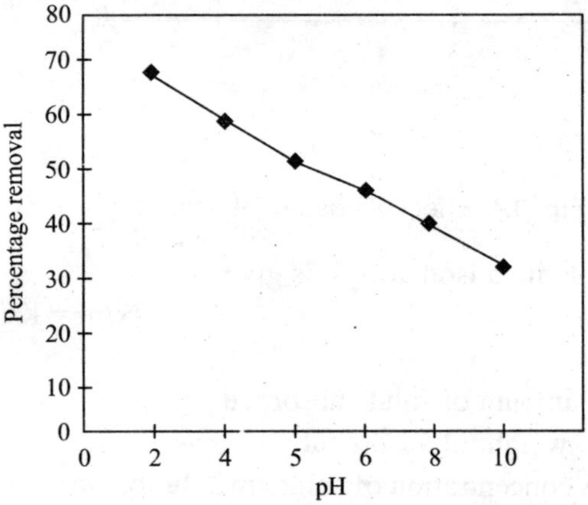

Fig. 3.7. Batch sorption studies at various pH.

Effect of time

It was seen that after some time the concentration of the residual solution becomes constant. Figure 3.8 shows the behaviour of the same.

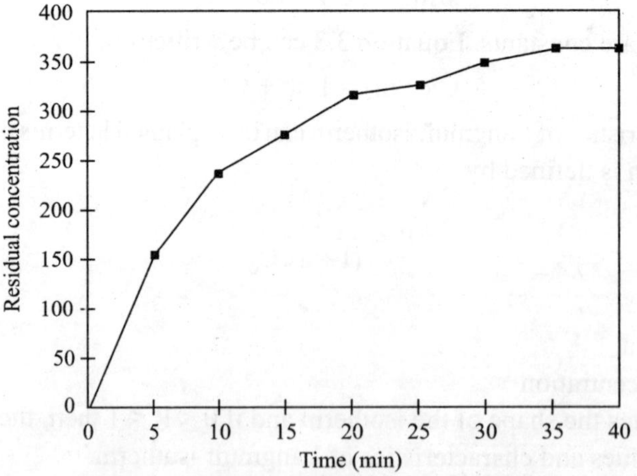

Fig. 3.8. Effect of time on phenol removal.

From the above, we can sum up the following:

1. The coconut shell can be used in the preparation of activated carbon and removal efficiency is high.
2. From the adsorption studies, it is found that adsorption of phenol obeys Freundlich and Langmuir isotherms.
3. From the studies it is seen that after some time the concentration of residual phenol solution becomes constant.
4. The constant value R is the Langmuir isotherm gives an indication of favourable isotherm.
5. Analysis of coconut shell activated carbon shows that chemical activation increases surface area.
6. Activated carbon derived from coconut shell have high removal potential at high concentrations and hence it will give higher removal efficiency for low concentrations.
7. This method of removal can be used in various industries which discharges phenol in their effluents like oil refineries, steel industries, pharmaceuticals, resin industries and insecticide industries.
8. On large scale, this process is economical as the activated carbon can be regenerated and re-used.
9. Here the study is for batch process, but this activated carbon can be also be used for continuous process.

Thus coconut shells can be gasified with air (with or without steam) into producer gas with a net calorific value of about 130 Btu per cubic foot. The raw producer gas together with tar can be burned efficiently and the absence of sulphur makes it most suitable as fuel for copra dryers and other mechanical dryers for foodstuffs. The coconut shells recovered from copra making are more than enough to meet the fuel requirements of the copra dryer. The surplus shells can be converted into charcoal, for which there is a growing market abroad. To improve charcoal recovery from coconut shells, integrated systems of copra drying and charcoal making are recommended, whereby the coco gas by-products from the carbonisation retorts can supply the fuel requirements of the copra dryers.

Use of coco gas and producer gas from coconut shells (instead of coconut husks) for copra drying, would make available more raw material for the coir fibre industry. This industry, which is at present supplying less than one per cent of the world demand for coir fibre, has the potential to secure the lion's share of the world market. A solution to the problem of disposal of its waste coir dust (the by-product of the defibring process) is gasification of coir dust and the utilisation of producer gas to meet fuel requirements for coir fibre drying.

4

Utilisation of Biomass

INTRODUCTION

Energy is a key input not only for daily household needs, but also in all developmental activities. As on today, the main primary source of energy is out of fossil fuels which are rapidly dwindling. But the demand for energy is ever increasing due to industrialisation, economic development and population growth. The total recoverable world reserves of the main fossil fuels namely coal, natural gas and oil applying current technology and assuming continued consumption of present day rates, have been estimated to be about 1000 years, 35 years and 20 years respectively. Modern industry is almost totally dependent on these sources. Approximately 93 per cent of fossil fuel is consumed for energy production worldover, with only 7 per cent being used by industry for the production of organic chemicals, plastics and solvents. The continued depletion of global fossil fuel has generated an urgent need to seek out alternate sources of energy. These have so far included harnessing of hydro, tidal, wave and wind power, capture of solar, geothermal energy supplies and also nuclear power. With all these systems, there is yet no definite answer on energy front for successful development strategy. However, it can not be doubted that the fossil fuels will be exhausted completely in not too distant future.

On this backdrop, more and more utilisation of renewable energy sources warrants top-most priority. It is heartening that there is growing consensus all over the world favouring the use of renewable energy. But this growing awareness should be further strengthened with the development of more advanced technologies for the exploitation of alternate renewable energy sources. Renewable energy sources are must if the economy of the world is to be carried forward.

BASIC REACTION

The basic reaction involved in biomass is photosynthesis, which is a process of fixation of carbon by living plants from atmospheric carbon dioxide with water by chloroplasts in presence of sunlight, releasing oxygen back to the atmosphere. Biomass is nothing but recycled carbon dioxide with solar energy in the most versatile useful form. Photosynthetic organisms, both terrestrial and marine, can be considered as continuous solar energy converters and are constantly renewable. Plant photosynthesis alone fixes about 2×10^{11} tonnes of carbon with an energy content of 2×10^{21} joules, which represents about 10 times the world's annual energy consumption and 200 times our food energy consumption. The magnitude and role of photosynthesis has gone so far largely unappreciated. Let it not be forgotten that photosynthesis which occurred in the remote past provides all the present fossil carbon sources. The fossil fuels, we are retrieving today, are made out of biomass buried inside mother earth millions of year ago.

Photosynthetically derived biomass exists in many available forms in the environment. These can be transformed into storable fuels and chemical feedstocks such as ethanol and methane gas. The actual efficiency of capturing solar energy by green plants may be around 3–4 per cent, the more effective photosynthetic plants like maize, sorghum and especially sugarcane being the most productive.

Biomass as a Renewable Sources

Biomass can be considered as a renewable energy source, and can be converted into either direct energy or energy-carrier compounds by:

1. Direct combustion.
2. Anaerobic digestion systems.
3. Destructive distillation.
4. Gasification.
5. Chemical or biochemical hydrolysis.

Biomass supplies may be achieved by three ways:

1. Cultivation of energy crops.
2. Harvesting of natural vegetation.
3. Utilisation of agricultural and other organic wastes.

The conversion of resulting biomass to usable fuels can be accomplished by biological or chemical means or by a combination of both, which are summarised in Fig. 4.1.

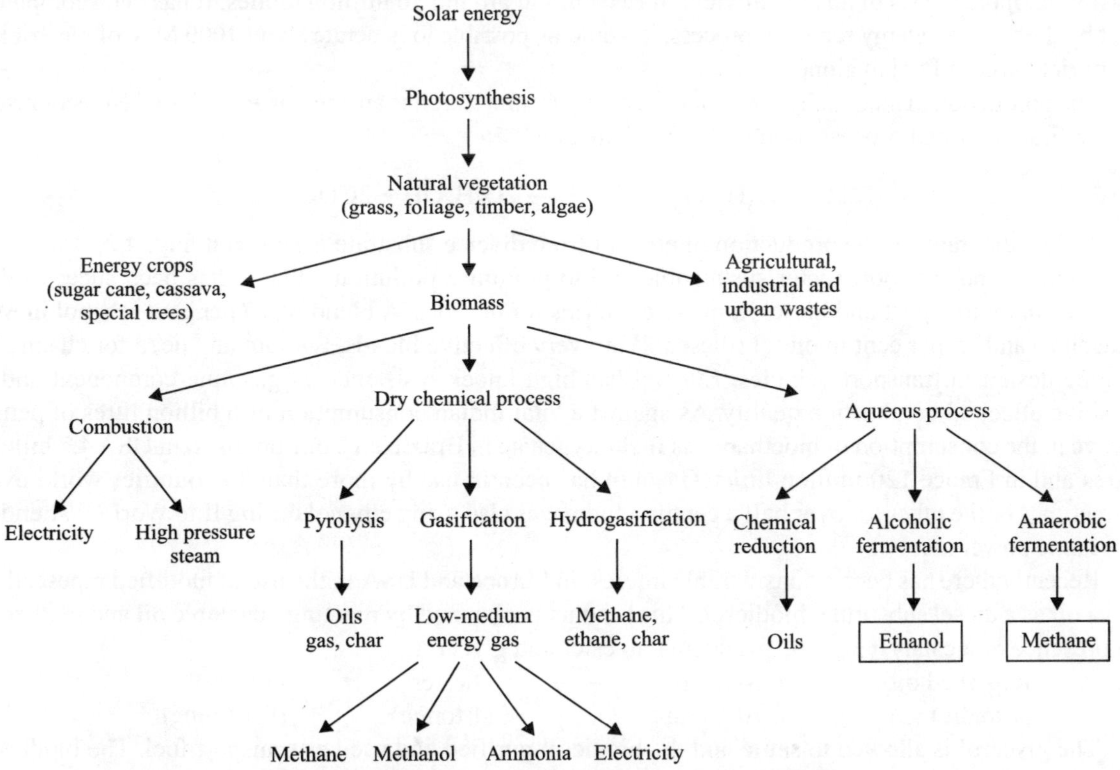

Fig. 4.1. Options for the conversion of biomass to energy.

The technical processing of the biomass depends on many factors, including moisture level and chemical complexity. Material with a high water content normally undergo aqueous processing, which avoids the need for substrate drying.

Alcoholic Fermentation

Alcoholic fermentation to ethanol, anaerobic digestion to methane, as well as chemical reduction to oily hydrocarbons, are all possible. Low moisture level materials such as wood, straw and bagasse can be burnt to generate heat or to raise steam for electricity generation. These may also be subjected to thermo-chemical processes such as gasifications and pyrolysis to produce energy-rich compounds such as gaseous oil, charcoal and eventually methanol and ammonia.

From time immemorial, biomass has been in use as a reliable source of energy all over the world. Burning of cowdung and wood, provides one-third of energy used by developing world. In fact, one-seventh of the energy consumed in the world is derived from biomass, which is equivalent to about 3 million tonnes of oil per day. The non-commercialised forms of energy like wood, charcoal, dried cowdung etc., make up over 85 per cent of the energy consumption in rural areas worldover. In India, even many industrial units in rice producing areas are using rice husks as energy inputs. Rice straw and rice husks are available in plenty in rice growing countries.

These can be used to generate power. About 4500 kg of milled rice produce 900 kg of husk with energy content equivalent to 1900 kg of wood. Few rice husk fired power plants are in operation in South East Asia. India boasts of an annual yield of rice husk at around 36 million tonnes. It has been computed that by the thermal energy recovery process, it would be possible to generate about 1000 MW of electricity from rice husk in Punjab alone.

The production of alcohol by fermentation of sugars and starch is an ancient art and is often recognised as the first microbial process used by human beings.

$$C_6H_{12}O_6 \longrightarrow 2C_2H_5OH + 2CO_2$$

A flow diagram for the production of ethanol from diverse substrate is shown in Fig. 4.2.

To make the transport sector sustainable and to minimise pollution, ethanol has been successfully blended in motor spirit and diesel in many countries for decades. A blend of 10 per cent ethanol in MS (gasohol) and 15 per cent in diesel (diesehol) are very effective blends, without any need for change in engine design in transport vehicles. Ethanol has high knock resistance as gasoline component and a positive effect on combustion quality. As against a total Indian consumption of 6 billion litres of petrol per year, the consumption of bioethanol as fuel/oxygenate in Brazil is 12 billion litres, in USA 4.7 billion litres and in France 120 million litres. Gasohol has been in use by more than 35 countries world over sometimes or the other for over half a century. India was also using ethanol during IInd World War under the name power alcohol.

Recently there has been a considerable interest in Europe and USA in the use of modified rapeseed or soya oil as a diesel substitute: biodiesel. This product is obtained by reacting vegetable oil and methanol in presence of a catalyst at 50°C producing an ester and glycerol:

Rapeseed oil	+	Methanol	$\longrightarrow$	Diester	+	Glycerol
(1 tonne)		(0.1 tonne)		(1 tonne)		(0.1 tonne)

The glycerol is allowed to settle and the biodiesel purified and used as transport fuel. The biodiesel has physical and chemical properties close to diesel and heating oil. It has been reported that the use of biodiesel does not require any specific engine modification. Biodiesel is non-toxic and more than 98 per cent biodegradable. Its contribution to the green house effect is three to five times less than that of diesel and above all, it is renewable.

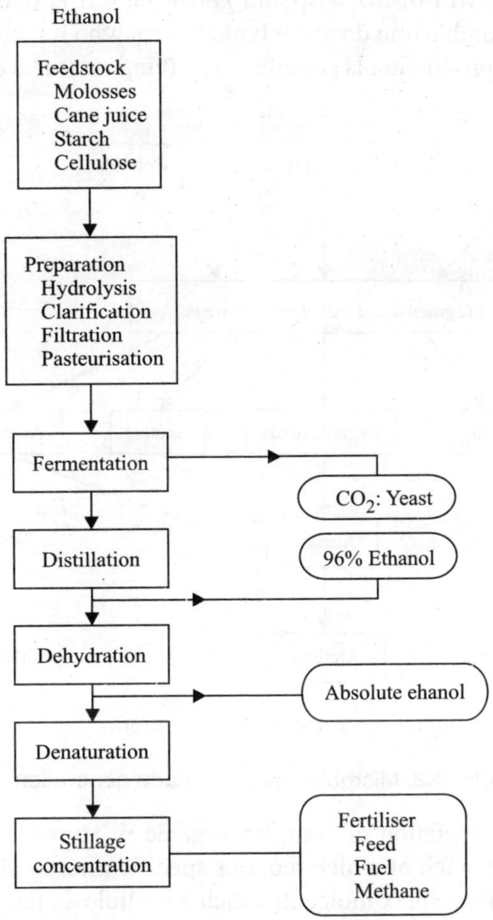

Fig. 4.2. Flow diagram for the production of ethanol.

France, Italy, Germany and USA have been the leading proponents of biodiesel only for environmental reasons. Biomass gasifiers convert solid biomass materials such as wood, agricultural wastes etc., into a producer gas through a thermochemical gasification process.

Producer Gas

The producer gas so produced, could be burnt directly for thermal applications. Methane gas exists in the atmosphere and is derived mainly from microbial action in natural wetlands, paddy fields and enteric fermentation in animals contributing about 20 per cent, 20 per cent and 15 per cent respectively, to the total methane flux. Domestic cattles are the major contributors, producing about 75 per cent of all animal emissions, where as humans produce about 0.4 per cent. After CO_2 methane is considered to be the most important green house gas. Methane 'fermentation' or biomethanogenesis, is an ancient process for the conversion of biomass into energy. It was discovered in 1776 by *Volta*, who demonstrated the presence of methane in marsh gas. When methane is produced by fermentation of animal dung, the gaseous products are usually referred to as 'biogas' and the installations called biogas plants (also called Klar-gas, refuse-derived fuel, RDF, will-of-the-wisp and gobar gas). It is odourless and burns with a blue flame without smoke and its combustion does not bother users who habitually burn firewood and dung. The microbiology of methane production is complex, involving mixtures of anaerobic micro-organisms (Fig. 4.3).

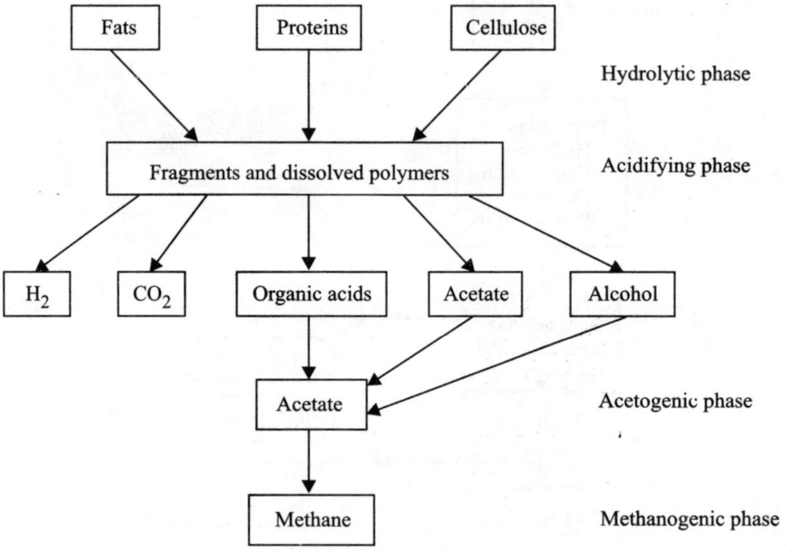

$$C_6H_{12}O_6 \longrightarrow 3CH_4 + 3CO_2 + 20 \text{ kcal}$$

Fig. 4.3. Microbiology of methane generation.

In principle, anaerobic fermentation of complex organic mixtures is believed to proceed through three main biochemical phases, each of which requires specific microbiological parameters. The initial stage requires solubilisation of complex molecules such as cellulose, fats and proteins, which make up most raw organic matter. The resultant low molecular weight soluble products are then converted to organic acids. In the final phase of microbial activity, these acids (primarily acetic) are specially decomposed by the methonic bacteria to methane and CO_2.

Biomethanation is a self-regulating symbiotic microbial process operating under anaerobic conditions, and functions best at temperature around 30°C. The organisms involved are found naturally in ruminant manures. Under ideal conditions, 10 kg of dry matter can produce 3 cubic metre of biogas, which can provide 3 hours of cooking. Biomass is renewable and ever replenishing and derived from inexhaustable source—the sun. In the age of diminishing natural resources, biomass will sure to be the ultimate energy solution of human civilisation.

REMOVAL OF COLOUR FROM INDUSTRIAL EFFLUENTS BY USING WASTE BIOMASS AND REAGENT LOADED POLYURETHANE FOAMS

Problem of colour removal is mainly confined to industrial waste-waters in general. Use of activated carbon is common in such cases but proves to be expensive. It will be useful if alternative adsorbents that are less expensive can be found. The present study is one such attempt in search of inexpensive adsorption materials.

Such a material is found in the continuously accumulating fungal biomass from *Aspergillus niger* used in the manufacture of citric acid. The dead biomass is dried in an oven to remove all moisture and sized into different particles sizes.

Adsorption studies were conducted on two basic dyes i.e., basic blue and malachite green and two acidic dyes i.e., fast red and congored. Among the basic dyes, good removals of more than 7.4 per cent w/w (98.8 per cent with 1500 ppm of initial concentration of dye) are obtained with basic blue while malachite green gave only 0.14 per cent w/w (98.6 per cent with initial concentration of only 300 ppm). Among the acidic dyes, fast red gave very good results with 6.0 per cent w/w removals (98.7 per cent with initial concentration of 1200 ppm) while congored gave only 2.0 per cent w/w (85 per cent with only 500 ppm).

Another recently developed low cost adsorption material is reagent loaded polyurethane foams which showed considerable promise. Loading the polyurethane foam with methyl-tri-octyl ammonium chloride gave reasonably good results with respect to the two acidic dyes with about 2.0 per cent w/w removals (around 87–88 per cent removals with initial concentration of 500 ppm).

Results are presented on studies on both biomass and loaded polyurethane foam with varying parameters generally used in all normal adsorption studies.

Large number of textile industries were established in several developing countries like India in the last two decades, which can lead to significant problems in discharging coloured aqueous wastes. The treatment processes like settling and clarification to remove dissolved dyestuffs and other chemicals, involve exposed surface areas for long periods in a hot climate incurring substancial evaporation losses of water. Adsorption processes produced high quality treated effluents but most systems use activated carbons, which are expensive and require additional costly regeneration plant. So there is a need to search low cost adsorbents for dye removal.

Preparation of Adsorbents

Biomass

Dead biomass (*Aspergillus niger*) was collected from citric acid plant, Biocon India (Pvt.) Ltd., Bangalore. The biomass was dried in sunlight and were powdered using a grinder. The powder was sieved according to the particle size ranges of 300–600, 600–710 and 710–850 μ.

Aliquat loaded polyurethane (AQ-PUF) foams

The rigid polyurethane foam was obtained from 'U-foam' India Pvt. Ltd., Hyderabad. The rigid polyurethane foams were made into small pieces of about 2 mm size and soaked in 2 M HCl for 2 hours, then washed with distilled water for several times until the excess chlorides were removed and finally washed with acetone and dried in air. To 5 gm of this polyurethane foam pieces, 1 ml of aliquat (cationic surfactant in 49 ml of chloroform solution) was added and then dried in air until all the chloroform was evaporated.

Coloured organics

The concentration of the colouring matter was determined by UV-240 Shimadzu Spectrophotometer, at wavelength of maximum absorbance for each dyestuff.

Procedure for Batch Mode Adsorption Studies

Aqueous solutions containing the dye were prepared in distilled water. Effects of agitation time, initial dye concentration and particle size on adsorption of dye were studied by taking 50 ml of 100 ppm dye solution and 1g adsorbent (waste biomass or reagent loaded PUF) in each glass bottle and agitating for pre-determined time intervals at 200 strokes/min, using a Remi-shaking water bath. The adsorbent and adsorbate were separated by centrifugation and the amount of dye adsorbed was measured spectrophotometrically and Langmuir isotherms were determined from the equilibrium data. The batch experiments were carried out with two basic dyes (basic blue and malachite green) and acidic dye (fast red and congo red) each with three different particle sizes (300–600 μ, 600–710 μ and 710–850 μ) of biomass. In the case of AQ loaded PUF, since it has cationic surface only acidic dyes (fast red and congo red) were examined for their sorption process.

Results and Discussion

Effect of initial concentration

The sorption capacities of waste biomass with both acidic and basic dyes and aliquat loaded PU (AQ-PUF) foam with acidic dyes were tested under different conditions. The equilibrium dye concentration varied between 10 ppm to 500 ppm while the absorbed capacities varied from 0.325 to 74.1 mg/g, with both the sorbents. The convex shape of the isotherms indicate the favourable sorption nature of the process. The adsorption capacity of biomass is found to be for congo red, 20.87 mg/g, for fastred, 59.25 mg/g, for malachite green, 14.80 mg/g and basic blue, 74.1 mg/g which is found to increase with decreasing particle size due to the increase in surface area of the adsorbent. With AQ-PUF, it was found to be 21.9 for fast red and 22.05 for congo red (Figs. 4.4a and 4.4b).

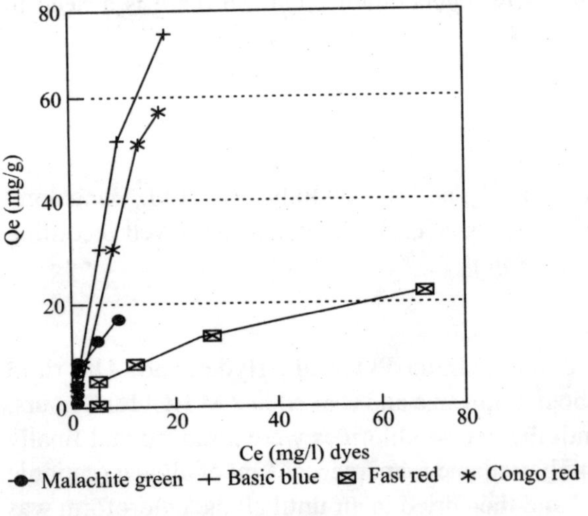

Fig. 4.4a. Adsorption capacity of two different acidic and basic dyes by biomass.

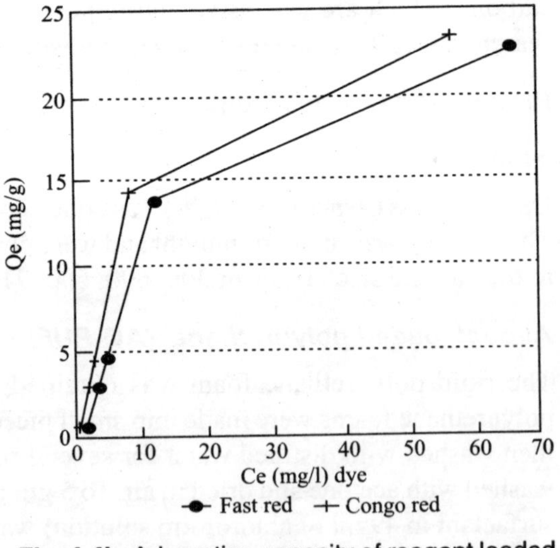

Fig. 4.4b. Adsorption capacity of reagent loaded PU foam for acidic dyes.

Langmuir isotherms

The monolayer nature of the adsorption process with biomass is indicated when 1/Ce Vs 1/Qe was plotted (Figs. 4.5a and 4.5b) as per langmuir equation (4.1).

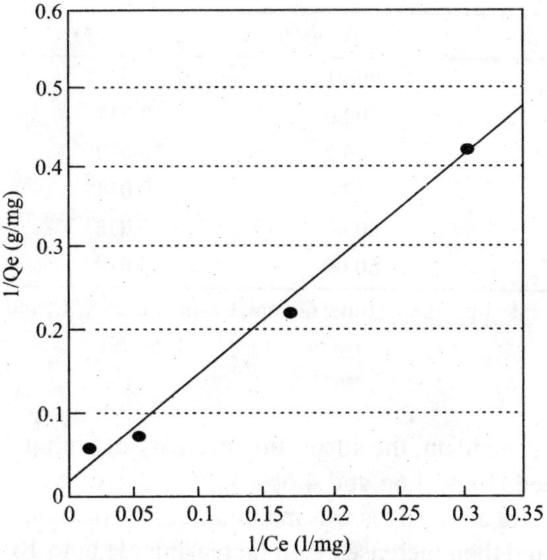

Fig. 4.5a. Langmuir adsorption isotherm for the adsorption of fast red by PU foam.

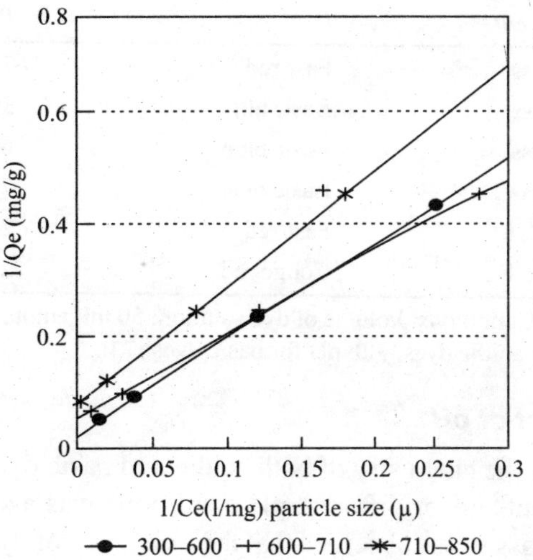

Fig. 4.5b. Langmuir adsorption isotherm for the adsorption of congo red by biomass of different particle size.

$$\frac{Ce}{Qe} = \frac{1}{Qo\,b} + \frac{Ce}{Qo} \quad \text{... (4.1)}$$

Qe = amount of adsorbate in mg adsorbed per gram of adsorbent.
Ce = equilibrium dye concentration present in liquid phase.
Qo = Langmuir constant (adsorption capacity).
b = Langmuir constant (energy of adsorption).

The corresponding Langmuir constant Qo (adsorption capacity) and b (intensity of adsorption) values were calculated for each process and presented in Table 4.1. The dimensionless separation factor R (equilibrium parameter) which indicates the favourable and unfavourable nature of the sorption process was also tested from the Langmuir isotherm data and these values are in between $0 < R < 1$, indicating high favourability of the sorption process for both the adsorbents.

Table 4.1. Langmuir constants for the adsorption of dyes by biomass and AQ-PUF.

Adsorbent	*Dye*	*Particle size (μ)*	*Qo (mg/g)*	*b (l/mg)*
Biomass	Congo red	300–600	34.48	0.018
Biomass	Congo red	600–710	25.64	0.026
Biomass	Congo red	710–850	16.94	0.028
Biomass	Malachite green	300–600	33.33	0.300
Biomass	Malachite green	600–710	26.00	0.333
Biomass	Malachite green	710–850	25.00	0.100
Biomass	Fast red	300–600	200.0	0.025
Biomass	Fast red	600–710	133.7	0.024

Adsorbent	Dye	Particle size (μ)	Qo (mg/g)	b (l/mg)
Biomass	Fast red	710–850	90.91	0.025
Biomass	Basic blue	300–600	200.0	0.034
Biomass	Basic blue	600–710	166.7	0.030
Biomass	Basic blue	710–850	143.0	0.044
AQ-PUF	Fast red	2 mm	20.00	0.038
AQ-PUF	Congo red	2 mm	50.00	0.019

Note: Conditions: Volume of dye solution: 50 ml; amount of absorbent : 1 g; contact time: 60 min; temperature: 30°C; and pH for acidic dyes: 3.0; pH for basic dyes: 7.0.

Effect of pH

Since the chemistry of both acidic and basic dyes are pH dependent, the adsorption capacity of both the sorbents under different pH conditions was also examined (Figs. 4.6a and 4.6b). In the case of waste biomass it was observed that the efficiency of dye removal for acidic dyes was maximum in the pH range 1–3, and then decreases with increasing pH upto pH 6.0 and then increases with increasing pH upto 10. In the case of basic dyes, there is a continuous increase in the removal of the dye with increase in pH, and the maximum removal was observed in the pH range 6–10.

The mechanism of the adsorption process can be explained considering the role played by the binding sites of the chitin/chitosan components of the fungal cell walls of the biomass in which the amine group with a lone electron pair will be available for co-ordination. Representing the chitin/chitosan monomer unit as CN:, the dissociation equation of the amine can be written as:

$$CN: + H_3O^+ \longleftrightarrow CN^- H^+ + H_2O \qquad \text{... (4.2)}$$

and that for acidic dye would be:

$$HA + H_2O \longleftrightarrow H_3O^+ + A^- \qquad \text{... (4.3)}$$

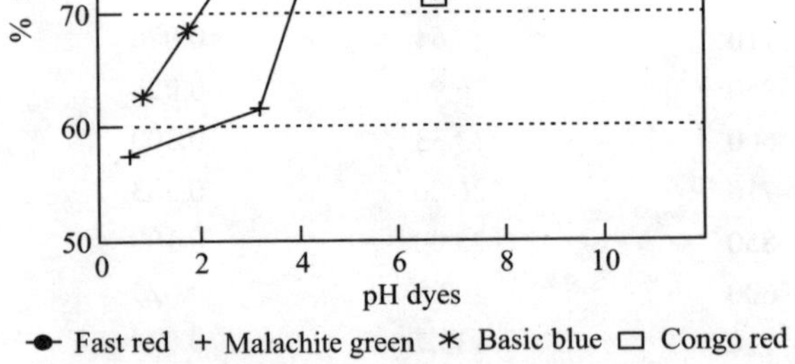

Fig. 4.6a. Effect of pH for the removal of two acidic and basic dyes by biomass.

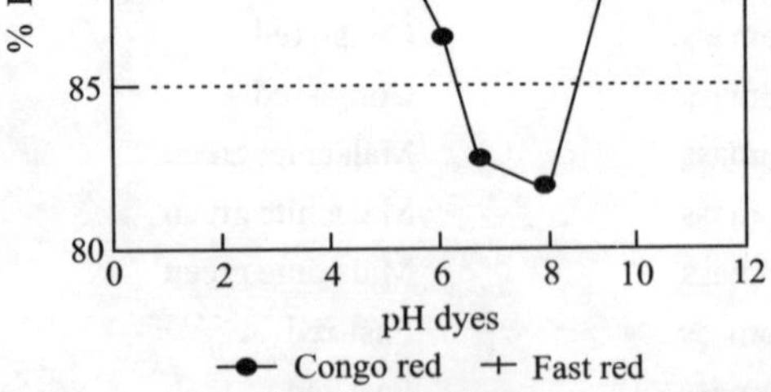

Fig. 4.6b. Effect of pH for the removal of acidic dyes by AQ-PUF foam.

Increasing pH shifts the equilibrium of the first equation to the left and that of the second equation to the right. This would lead to increased concentrations of A^- and CN:, both of which are negatively charged; the repulsive forces that result are reflected by the diminished uptake of acidic dye, with increasing pH. However, in the case of basic dyes, at higher pH value, the association of dye cations, with negatively charged oxide surfaces of metal ions like calcium, silica, aluminium, iron etc., likely to be present in the fungal cells may take place as shown below:

$$-\text{MOH} + \text{OH}^- \longleftrightarrow \text{MO}^- + \text{H}_2\text{O} \quad \text{... (4.4)}$$

$$-\text{MO}^- + \text{Dye}^+ \longleftrightarrow \text{M–O Dye} \quad \text{... (4.5)}$$

In addition to the oxides of these metal ions, the physical adsorption, by cell wall may also be playing a major role with both acidic and basic dyes above pH 6.0.

With AQ-PUF the concentration of dye removal was found to be maximum in between 2–4 pH and slightly decreases in the pH range 4–8 for both the adcidic dyes (fast red and congo red) and then increases. The dissociation of acidic dyes can be respesented by equation 4.6, while the nature of the surface of AQ–PUF can be shown as equation 4.7.

$$\text{HA}^+ + \text{H}_2\text{O} \longleftrightarrow \text{H}_3\text{O}^+ + \text{A}^- \quad \text{... (4.6)}$$

$$\text{RNH}_4\text{Cl} \longleftrightarrow \text{RNH}_4{}^+ + \text{Cl}^- \quad \text{... (4.7)}$$

In the between 2–4 pH, the equilibrium of the equations both 4.6 and 4.7, will be shifted more to right, resulting greater sorption of dye anion by the cationic amine surfaces offered by the AQ-PUF by simple electrostatic attraction. However, above pH 4.0, simple solubility of the dye molecules in the organic reagent quoted surface of the PU foam, may be taking place by simple solvent extraction mechanism, resulting less influence of pH, of the sorption of the acidic dyes.

Effect of contact time

The adsorption capacity of each sorbent was followed by varying contact time was studied. With AQ-PUF, maximum sorption capacity reached within five minutes, whereas with biomass, minimum 60 minutes is required, indicating the chemisorption is the main mechanism of the sorption in the former case, whereas in the latter case, both surface adsorption and chemisorption may be working.

Effect of temperature

There is no change in sorption capacity with the change in temperature from 30°C to 60°C for AQ-PUF, indicating stronger binding nature of chemisorption nature, where as with biomass, the sorption capacity decreased with increasing temperature from 30°C indicating the endothermic nature of the adsorption process. The thermodynamic parameters were calculated and presented in Table 4.2. For both acidic and basis dyes positive values of H° were obtained indicating the endothermic nature of adsorption process, while negative values were obtained for G° indicating the fastness of the adsorption process. The positive values of S° show increase randomness at the solid-solution interface during the adsorption of dyes on biomass.

APPLICATION OF WASTE BIOMASS FOR THE REMOVAL OF COLOUR FROM TEXTILE AND PAPER MILL EFFLUENTS

Textile industry with an annual production of about 3960 million metres of cloth is one of the biggest in our country. The waste-waters resulting from these mills contain considerable amounts of different pollutants due to which it is not possible to discharge the untreated waste either into a waste course or into a municipal sewer without causing damage. Many dyes used in the textile industry are particularly difficult to remove by conventional waste-water treatment methods, because they are stable to light and oxidising agents and are resistant to aerobic digestion.

Table 4.2. Thermodynamic parameters for the adsorption process using waste biomass.

Dye	*Temperature °C*	K_c	G^o *(kJ/mol)*	H^o *(kJ/mol)*	S^o *(J/mol/k)*
Malachite green	30	19.10	–7.43	0.640	373.400
Malachite green	40	19.20	–7.69	0.640	373.400
Malachite green	50	19.32	–7.95	0.640	373.400
Malachite green	60	19.49	–8.24	0.640	373.400
Congo red	30	11.20	–6.09	0.014	30.444
Congo red	40	14.23	–6.92	0.014	30.444
Congo red	50	15.66	–7.40	0.014	30.444
Congo red	60	27.57	–9.19	0.014	30.444
Basic blue	30	165.70	–12.90	0.253	1.005
Basic blue	40	110.10	–12.23	0.253	1.005
Basic blue	50	89.91	–12.08	0.253	1.005
Basic blue	60	70.43	–11.80	0.253	1.005
Fast red	30	99.00	–11.60	0.132	2466.630
Fast red	40	82.33	–11.50	0.132	2466.630
Fast red	50	64.80	–11.20	0.132	2466.630
Fast red	60	52.20	–10.95	0.132	2466.630

Note: Conditions: Amount of adsorbent: 1 g; contact time: 60 min; particle size 300–600 μ; volume of dye solution 50 ml; initial concentration of congo red: 50 ppm; initial conc. of basic blue 100 ppm; initial concentration of malachite green: 50 ppm; initial conc. of fast red 50 ppm; pH for acidic dyes: 3; and pH for basic dyes: 7.

Most of the paper mills in the India are integrated pulp and paper mills and employ sulphate process of pulping. The sources of effluents are wood washings, digestion process, pulp washings, chemical recovery and paper machine of which the pulp mill effluent is the strongest. In this section applications of waste biomass (*Aspergillus niger*) as an adsorbent for the removal of colour/COD from textile and paper and pulp mill effluents are discussed. The adsorption of the coloured material in the effluent on the waste biomass is monitered by following the change in the COD value of the effluent, by batch studies. Langmuir isotherms were plotted from the data obtained by batch process, and the constants are given in Table 4.3. The effluent characteristics before and after treatment were given in Table 4.4.

Table 4.3. Langmuir constants.

Effluent	*Qo (mg/g)*	*bl/mg*
Textile effluent (with biomass)	2.9422	0.07553
Pulp mill effluent (with biomass)	0.764	0.5275
CETP effluent (with AQ-PUF)	1666.7	0.00041

Note: Conditions: Volume of effluent: 50 ml; temperature: 30°C; pH: 10; initial COD of textile effluent: 1400 ppm; COD of paper and pulp mill effluent: 800 ppm; and COD of CETP effluent: 2400 ppm.

Table 4.4. Characteristics of textile and paper and pulp mill effluent before and after treatment.

Parameter	*Textile effluent*		*Paper and pulp mill effluent*	
	Before treatment with biomass	*After treatment with biomass*	*Before treatment with biomass*	*After treatment with biomass*
pH	10	8.5	8.5	6.8
Suspended solids	950 mg/l	260 mg/l	1115 mg/l	100 mg/l
BOD	820 mg/l	30 mg/l	239 mg/l	30 mg/l
COD	1400 mg/l	250 mg/l	1087 mg/l	210 mg/l

M/L Values for Batch Adsorber

In any batch adsorber, the effluent to be treated and the adsorbent are intimately mixed in the treatment tank for a said time to enable the system to approach equilibrium, following which the thin slurry is filtered to separate the solid adsorbent and adsorbate from the liquid. The equation (4.8) permits calculation of the adsorbent-solution ratio for the given change in effluent concentration, Co-ce.

$$M/L = \frac{\text{Co-Ce}}{\frac{\text{K Ce}}{1+\text{aCe}}} \quad \text{... (4.8)}$$

The M/L values for the reduction of the initial COD value of textile effluent of 1400 ppm and pulp mill effluent of 1087 ppm, to different lower levels are calculated and plotted in Fig. 4.7, which can be used to predict the design parameters of single stage batch reactor.

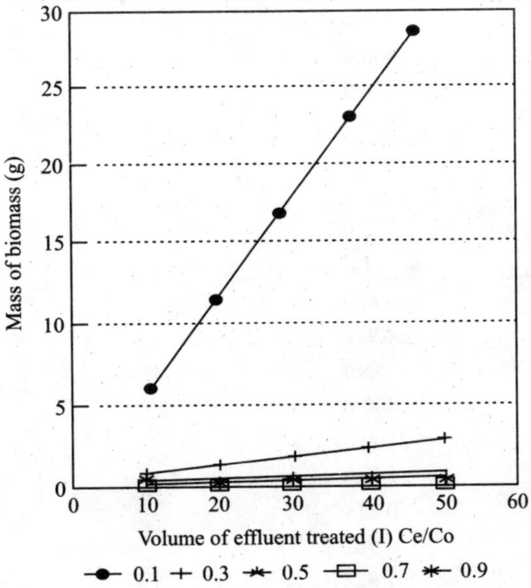

Fig. 4.7. M/L values for the removal of COD from paper and pulp mill effluents.

Application of AQ-PUF for the removal and COD/colour from finally disposed effluent of common effluent treatment plant

The small scale industries find it difficult to treat the waste themselves due to shortage of land and finance required for construction of individual treatment plant. In view of this a common effluent treatment plant 'Jeedimetla Effluent Treatment Limited' (JETL) was designed and implemented in this industrial estate.

Though considerable reduction in BOD values of the effluents is achieved in treatment plant, it is observed that the finally disposed effluents posses high colour and the applications of AQ-PUF for the removal of COD/colour from the finally disposed effluent was tested by batch studies.

The maximum removal of COD was achieved in the pH range from 3.5–4.5. However, even at 8.0 pH and above, 80 per cent removal of colour was observed and all the further experiments were carried out at 8.0 pH without any further treatment to the effluent. The analysis of equilibrium data for the sorption of the colour from finally disposed effluent was examined in the light of Langmuir isotherm model and the Langmuir constants were given in Table 4.3.

The M/L values for the reduction of the initial COD values of 2400 ppm for finally disposed effluent of CETP to different lower levels were calculated and plotted in Fig. 4.8, which can be used to predict the design parameter of single stage batch reactor. The characteristics of the CETP effluent before and after sorption studies were given in Table 4.5.

Table 4.5. Characteristics of JETL effluent after treatment.

Parameter	*Value in mg/l except pH*	
	Before treatment with AQ-PUF	*After treatment with AQ-PUF*
pH	8.0	7.00
Suspended solids	400.0	100.00
Hardness as $CaCO_3$	720.0	220.00
Chlorides	160.0	40.00
Sulphates as SO_4	1780.0	80.00
Nitrates	200.00	20.00
Phosphates	13.0	3.00
Fluorides	1.0	0.30
BOD (5 days at 20°C)	800.0	30.00
COD	2400.0	40.00
Dissolved oxygen (DO)	Nil	3.00
Acidity as $CaCO_3$	400.0	40.00
Alkalinity	2800.0	80.00
Sodium	840.0	40.00
Lead	0.319	0.09
Cadmium	0.902	0.02
Chromium	1.824	0.08
Copper	1.784	0.04
Zinc	1.885	0.89

(Contd...)

Parameter	*Value in mg/l except pH*	
	Before treatment with AQ-PUF	*After treatment with AQ-PUF*
Nickel	1.818	0.18
Cobalt	0.174	0.04
Iron	4.960	0.60
Bismuth	9.702	0.70
Antimony	2.280	0.28
Manganese	0.849	0.05

Values are an average of three determinations.

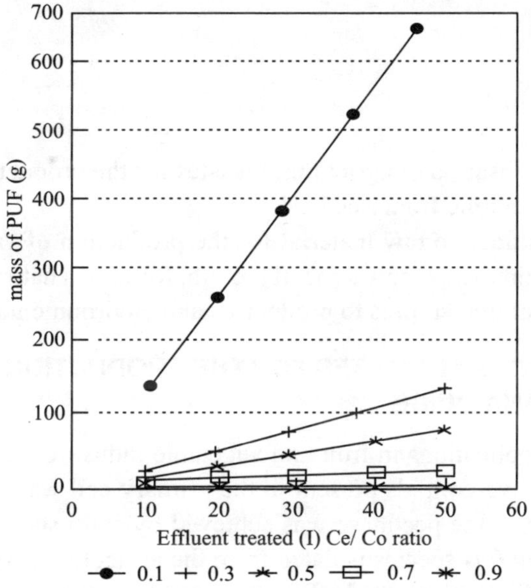

Fig. 4.8. M/L values for the design of batch reactor for the removal of COD by AQ-PUF foam.

Cost Analysis

Many factors like pH, temperature, initial dye concentration influence the rate and extent of dye removal by various adsorbents. Consequently a comparison relative economic evaluation of the process with different adsorbents is a complex problem. However, a brief comparison can be made between biomass, AQ-PUF and the activated carbon, the most widely used adsorbent, by performing isotherm studies under similar conditions. The adsorption process using biomass is much cheaper than activated carbon. Although activated carbon may be re-used four to five times after regeneration, biomass would still prove to be cheaper without regeneration, because dead biomass is a waste product obtained from food industries. Though the cost of AQ-PUF and activated carbon are of the same order, the reagent loaded PUF can be re-used upto six times with high flow rates by column process, without recharging and the recharging process is also easy. Whereas the desorption with activated carbon is very complex in nature. With these advantages, aliquat loaded PUF finds potential application for the efficient treatment of industrial effluents to reduce to pollutant levels to the desired values.

5

Enzyme Pectinase from Agricultural Waste and Ceramic from Rice Husk

INTRODUCTION

This chapter highlights the utilisation of agricultural wastes for the production of enzyme pectinase for extracting banana juice and ceramic from rice husk.

Agricultural wastes are attractive raw material for the production of value added products as they contain all the necessary requirements for microbial growth, forming a natural means of waste recycling. This approach puts the agricultural surplus to productive and environmentally safe use.

UTILISATION OF AGRICULTURAL WASTES FOR THE PRODUCTION OF ENZYME PECTINASE FOR EXTRACTING BANANA JUICE

Pectic enzymes have wide applications in fruit and vegetable industries. They are utilised to eliminate pectin and pectin like protective colloids present in the primary cell wall of the most plants and plant productions. Production of enzyme pectinase was achieved by solid state fermentation in presence of 0.1 per cent pectin by *Aspergillus* species isolated from the agricultural wastes. Isolation of the native organisms was achieved by *Blotter's* methods. Effect of supplementation of molasses, pectin and yeast extract on the effluent for maximal enzyme production was studied. 0.1 per cent pectin was found to be optimum to induce enzyme production. The crude extract showed a pH optima of 5.2 and a temperature optima of 40°C. Crude extract of enzyme could produce 2.5 litres of crystal clear banana juice having 22 per cent sugar content starting from 3.5 kg banana pulp.

The interest which now-a-days exists for environmental biotechnology, as well as the belief that its development is of crucial relevance for the entire technological development is based on several factors: the use of surplus renewable sources, economical processing etc. The use of agricultural crop wastes is of great importance as it is related to solving some ecological problems and hence protecting the environment. The agricultural wastes form a natural medium for the growth of various organism under solid state bio-processing conditions. Solid ground is the source of various mineral salt and trace elements needed for the development of micro-organisms. Low water content of such substrate is a selective factor for the growth of various moulds. Some of the moulds that grow under such conditions produce certain useful enzymes. Some of the enzymes produced (e.g., pectinase) find use in food industry. Most pectin in unripened fruits is present as protopectin. During maturation and ripening this protopectin is transformed into soluble pectin, leading to fruit softening. Protopectin and other partially soluble pectic substances confer cloudy properties to the fruit extracts. Pectinases break the complex acidic polysaccharides pectin and pectin like colloids present in the cell walls of most plant and plant products. This study reports the isolation of pectinases from *Aspergillus* grown on agricultural wastes under solid state bio-processing conditions.

Materials and Methods

Strain

The organism used for pectinase production was *Aspergillus* species isolated from the agricultural wastes. The native organism was isolated by Blotter's method.

Culture medium

About 25 gm of crop waste chopped into 5–10 mm particles after removing the skin and leaves and mixed with 12 ml of mineral solution. (Composition of mineral solution was 0.007 per cent of $ZnSO_4.7H_2O$, $FeSO_4.7H_2O$ and $CuSO_4.5H_2O$ in 0.2 N HCl). It was also supplemented with 0.1 per cent pectin to induce the enzyme formulation and various concentrations of nutrients like molasses and yeast extract. The moisture was maintained at 50 per cent. It was measured gravimetrically as a difference between a fresh weight and after 24 hour of sample dried at 105°C.

Experimental conditions

The raw materials are inoculated with the native *Aspergillus* strain and incubated at 40°C for 7 days under solid state processing conditions. It was then dried at 60°C in an oven and ground into a paste using distilled water. The enzyme was extracted with distilled water and filtered under vacuum. The crude enzyme complex was then applied to extract juice from banana.

Fruit juice extraction

3.5 kg banana fruit was washed, deskinned and then crushed for processing. The mash was heated to 60°C and 1.05 grams potassium metabisulphite was added to stop the oxidation and browning of the fruit extract. The extract was cooled to 40°C. 100 ml of the enzyme was added and incubated for 48 hours. The banana pulp was pressed to recover the juice which was centrifuged to remove the suspended particles. The juice was then decolourised with activated charcoal treatment and then filtered to recover at 2.5 litres of crystal clear juice (Fig. 5.1).

Results and Discussion

Biosynthesis of *Aspergillus* pectinase enzyme by solid state fermentation from agricultural wastes is a simple and low energy consuming process. It is based on the utilisation of cheap substrates, crop wastes combined with simple salt solution and growth nutrients. Crude enzyme extract showed an activity of 280 U/ml, when supplemented with 0.1 per cent pectin, 1 per cent molasses and 0.5 per cent yeast extract. Further purification of the pectinase enzyme from the crude extract was not attempted before its application on the fruit extract as different activity ratios of other contaminating enzymes such as cellulases and hemi-cellulases were also observed. Addition of 0.1 per cent pectin was sufficient to induce the production of enzyme. Supplementation with molasses and yeast extract aided in the growth of *Aspergillus* species pectinase showed its maximal activity at pH 5.2 and temperature 40°C. For high production of enzyme, it was observed that the process moisture content has to be less than 50 per cent. Solid state fermentation on agricultural wastes is therefore advantageous due to usage of less water in the process.

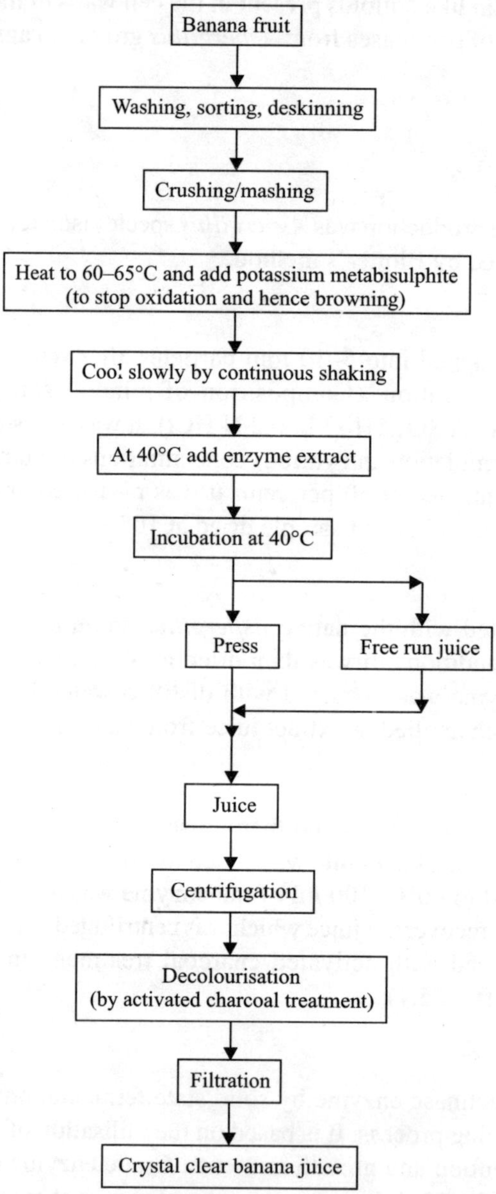

Fig. 5.1. Extraction of juice from banana fruit.

Hence, lower amounts of liquid waste is present, so lesser disposal treatment problem. The crude enzyme extract when applied to 3.5 kg banana pulp yielded 2.5 litres of banana juice. The juice had 22 per cent sugar content which is quite high in comparison with the sugar content of sugarcane juice which was found to be 10 per cent. When banana fruit is crushed during processing, some of the pectin appears in either soluble or insoluble form in the juice and this affects the quality or yield of the juice which is recovered. Pectinases modify this pectin material. This is done with a view to improve both the yield of juice extract and its quality. The addition of this enzyme resulted in rapid reduction of the viscosity of the juice and simultaneously removal of turbidity.

Thus the state bio-processing of the agricultural crop wastes form an alternative and economical process for the production of enzyme pectinase. It serves dual function: the removal of unwanted substances and production of value added compounds. The enzyme produced by the native *Aspergillus* yielded about 200 ml juice from three bananas. The juice was found to be highly nutritious having 22 per cent sugar content with essential minerals and vitamins. Extraction of crystal clear juice from banana is possible only through this enzymatic procedure.

CERAMIC FROM RICE HUSK

Pottery making is one of the most interesting ancient industries developed by man. The ceramic industries are also called silicate industries.

The white-ash produced by burning rice husk contains about 95 per cent silica since silica is one of the basic raw materials in refractory, glass and ceramic industries. Silica used in ceramic industry is obtained from quartzite rocks, which are pulverised to fine powder. The temperature required for firing various ceramic bodies is in the range of 1000 to 1300°C. The main advantages of using rice-husk ash as a source of silicon are as follows:

1. It contains 95 per cent silica along with some minor ingredients like Al_2O_3, CaO, P_2O_5, Na_2O, K_2O, etc. It is obtained in finely divided form and does not require any further grinding.
2. As amorphous rice-husk ash (silica) is reactive in nature, firing temperature will be lower than those used in ceramic industry.
3. Production of ceramic bodies from rice-husk ash is expected to be cheaper.

It is now understood that sodium oxide reduces the melting point of silica, boric oxide provides high electrical resistivity, alumina increases the mechanical strength and electrical resistivity and calcium carbonate provides chemical durability.

Production of Ceramics

The process flow chart for making ceramic materials is shown in Fig. 5.2. The process developed on a laboratory scale is described as under:

$$V_m = \text{Volatile matter}$$

Mixing of raw materials

White ash (obtained from rice husk) was mixed with other desired raw materials in powder form according to the required quality of the ceramics. Mixing was done thoroughly to get a homogeneous mixture as far as possible to avoid non-uniform fusion while heating.

First firing

Each of the mixture composition taken in ceramic crucible was heated in a chamber-type electric furnace, at different temperatures ranging from 800 to 1000°C for 1 to 4 hours. The purpose of this firing study was to get a suitable combination of time and temperature for each composition, so that all the volatile gases like CO_2, H_2O, etc., are released, the mixture is partially fused and the shrinkage in final firing is also reduced.

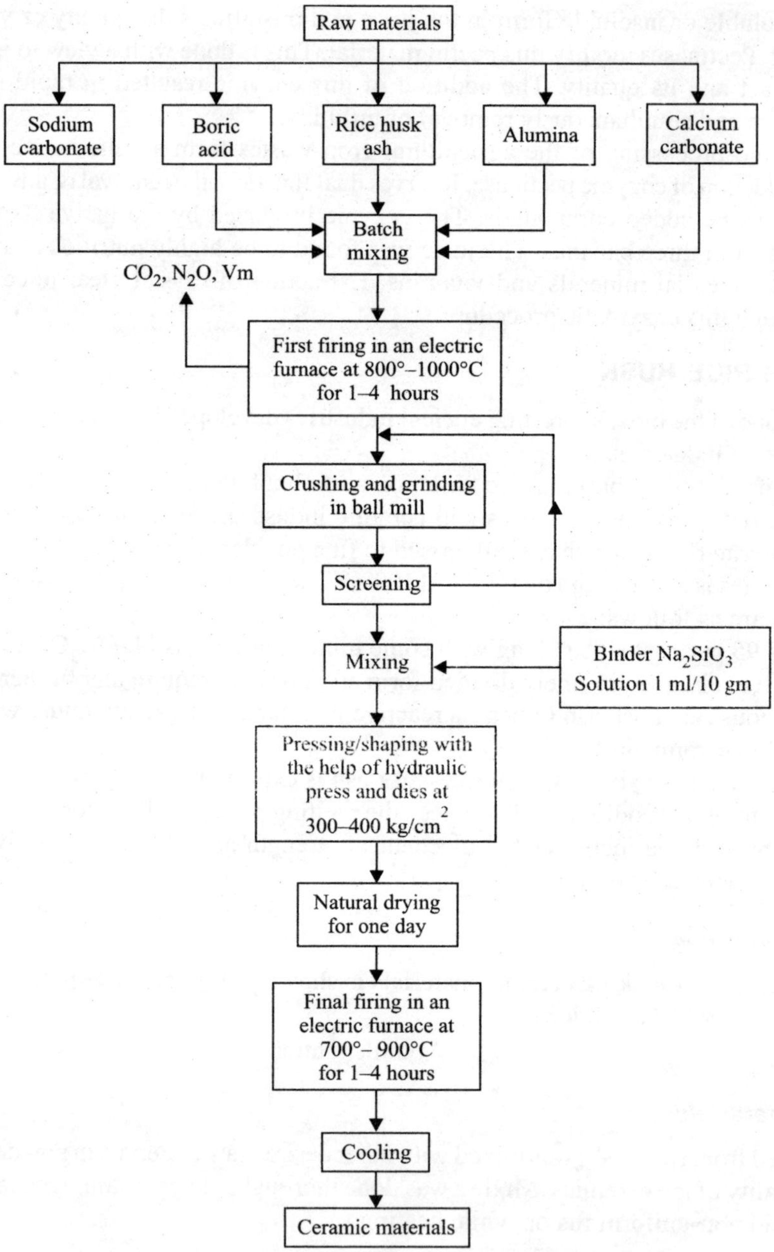

Fig. 5.2. Process flow chart for making ceramic bodies from rice-husk.

Crushing and grinding

Partially fused material obtained from first firing was subjected to grinding in a ball mill. If the fused material contains lumps of more than 0.5 cm size, then it is necessary to reduce first to 0.5 cm size.

Sieving

Ground materials were sieved through 212 micron sieve and particles less than 212 microns size only were used.

Mixing of binder

Sodium silicate solution of density 1.225 g/ml was added to the material already sieved, at a ratio of 1 ml/10 g of the material. It was mixed thoroughly to make it homogeneous.

Shaping

The mixture with the binder was pressed and given shapes with the help of hand operated hydraulic press and dies. The pressure applied was 300–400 kg/cm^2.

Drying

After giving the shapes, the specimens were dried in the sun for one day.

Firing of body

The bodies were fired in a closed electrically heated chamber at different temperatures from 700 to 900°C for 1, 2 and 4 hours.

Cooling of fired body

After firing the furnace was switched off and bodies were left in the furnace to cool down gradually.

Properties of the ceramics

Ceramic bodies using the rice-husk ash can be sintered at a much lower temperature of 700 to 900°C compared to 1100 to 1300°C required for conventional porcelain.

The densities of the prepared samples were between 1.3 and 2.1 g/cc compared to 2.2 to 2.5 g/cc for commercial porcelain.

Modulus of rupture of the samples were from 215 to 360 kg/cm^2, which were slightly lower than those of the industrial ceramics (200 to 600 kg/cm^2).

Electrical resistivity of the samples containing no Na_2O was extremely high, which was in the order of 7×10^{15} Ω-cm at room temperature and the material was very much suitable as an electrical insulator.

The material also had high chemical resistance towards water as well as mineral acids.

Considering the low temperature of sintering and other properties of the ceramic specimens prepared, it is claimed that the ceramics produced from rice-husk ash can be used for electrical insulators. It is also expected that tiles, potteries and white wares can be produced economically.

6

Paper, Paper Board and Furfural from Agricultural Waste

INTRODUCTION

Hand-made and small-scale paper industry can play a vital role in easing the growing shortage of paper. Formerly when one thought of manufacturing paper, one had to think of vast investments, involving crores of rupees. It is no longer so. Technology is now available for making paper in a small way and economically. India does not have vast forest resources which can supply raw materials for big paper mills. But large quantities of agricultural wastes are available from which paper pulp can be produced. It may not be economical to use these raw materials by sending them to big paper mills over long distances, but it is certainly possible to start small-scale and hand-made paper units based on them.

PAPER

Hand-made/Small-scale Paper

Though commonly known as hand-made paper industry, the processes involved in the manufacture of hand-made paper are not entirely based on manual labour. Manual labour is limited only to the formulations of sheets; in most other processes electrical power is used. All the machines used in unit operations are scaled down models used in big paper mills. If power is used for preparing sheets also, it is called a small-scale paper unit.

The escalation of prices on investment for setting up big paper mills and the acute shortage of raw materials have made mill-produced paper very costly. On the other hand, the low investment per annum, cheap raw materials (depending upon the local conditions in scattered areas), the availability of comparatively cheap labour in rural areas, and recent technological developments in machinery employed by modern hand-made and small-scale paper units have brought down the cost of production sharply. These factors can now play a significant role and hand-made and small-scale industry can stand very well in competition with big paper mills in quality and price in certain varieties of paper boards. Apart from these varieties, there are certain special varieties of paper which hand-made/small-scale industry alone can manufacture.

Some of them are the art and engineering drawing paper, permanent document paper, watermarked certificates for universities and commercial firms, high grade stationery, card sheets, greeting cards, album papers, filter papers, blotting paper and electrical insulation paper.

Agricultural Wastes as Raw Materials

Studies conducted at various research institutes in India have shown the possibility of using various agricultural wastes in combination with rags, waste paper, tailor cuttings, cotton wastes, etc., for manufacture of different types of papers and paper boards. Some of the agricultural wastes are given in Table 6.1.

Table 6.1. Agricultural wastes which can be used for manufacture of paper and paper boards.

Agricultural waste	*Remarks*
Jowar chaff	Not useful as feed for cattle
Cotton stalk	Not decomposed easily
Maize cobs after removal of grains	Not decomposed easily
Sunflower stalk	Difficult to decompose but with hard fibre components
Mature bajra stalk	Not useful as animal feed
Tapioca stalk	Thrown off as waste and not easily decomposed
Banana pseudostem	With valuable fibre but not used in industry
Bhendi stalk	Over matured ones are too fibrous to decompose
Papaya stem	Thrown away as waste, not easily decomposed
Mature napier grass stem	Not useful as cattle feed
Glyricidia pods	Available in plant after green mature salvage
Lantana weeds	Noxious thorny plant, difficult to eradicate
Mesta (defibred) stem	Cannot be used for further fibre extraction
Prosopis juliflora	Noxious thorny weed posing eradication problem
Parthenium hysterphorus Linn. (Congress weed)	Rapidly growing weed, difficult to eradicate
Sugarcane bagasse	If available as waste
Various cereal straws	Surplus available as waste

There are several processes used for making paper and paper boards by different industries. However, the 'soda process' is the most commonly used process in hand-made/small scale paper units. The process is shown in Fig. 6.1. It consists of the following steps.

Chipping and size reduction

The raw material is first dried and then cut into pieces. It is further cut into very small bits or ground in some grinding mill or pulveriser.

Digesting and washing

Digestion is generally done by cooking of the ground raw material. The digester is either electrically heated or is a boiler type using organic fuel. While cooking 2 to 3 per cent caustic soda is added. The cooking time varies with the type of raw material used.

Pulping and washing

The digested material is converted into pulp by beating in a Hollander type beater. The cooked material is washed well before being fed into beater.

While the charge is fed, waste paper or suitable good quality cellulosic material is also added in proper form and in suitable proportion. The beating operation lasts about 6 hours.

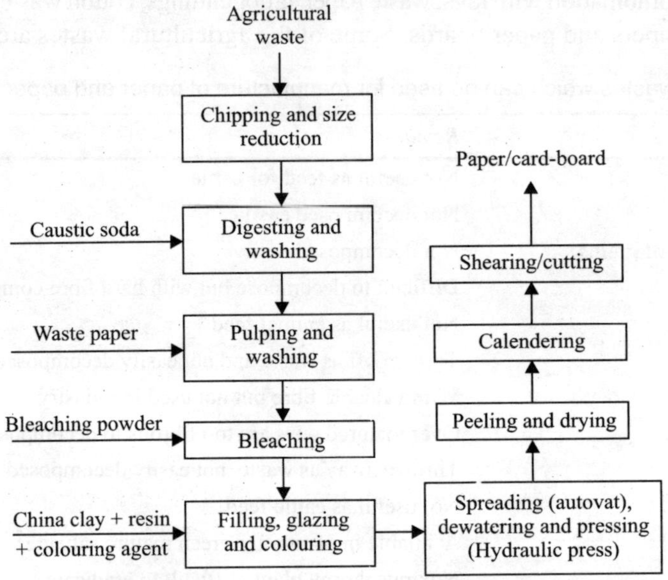

Fig. 6.1. Soda process for making paper and paper boards.

Bleaching, filling, glazing and colouring

The charge is thoroughly washed and then bleached by adding bleaching powder. For proper filling and glazing, chemicals like China clay, resin, alum, etc., are added and again beating is carried out for another 2 hours period. For coloured papers some colouring agent is added before further beating.

Lifting and couching

The pulp from the beater is transferred to a trough where the pulp is mixed with a sufficient quantity of water to make the suspension uniform and free from clogging. According to weight required for each sheet of paper, suspension is diluted with water and used. Autovats are used to prepare papers. A wire-mesh frame forms the base of the autovat. The pulp when poured over the mesh gets distributed evenly and excess water drains off. A cloth is spread over the evenly spread pulp and it is transferred to hydraulic or mechanical press.

Pressing

A number of layers are pressed in a mechanical or hydraulic press. Here the drainable water is squeezed out and the wet papers are ready for drying.

Drying and sizing

The sheets formed are dried along with the backing in the sun and are then peeled off from the cloth. For boards, they can be peeled off even under wet condition.

Calendering

The sheets are sandwitched between polished metal sheets and then rolled to and fro between calender rolls for removing the wrinkles and getting the desired smoothness.

Shearing

The papers/paper boards are finally cut to the desired sizes.

Recent innovation

A new vat for formation of sheets in place of existing autovats has been introduced in which more than 30 such machines are in use in India.

The vat consists of a cylinder mould rotating in a cement concrete tank on ball bearings. When the mould rotates, a thin layer of a continuous sheet is formed which is picked up by a woollen felt from the couch roll. The wet sheet is carried by woollen felt over a series of rollers, to the press part and passed between a top metal roller and a rubber roller, when the sheet gets wound over the metal top roller. The sheet is removed from the top metal roller and sent for drying.

The vat is manually operated or it can be run by a 2 HP motor. Any variety of paper/paper boards can be produced with this vat. In case of autovats, one has to depend on skilled and experienced persons, whereas semi-skilled persons can operate the cylinder mould vat to get uninterrupted production. The sheets are uniform and thickness variation is negligible. All edges are intact and the damages and wastages are minimum. The operation of the cylinder vat is easy and less labourious.

FURFURAL

Furfural is a very useful versatile aldehyde. It is used as a selective solvent for separating saturated compounds from unsaturated compounds for certain petroleum products. Furfural can be used as a basic raw material for the synthesis of furfuryl alcohols, nylon, butadiene, etc. Furfural can be produced from any agricultural residues, wastes and by-products, which are rich in pentosans such as bagasse, rice husks, cottonseed hulls, oat hulls, corn-cobs, flax shieves and other cellulosic wastes.

Methods of Producing Furfural from Agricultural Wastes

Zakoshchikon and Abramyants method

This method of production of furfural from rice husk is based on the use of more concentrated solution of H_2SO_4 (20–40 per cent). The ratio of the solid to acid solution is 4:8. The digestion process is carried out under atmospheric pressure. After the production of furfural, the sulphuric acid is regenerated. This method of furfural production is also suitable for sunflower hulls, cotton hulls, corn-cubs and some other cellulosic materials.

Fold's method

In this method, rice husks are hydrolysed in autoclaves, fitted in a thermostatically controlled rotor digester. The heating time is adjusted to 1.5 to 5 hour. The ratio of acid solution to husk is kept at 5:1. The optimum conditions for getting the maximum yield of furfural (which is about 53.4 per cent of pentosan content of rice husk) are as follows:

Concentration of sulphuric acid	=	6 per cent
Reaction temperature	=	150°C
Reaction period	=	2 hour

Regional research laboratory (RRL), Jorhat

The RRL (Jorhat) has developed a process for the production of furfural from the agricultural wastes, namely sawdust, corn-cobs, paddy husks, etc. The flow chart of the process is shown in Fig. 6.2. In this process, along with furfural a small quantity of methanol is obtained as a by-product and acetic acid is left in the residue.

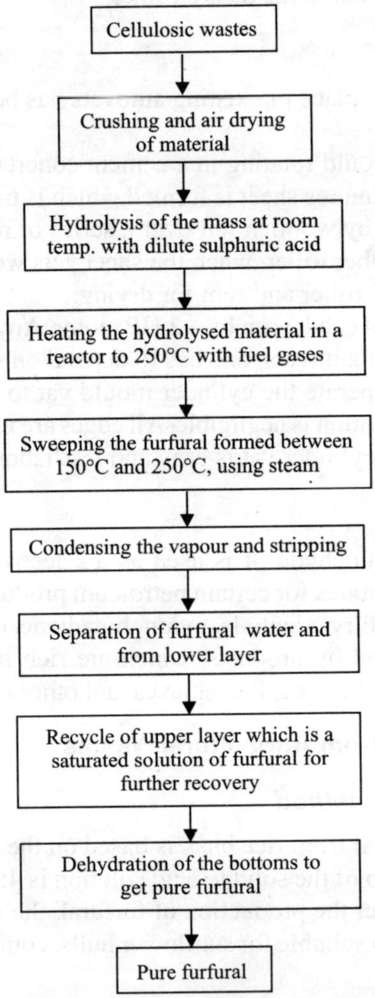

Fig. 6.2. Flow chart of RRL (Jorhat) process.

SECTION II

Food Processing Waste

7

Fruits, Vegetables, Plantation and Forest Products

INTRODUCTION

In the preparation of fruits and vegetables for processing purposes, large amounts of wastes are accumulated in the factories. Such wastes may be divided into solid and liquid wastes, the former consisting chiefly of peels, cores, seeds, trimmings, pits, while the latter of floor washings from storage and processing containers. Further, the above large volume of water used for washing, cooling and blanching, etc., which is also a potential source for bacteria, has to be disposed off.

The quality and characteristics of fruit processing wastes are diverse. Depending upon the raw materials, these wastes vary from liquid wastes to liquid slurries, from continuous to variable flows and concentrations, from harmless to nutrient excess wastes, and from seasonal to year-round waste loads. Geographical location, availability of land and proximity to ground and surface waters, are factors to be considered.

WASTE UTILISATION FROM FRUIT AND VEGETABLE INDUSTRY IN INDIA

The manufacturing units of fruit and vegetable products in India produce major wastes from these industries, such as citrus peel, mango peel and stones, pineapple cores, apple pomace, and various seeds, which are not utilised in India in an organised manner, even though machinery to do so is available in the country.

It has been found that the citrus waste could be converted economically to an animal feed that is nutritionally comparable to snapped corn.

Available product from citrus waste is peel oil. Recovery by cold pressing and distillation fluctuates from 1.85 to 9.7 lb per tonne of peel. Citrus oil finds varied application in the food, perfume, cosmetic, insecticide, rubber and textile industries.

Fatty oil from citrus seed is another product suitable for edible purposes either in refined or hydrogenated form.

Citrus waste is an abundant source of pectin. About 5 million pounds of pectin is recovered for use in food products in the USA.

Cannery residues, cured with lime and dried, are a good cattlefeed, and recovery is 25 to 27 per cent. Orange peel waste is used for extraction of natural colour, 402 to 669 mg/kg being obtained as a colour concentrate. Peels from malta orange and lime are used in candying.

Apple pomace is dried and used for manufacture of high-quality pectin, which is diluted with sugar for marketing as 200-grade pectin. After extraction of pectin the wastes are dried and used as cattlefeed.

Temperate fruit kernels, like those of the apricot, peach and cherry, yield 33, 25 and 30 per cent of edible oil on a dry basis.

Stones and peels are the main waste from mango. Some 10 per cent oil is recoverable from the kernel stones and is used as a substitute for cocoa butter. The residual cake, rich in carbohydrate and high in protein, is a good cattlefeed.

Feeding of dried pineapple waste is reported to increase yield and quality of milk.

NEW PRODUCTS FROM FRUIT AND VEGETABLE PROCESSING WASTES

Mango Wastes

Mango wastes constitute 40–50 per cent of the fruit:

1. Peels (12 to 15 per cent), on treatment with pectic enzymes and pressing, yield 50 to 55 per cent juice (15–16° Brix). This can be used to make wine or vinegar, or in a concentrated form (70° Brix), obtained by refining with hydrated lime or gelatine and evaporation, as a syrup with many outlets. The juice can be solvent-extracted to yield concentrated carotenoid pigments (14 to 17 mg/100 g) for food colourant use. Pectin in the peels (13 to 18 per cent, dry basis) is of jelly grade 200–240.
2. Coarse, fibrous pulper waste (5 to 10 per cent) is also a source of wine, vinegar and pigment concentrate.
3. Kernels (15 to 20 per cent) yield tannins (8 to 9 per cent) on spirit extraction; fat (7 to 9 per cent) on hexane extraction that finds use in soapmaking, margarine preparation and cocoa-butter substitution; and glucose syrup (70°C Brix) by enzyme or acid hydrolysis of kernels that have been freed of tannins and fat. Thus proper processing of root of mango wastes should yield sugar solids 36 tonne, tannins 4 tonne and fat 3.5 tonne, besides carotenoid concentrates in considerably amount.

Citrus Wastes

Citrus wastes constitute 50 per cent in oranges and 60 per cent in acid limes:

1. Peels on cold pressing yield the essential oil (0.05 per cent from lime peels, 0.5 per cent from lemon peels); the press-cake can be steam distilled to recover residual essential oil. Pectin of jelly grade 225–250 can be obtained either from fresh, or blanched and dried peel (yield 2.0 to 3.5 per cent on fresh weight). Brined peel preserved with 500–600 ppm SO_2 finds ready use in marmalade or candy manufacture.
2. Cull fruits, or slightly damaged fruits, are pressed for juice which is fermented, boiled, filtered, treated with lime and calcium carbonate and boiled to precipitate calcium citrate, from which citric acid can be made.
3. Flaked citrus seeds on solvent extraction yield an oil used in leather and textile industry.

Apple Wastes

Apple wastes make up 20 to 30 per cent of the fruit:

1. Pomace may be dried for cattlefeed uses; or boiled with 4–5 parts of water to yield a juice (4 to 5° Brix, 0.15 to 0.20 per cent acidity) which can be worked up after addition of sugar and a food

acid to yield an acceptable soft drink (12° Brix, 0.4 per cent acidity), or fermented to wine or vinegar.

2. Pectin of jelly-grade 250–275 can be extracted from the blanched pomace, as such or after drying, in 2.5 to 3.4 per cent yield (wet weight).

Pineapple Wastes

Pineapple wastes (peel, core, trimmings and shreds, 30 to 50 per cent) on shredding and continuous screw-pressing yield 70 per cent juice (12 to 14° Brix, 0.4 to 0.8 per cent acidity), which, after ion-exchange refining, can be used either as a mix in pineapple canning syrup, or in squash, or for citric acid recovery (as from citrus cull fruits). The press-cake, after rotary drying, yields pineapple bran, a cattlefeed.

Grape Processing Wastes

Grape processing wastes (20–30 per cent) consist of several components:

1. Stems yield 2 per cent cream of tartar on boiling, filtering, concentration and cooling.
2. Pomace on drying constitutes cattlefeed (protein 12, crude fibre 36, fat 5 and ash 6 per cent respectively), while aqueous extraction yields calcium tartrate and tartaric acid.
3. The sediment formed on ageing of wine, called argol, on dissolving in a minimum of boiling water and cooling yields pure cream of tartar.
4. Grape skins, when obtained free of seeds, yield food colourants by extraction and concentration.
5. Grape seeds on screw-pressing yield an oil that is used in soapmaking and paint formulation, or for edible purposes after refining; and a cake (protein 14, crude fibre 43 and carbohydrates 30 per cent) used as animal feed.
6. Seed hulls, obtained by decortication, yield on aqueous extraction a tannin concentrate (15 per cent) suitable for tanning hides.

Tomato Waste

Tomato waste (20 to 30 per cent; seeds, peels and cores), dried as a whole, constitutes a good animal feed (protein 8-10, fat 10–12 per cent):

1. Peels on hot caustic extraction yield a resin (50 per cent), which on refining and polymerisation functions as a food can lacquer.
2. Seeds yield oil by expelling (20 per cent) or solvent extraction (33 per cent) for soapmaking, edible or semi-drying use. The meal (protein 37, crude fibre 28 and minerals 7.4 per cent) is a useful cattlefeed or fertiliser.

UTILISATION OF FRUIT CANNING WASTES AS CATTLEFEED

Estimates of daily availability of feeds and fodders for various classes of livestock indicate considerable. degrees of underfeeding.

Many fruit and vegetable canning industry wastes are good sources of crude protein, such as tomato waste, cabbage leaves and red onion waste. Others like mango seed kernels and fruit peels, orange peels, pulp and seed, and red onion peels are rich in energy. Calcium-rich sources are orange peels, orange pulp, cabbage leaves and wastes, and red onion peels and slice waste.

Indian Council of Agricultural Research set up studies at 8 centres to examine the use of forest and industrial by-products to augment feed resources. Work on tomato wastes and mango seed kernels at the Anand (Gujrat), centre is discussed.

Tomato Waste to Milch Cows

1. Four lactating Kankrej cows served as controls, receiving (as percentages): wheat bran 20, maize bran 20, guar bhardo 10, groundnut cake 10, maize cake 17, rice polishings 20, salt 2 and bone meal 10. Four other cows received 8 per cent tomato waste, with a one per cent reduction each in the proportion of maize bran and guar bhardo, and a 2 per cent reduction each in groundnut cake, maize cake and rice polishing. Over a 12-week period, average yields per day were: milk 6.94 kg and fat-corrected milk (FCM) 7.27 kg for the controls, and 7.57 kg and 8.27 kg in the treated group.
2. In a second 12-week feeding trail with 3 cows in each group, replacement of 20 per cent of the crude protein concentrate with 16 per cent tomato waste resulted in the following daily average yields of milk and FCM: control 6.38 and 6.22 kg, treated 6.26 and 6.66 kg.
3. Tomato waste had no adverse effects; thus it can be used to augment normal feed concentrates, but should be well ground to ensure that the seeds do not escape digestion.

Digestibility and Balance Trials with Mango Seed Kernels on Adult Bullocks

Five Kankrej bullocks were fed daily with 3 kg of mango seed kernels, one kg of groundnut cake and pasture hay *ad lib*. The mango seed kernels contained (as percentages on a dry basis): crude protein 6.65, ether-extract 12.84, crude fibre 0.14 and nitrogen-free extract (NFE) 76.8. The digestibility co-efficients indicated that protein was digested to be extent of 15.7 per cent, ether extract 58 per cent and NFE 72 per cent.

Mango seed kernels provided 73.7 kg of total digestible nutrients over the feeding period indicating that they were a good source of energy. The N, P and K balances, all positive, were: 22 ± 1.59, 1.3 ± 0.19 and 12 ± 1.19 respectively, expressed in gm per animal per day.

Feeding Mango Seed Kernels to Growing Calves

Calves (8 Kankrej and 12 Surti) were divided on the basis of species, sex, bodyweight and age into two groups. A normal concentrate mixture was fed (according to the requirement of digestible protein) to the control group; the other group received 20 per cent of mango seed kernels but the same digestible crude protein and total digestible nutrients as the controls. Each animal also received daily 2 kg of guinea grass and mature hay *ad lib*. Over a 12-week period, the gain in body weight was: control 40 and treated 37 kg. Only male animals were used for the digestibility and balance trials, which yielded the following results for the control and treated animals: total digestible nutrients 67.9 and 67.2 kg; crude protein content 28.52 and 28.74 per cent; digestible protein during the feeding period 21.97 and 21.59 kg; digested protein 77 and 75 per cent; digestibility co-efficients both for the ether extract and the NFE 70 and 70 per cent; calcium balance +20.3 and +20.1 g; nitrogen balance +62.1 and +55.6 g; and phosphorus balance +6.4 and +5.6 g.

Feeding Mango Seed Kernels to Milch Cows

Twelve Kankrej cows were divided into two groups; one was fed with the control concentrate and the other with feed containing 10 per cent mango kernels. Average yields in kg/day of milk and FCM were: controls 7.59 and 8.80, treated 7.70 and 9.65. Use of defatted mango seed kernels at 30 per cent levels in the concentrate mixture had also no adverse effect on milk production. All these experiments show that mango kernels obtained by decortication would be suitable as livestock feed.

UTILISATION OF WASTES FROM FRUIT AND VEGETABLE PROCESSING INDUSTRIES WITH SPECIAL REFERENCE TO THE ISOLATION OF BIOFLAVONOIDS AND THEIR UTILISATION AS ANTIOXIDANTS

Bioflavonoids find use in different fields: for treatment of hypertension, radiation damage and allergic diseases; in bacteriology for their bacteriostatic activities against *Staphylococci*, *Typhi* and *E. coli*; as food additives for their antioxidant property; and recently in therapeutics for their anticarcinogenic activity.

Flavonoids are present in large amounts in the peels of citrus fruits and onions, enormous quantities of which are being thrown out as waste by food industries. Methods are already available for the extraction of flavonoids like quercetin from onion peel and hesperidin from orange peel.

The antioxygenic properties of crude citrus bioflavonoids (CBF) were compared with phenolic antioxidants like propyl gallate (PG), butylated hydroxyanisole (BHA) and butylated hydroxytoluene (BHT) by incorporating them at a level of 2 mg/100 gm into ghee, refined safflower oil and whole milk powder. In another series, potato chips were fried both in refined groundnut oil and in hydrogenated oil carrying CBF, PG, BHA and BHT.

Ghee and safflower oil were stored in beakers at 37 ± 0.5°C. Whole milk powder was packed in paper/Al foil/polyethylene laminate pouches, and fried potato chips in cello-poly pouches, and both were stored at 45 ± 0.5°C. The rate of oxidation was measured by determining the peroxide value and malonaldehyde content, and by organoleptic evaluation initially and at periodic intervals.

It was found that CBF has limited antioxidant potency in safflower oil. In ghee it shows considerable potency, comparable to that of PG. In whole milk powder and fried potato chips, the antioxidant activity of CBF was comparable to that of BHA and BHT, but inferior to PG.

PRODUCTS FROM CITRUS WASTES AND THEIR UTILISATION IN FERMENTATION INDUSTRIES

Small industries discard their wastes, and even in bigger units, utilisation is confined to the extraction of citrus oil from the peels, or their conversion into manure by natural anaerobic digestion. Effluents are in generation too high in acidity and BOD to be handled by municipal sewerage systems.

Wastes consist of citrus peels, citrus seeds and marc:

1. Citrus peel (20 to 60 per cent of fruit) has 75 to 80 per cent water, and the solids contain (as percentages): sugar 44 (glucose; fructose: sucrose, 63:21:16), cellulose (including lignin and pentosans), 33, and pectic substances (which adhere tenaciously to the sugar moieties) 20. Removal of essential oil gives citrus peel pulp, and animal feed, an peel juice, which is concentrated to molasses. Other uses are in preparation of boards, plastic materials, pigments, pectin and pectic acid, and as dried, dehydrated, brined and ensilaged peel.
2. Citrus seeds when dry contain 30 to 35 per cent of a fatty oil rich in oleic and palmitic acids, and about 16 to 20 per cent protein; the presence of limonin imparts a bitter taste. Immediate drying of seeds is necessary to prevent mould growth.
3. Citrus marc is obtained only in small proportion, and is usually added to peel waste prior to its processing for cattlefeed or pectin manufacture.

Utilisation in Fermentation Industries

1. For acetone-butanol production, exhausted citrus peels are fermented using *Clostridium acetobutylicum*. From every tonne of sugar in the peel 50 cu.m of gas (45 hydrogen to 55

carbon dioxide) is evolved, while 500 parts of peel yield 12 p of butanol, 5 of acetone and one of ethanol, besides methanol in small amounts.

2. From peel juice or citrus molasses, enrichment with mineral salts, pH adjustment, 24 hour fermentation and distillation yields ethanol of 95 per cent strength. The crude distillate contains both peel oil and fusel oil, separable in a special decanter, and methanol (derived from pectin hydrolysis), whose removal is difficult; hence citrus alcohol is best used as fuel spirits or in industrial outlets. The distillation residue is a source of vitamins C and P. One tonne of alcohol is obtained from 25 tonne of waste peel juice, and recovery costs are about half that from blackstrap molasses.
3. Using *Torula utilis*, yeast in a yield of 46 per cent of total sugar present can be produced from citrus peel juice or molasses by dilution, mineral salt addition, pH adjustment and fermentation. Use of macerating enzymes by solid-state fermentation techniques, recently developed in Japan, yields citrus peel sugars; these could support production of SCP by such enzymes as *Debaryomyces hansenii* and *Candida utilis*.

CFTRI has examined citrus peel utilisation. Exhausted peels constitute a disposal problem, and could be a source of pectin. The pulp left thereafter contains sugars which are extracted to give peel juice for further fermentation or SCP productions. The solid residue can be converted into a nutritionally-balanced cattlefeed after pH adjustment, treatment with pectin-degrading enzymes (themselves manufactured in India using CFTRI technology), dewatering and drying.

IMPROVED METHOD OF EXTRACTION OF OIL FROM ORANGE PEELS USING FUNGAL XYLANASE

Citrus peel, an important by-product of the citrus fruit processing industry, is commercially important in its many applications in food, pharmaceutical and cosmetic industries. The bulk of the aromatic essential oil is lodged inside oil sacs located in the flavedo area of the peel. Oil is obtained by cold pressing, distillation or solvent extraction, none of which leads to efficient oil liberation.

To this end, enzymic digestion of the oil sacs was examined. The wall of the sac is composed of cellulosic, hemicellulosic and pectic materials, and the fungal enzyme employed carried considerable xylanase, along with amylolytic, cellulolytic and pectinolytic activities. As a quick evaluation method to judge the efficiency of enzyme treatment, estimation of limonene, the major component of the orange peel, was chosen.

Cut pieces of orange peel were treated with crude *Monilia sp.* enzyme with 725 units of xylanase activity per ml, and maintained for 2 hour at temperatures of 30, 40, 50 and 60°C. The reaction mixture was then filtered and extracted with ether, and the extractives were subjected to GLC analysis. The limonene content was several-fold higher at 30°C than for the control not treated with enzyme.

The much lower limonene content obtained at higher temperatures was presumably due to evaporation of the oil. Although 55°C is the optimum temperature for xylanase activity, some 30–40 per cent of the activity was still retained at 28°C, and this was adequate to act upon the sac tissue and liberate oil. This observation is significant, since processes that can be worked at lower temperatures are desirable with volatile oils.

SOME USEFUL PRODUCTS FROM LIMONENA, A BY-PRODUCT OF CITRUS INDUSTRY

Deterpenated oils constitute only 2–10 per cent of citrus essential oils and find their use in soft drink industry because of their increased stability and desirable flavour components. Residual terpenic hydrocarbons (80–96 per cent) go largely unutilised. Means of transforming α-limonene, the terpenic hydrocarbon fraction of orange peel oil, through synthesis of flavouring compounds have been explored.

CFTRI has developed process for conversion of limonene into α-terpineol in about 70 per cent yield by passing a solution of limonene in acetic acid through a cation exchange column. Separation of α-terpineol leaves a mixture of limonene, α-terpinene and terpinolene; the latter could also be obtained directly from limonene by isomerisation using silica gel as the acid catalyst. Terpinolene could be separated easily from the mixture by treating it with peracetic acid under controlled conditions to give terpinolene oxide.

Terpinolene oxide is a key intermediate for several perfumery and flavouring chemicals. A method was developed for reduction of terpinolene oxide to a 3:1 mixture of 4-terpinenol and α-terpineol. Acid-catalysed rearrangement of terpinolene oxide gave karahanaenone, a constituent of hops used in flavouring beer. Limonene could also yield terpineol esters. Since conversion of a tertiary alcohol to its esters is difficult, α-terpinyl chloride was first obtained by HCl-treatment of limonene. Thereafter the chloride was quantitatively converted to α-terpinyl acetate by simply stirring with acetic acid in the presence of zinc oxide; apparently the latter became zinc acetate during the reaction and displaced chloride from terpinyl chloride, a new reaction of considerable theoretical interest. Not only acetates but other esters (formates, propionates, etc.), ethers and alcohols can be obtained by substituting acetic acid with other acids (formic or propionic), alcohol or water, as can any tertiary alcohol, ether or ester. Several of these are commercially important flavour and perfumery chemicals (e.g., myrcenol, ocimenol, hydroxycitronellal and linalool). Newer compounds prepared using this reaction are *cis*- and *trans*-dihydro-β-terpineols; these have a pleasant floral odour and should blend well with tobacco flavourants in cigarettes and cosmetics. Other products from limonene being investigated are L-menthol, some juvenile hormones like juvabione, etc.

DEVELOPMENT OF EDIBLE BY-PRODUCTS FOR NUTMEG WASTE (RIND)

The nutmeg (*Myristica fragrans*) yields two spices: nutmeg mace, the dried, reticulated aril, and nutmeg, the dried seed or kernel.

The waste from processing this spice is nutmeg rind, the outer fruit or pericarp. This is obtained as two peach-like halves, and constitutes 80 to 85 per cent of the whole fruit.

The rind is acidic and acrid to taste and carries a terpenic flavour. It contains, as dry weight percentages: pectin 12.75, total reducing sugars 14.36 and crude fibre 28.20. Acidity as citric acid is 21.4 per cent.

Because of the ovoid shape, abrasive peeling of the rind is not possible. Chemical peeling and blanching were explored, and a procedure worked out that involved soaking for 1 to 2 minutes in boiling 10 per cent sodium hydroxide solution.

Using the peeled rind material, a number of products were prepared including preserves, jelly, chutney, jam, pickle, canned fruit, canned mixed fruit (cocktail), squash, syrup and nectar (ready-to-serve beverage). Standard methods of preparation are employed for these products, with certain specific observations. To make preserves, the peeled material steeped in brine of increasing concentration (upto 8 per cent) for a total of about 8 days, followed by soak-washing and boiling in sugar syrup (containing ascorbic acid) of increasing concentration upto 70 per cent. Brine-steeping also serves to remove acrid components and to achieve softening prior to canning. A mixed fruit combination of nutmeg, pineapple, papaya and banana in a 4:2:2:2 ratio gave an excellent canned cocktail product as well as jam. Pectin extract for jelly-making was derived by boiling with 3 volumes of water containing 0.75 per cent citric acid. Pickle made in combination with sliced green ginger, whole green chillies, halved limes, etc., and chutney of a sweet-hot type with sugar, red chillies, onions, green ginger, garlic and vinegar. Squash and syrup prepared using the juice extracted from nutmeg pulp either in a press or by forcing through cloth, which was therefore thickened and preserved using potassium metabisulphite.

Storage trials of all the products indicated good shelf life. Taste evaluations using a panel of judges showed high degrees of acceptability. These products from waste nutmeg rind show promise of internal and export markets.

RECOVERY AND QUALITY OF PECTIN FROM NUTMEG WASTE

Nutmeg rind, comprising 80–85 per cent of the nutmeg (*Myristica fragrans*), is a waste. The present studies indicate that the rind can yield 11.1–12.5 per cent of pectin of high quality with a jelly-grade of 225–250.

The best recovery of pectin from nutmeg waste by using either 0.25 per cent HCl or 0.75 per cent citric acid, when the yield of pectin on a dry weight basis were 12.73 to 14.33 per cent and 12.54 per cent, respectively. A 1:4 ratio of solids to extractant was employed. Some 93 per cent of pectin could be recovered with dilute HCl in a single step, but with citric acid two extraction steps are necessary. The optimum time of extraction was 60 minutes at a temperature of 85–90°C.

The pectin obtained in 12 per cent yield was found to have an average methoxyl content of 8 to 12 per cent, an anhydrous uronic acid content of 75 to 80 per cent and a jelly-grade of 225. Obtaining pectin from nutmeg rind appeared to be a viable process.

BY-PRODUCTS FROM PLANTATION CROP PROCESSING INDUSTRIES

The diverse plantation crops in India earn Rs. 150 million in foreign exchange yearly, apart from meeting domestic needs. By-product utilisation could further improve their economics.

Arecanut

Arecanut (1.7 lakh tonnes from 175,000 ha) yields the following by-products:

1. Area husk (600–700 kg/ha), useful in making fibreboards, hardboards and insulating boards, brown paper if mixed with bamboo and jute pulp, and rubberised mattress material.
2. Leaf sheath and leaves (90,000 tonnes), now used as thatch, for broom-making, or as fuel, but with potential for preparation of box-lining material, or of moulded cups from the sheath.
3. Chogaru liquid or kali left after the tender nuts are boiled in water, which carries a sediment of high tanning content that could find use in the leather industry if a steady supply could be assured.
4. Arecanut fat (11 per cent of the nut), a white plastic solid of mp 38°C rich in lauric, myristic and palmitic acids (total 75 per cent), and useful in the soap industry.

Cashewnut

Cashewnut (1.8 lakh tonne) plantations yield three major by-products:

1. Some 10 lakh tonne of cashew apples are available annually, of which only 40,000 tonne are used for liquor ('feni') production in traditional or modern ways. The fruit can be converted to candy, syrup, canned fruit, chutney, curried vegetable, pickle, etc. The pressed juice, after removal of astringent and acrid principles, yields a variety of beverages, blended fruit drinks, concentrates and fermented liquor. Collection of fruit from small, scattered plantations is the problem.
2. Testa is the thin brown skin discarded during cashewnut processing. Now used as manure, it could yield excellent tanning extracts that are at present imported.

3. Cashewnut shell liquid (CNSL) is a valuable polymerisable material not yet adequately utilised within the country. The shells could also form a source of galactose (100 kg from 3 tonne of shells). Further, solvent extraction could double the yield of CNSL over the present oil-batch extraction procedure.

Coconut

While a number of coconut by-products are utilised, there are others that need attention:

1. Coir dust, a mixture of corky pith, short fibres and pericarp, accumulates to the extent of 1.6 lakh tonne near the coir-extraction centres and is dumped into the backwaters. It is an excellent soil conditioners, absorbing 8 times its volume of water and releasing it only slowly. Use in insulation boards, in moulded articles, as expansion joint fillers and furfural production have all been suggested.
2. Coconut shell (8.8 lakh tonne, of which half is easily collected) is now used as a fuel and for making fancy articles, but could find outlets for manufacture of charcoal, activated carbon, fine shell flour and furfural.
3. Coconut water from ripe nuts (100 litres from 200 nuts) has a lower sugar and high salt content, and is therefore unsuitable for either sugar or alcohol manufacture. Use as a rubber coagulant, growth media, source of potassium fertiliser or soft drink is hampered by the difficulty of large-scale collection, quick deterioration in quality and packaging problems.
4. Toddy is produced in a yield of 1400 ml/tree/day, and proper preservation and standardisation are the important problems.

Cocoa

Cocoa (1000 tonne, with a tenfold increase possible yields cocoa beans as the main product, and several by-products:

1. Pod wall of husk carries mucilage and sugars which could have feed or manurial value.
2. Cocoa shell has a high crude fibre content (17 per cent) but also 14 per cent protein, 7 per cent ash and 3 per cent fat. The fibre content militates against human use, and the theobromine level (0.75 to 1.30 per cent) against feed use; extraction of theobromine could be important for further conversion by methylation into caffeine. Present use of shell is as fuel (4000 cals/kg) or fertiliser (10–12 per cent phosphate in the ash).
3. Cocoa germ carries 3–5 per cent of a dark, odorus fat with a high content of unsaponifiable matter and of pro-vitamin A. The germ has 1.5 to 2.5 per cent of theobromine, sufficient to repay extraction.
4. Shell-nib mixtures carry from 10 to 38 per cent of fat; the residue after solvent extraction has 2 to 3 per cent theobromine.

Coffee

Coffee, a major plantation crop, generates three major by-products:

1. The pulp and husk (75,000 tonne) has been used as animal feed after drying or ensiling, and as a good organic fertiliser after natural fermentation in which biogas is evolved. Upto 30 per cent husk could replace chicory in its use as a coffee blend.
2. Coffee husks are now used as fuel in drying coffee beans, but an outlet as roughage in animal feeds is a possibility.

3. Dry spent coffee beans and surplus beans are used in Brazil as a source of oil that can be employed even without refining in soapmaking.
4. There are possibilities of obtaining furfural from coffee parchment (13.4 per cent yield) and coffee husk (5.4 to 6.9 per cent yield); tannins and organic acids from coffee residues by extraction or fermentation; and pectic enzymes from mucilage. Removing caffeine from coffee beans prior to roasting to give decaffeinated coffee has a distinct future in India.

Citrus Fruits

Citrus fruits like orange, lemon and lime yield 500 to 600 kg of waste peel for every tonne of fruit processed. Alcohol-soluble carbohydrates if extracted and concentrated yield molasses, which can be fed to cattle or fermented to alcohol. Processes for manufacture of pectin, especially from grape fruit peel (yield 12 per cent), have been developed by CFTRI, as as process for removing terpenes from the total essential oils to yield products of enhanced flavour value, stability and solubility. Citrus seeds yield 25 to 30 per cent of a fatty oil, from which the bitter limonin taste can be removed by alkali refining to give an edible fat. Hexane-extraction of wet peel yields pigments that find particular use in colouring citrus juices and concentrates.

Rubber Seeds

Rubber seeds are collected after dehiscence of the fruit, and must be heated immediately to destroy lipase activity. The kernel oil (50 per cent) is used in surface coatings, while the extracted cake is rich in protein (30 per cent) but carried cyanogenetic glucosides and hydrocyanic acid.

Spice

Spice oleoresin preparation generates 10 to 50 times the quantity of spent materials. These are now used as manure or fuel, but may find use either in cattlefeed, or as source materials for various purposes:

1. Extraction of colours, as from ripe pepper and paprika berries.
2. Recovery of flavouring components, as from green ginger peels; onion outer skins and leaves which yield quercetin; or tender leaves and twigs of cinnamon, which give an essential oil rich in eugenol that has wide uses.
3. Use in agarbathi manufacture of such materials as cardamom husk and spent seeds.

Tea

Tea production generates some 15,000 tonne of tea waste (fluff, stalks, dust and sweepings), most of which is today burnt. The caffeine content is 1.5 to 3.5 per cent, extractable with solvents, and a plant in Assam produces 20 tonne/annum. There is over ten times this potential, but collection and transport of tea waste is neither easy nor economical. The residue after caffeine extraction has cattlefeed potential, except for a 20 to 25 per cent tanning content. Waste tea could be a source of green, yellow and brown pigments, pectin (5 per cent), cellulose (20 per cent), or polyphenols for antioxidant use, and most profitably, a raw material for instant tea manufacture. Instant tea waste is a good source of the plant growth-promoter triacontanol. Tea seed (100 kg/ha if exclusively produced) carries 20 per cent of fatty oil of iodine value 89; the residual cake has 10 to 15 per cent of saponins which are a foam-additive in the beer industry and after saponin removal, the protein level (12.6 per cent) is high enough to suggest feed utility.

Tapioca

Tapioca or cassava, a tuber food crop, has edible leaves of good protein content. The rind from domestic or industrial use is starch-rich; it is increasingly being used for livestock feeding, but has potential for microbial protein growth and amylase production. Some tapioca varieties produce seeds that carry 30 per cent of a linoleic-rich oil and a high-protein cake.

BY-PRODUCTS FROM THE PLANTATION PRODUCTS INDUSTRY

Use of by-products of the plantation industries as fuel or materials do not exploit their specific properties, which must be considered in seeking other uses. These industries fall into three groups, namely spices, tree nuts, and other products (tapioca, rubber, etc.).

Spices long used as a whole, are now extracted, and the oleoresins obtained after concentration are utilised in food processing and pharmaceuticals. Also essential oils are recovered. Oleoresin residues comprise from 80 to 90 per cent of the material processed, and those from essential oil recover 95 to 98 per cent. Since flavouring materials are still present, use as insect repellants and fumigants (that function by burning) it suggested for pepper spikes, chilly stalks and ginger peels.

Solvent ethylene dichloride that is present (1000 ppm) in oleoresin residues suggest their usage as chlorinated insecticides or for bacterial control in latrines and sewage farms. Nutmeg rind could be a source of pectin (14 per cent), and an ingredient of pickles and chutneys.

Tree nuts include the coconut, cashewnut and arecanut:

1. Coconut (6,000 mill nuts annually) yields husk, coir, pith and shell (4.80, 0.78, 0.66 and 0.85 mill tonne) which now have varied uses. The high calorific values of coconut pith, shell and shell char (7437, 8630 and 13000 Btu/lb), and the very low contents of sulphur, phosphorus and ash suggest their use as excellent reductants for metallurgical applications; at the Regional Research Laboratory Trivandrum (RRLT), upto 95 per cent reduction was achieved of iron ore and mill scale to sponge iron of superior quality. Properties of these materials suggest other uses: husks as ingredients of hot-press particle boards; coir fibres, after blending with other fibres, as fillers in clay, cement and plastics for low-cost roofing, for rubberisation, and after carbonisation and graphitisation, as reinforcement for metals, plastics and clays. Further, from coconut leaves, long fibres can be extracted and the mid-ribs used as match stricks. Future expansion of the coconut depends on control of root wilt diseases.
2. Cashewnuts yield as by-products a brown cashew testa from which some 37 per cent of tannin is extractable, and cashewnut shell liquid (CNSL), a versatile source of resins, varnishes, paints, adhesives and other chemicals. Present oil-bath extraction at 180°C yields about 38 per cent of the CNSL present, followed by another 33 per cent from subsequent expelling, thus leaving 29 per cent; a good part of this could be obtained by solvent extraction but this is not as yet commercialised. The expeller residue, now burnt, could be used as a binder-filler in particle boards.

The tapioca has been bred to double its starch yield and increase its starch content; these new varieties could thus raise future production over the same area. Tuber are used to produce starch yielding as wastes tapioca rind or peelings and spent pulp or tipi. If rind were to be separated by simple winnowing into outer cellulosic and inner hydrolysable material, the latter could find use as cattlefeed. Spent pulp as obtained contains 90 per cent moisture. Where sun-drying is feasible, the dried mass can be used as cattlefeed but a product with 20 per cent solids can be derived by centrifuging, and used for hydrolysis to glucose and further fermentation to alcohol. The soluble sugars (2 to 3 per cent of tapioca) thus recovered could also be fermented to alcohol. Tapioca-derived power alcohol of dry tapioca could be an attractive petroleum substitute, and such an outlet would stimulate tapioca production. Other by-products of the tapioca plant are the steam, which have been hot-pressed with some binder into boards, and the leaves too fibrous for human consumption but potentially useful as cattlefeed or for leaf protein extraction.

UTILISATION OF AGRICULTURAL RESIDUES FOR PULP, PAPER AND BOARD

Rice straw on suitable cooking with caustic soda and subsequent calcium chlorite bleaching, was blended with 10 per cent bamboo pulp. On suitable beating, the papers formed met the requirements of writing and printing papers in terms of breaking length, tear factor and burst factor. The cold soda process applied to rice straw sufficed to yield grease-proof papers, and tough wrapping paper with an excellent folding endurance resulted from hot caustic soda digestion. Straw board and insulation fibre board of good quality could be had by lime digestion.

Wheat straw, by broadly similar treatments, likewise yielded writing and printing papers, grease-proof paper and strawboards of good quality. For production of corrugating medium, hot digestion either with caustic soda or with a lime-soda mix was employed; a 60:40 blend of wheat straw and gunny waste gave the best results.

From jute stricks, writing and printing papers could be made in high yield by a mechano-chemical process, and in lower yield but better quality by the hot sulphate process. The same was true of wrapping paper, of which a cheap grade could be made in high yield using cold soda digestion. Grease-proof paper of good quality could be had at pilot-plant level both from sulphate pulp and from neutral sulphite semi-chemical (NSSC) pulp. Of particular promise was the use of a mix of 60 per cent of the mechanical pulp made from jute stricks and 40 per cent of bleached sulphate pulp to make newsprint of excellent quality. Building boards have also been produced.

Conditions were worked out for using bagasse for the production of writing and printing paper, wrapping paper, rayon grade pulps, grease-proof paper, straw boards and fibre board. Newsprint prepared from a mixture of mechanical or groundwood pulp from eucalyptus hybrid (40 parts) and soda semi-chemical blenched pulp from bagasse (60 parts) compared very favourably with both imported and Indian newsprint.

From cotton stems, lime-cooking gave board of good quality. Wrapping paper could be had both by the sulphate and the soda processes, the latter in higher yield, but for writing and printing paper, quality was better by the sulphate process.

Other miscellaneous raw materials have been explored. Arecanut husk could not be pulped either by the cold soda or the NSSC process. Groundnut shells could be pulped by the chemical sulphate process, but the poor strength of the pulp rendered it unsuitable even for making brown wrapping papers. Particle boards could be made from the hulls using 10 per cent of several types of resins. Tea stems, either with or without bark, were suitable for the production of writing, printing and wrapping papers after caustic digestion. Castor stems were likewise suitable especially of 70 parts were mixed with 30 parts of unbleached bamboo sulphate pulp. Sunflower stalks by the sulphite process proved unsuitable for pulp and paper making. Arhar and jowar stalks yielded lime pulps of a strength higher than in ISI standards. Sugarcane leaves gave only low-grade fibre boards by the semi-chemical processes, but could be used for 30 per cent extension as filler in phenol-formaldehyde resins. Use of paddy husks and mustard stalks in building boards has been examined.

8

Animal and Poultry Products

INTRODUCTION

Phenomenal progress has been noticed in recent years in Indian meat industry. This is evidenced by construction of modern abattoirs, meat complexes, bacon factories and poultry processing plants. The Indian meat industry is based upon cattle, buffaloes, sheep and goat, pigs and poultry. At present, these animals are slaughtered in recognised municipal abattoirs apart from abattoirs, meat complexes and bacon factories. Also animal slaughter takes place on open grounds or backyards of the houses in rural or sub-urban areas.

The products, that emanate from such animal slaughter are meat, by-products, waste products and effluents. In recent years, the meat production has been boosted to about 1.5 million tonnes. Correspondingly, the quantum of by-products and effluents increased considerably. The economic losses incurred due to under utilisation or improper utilisation of animal by-products have been reported to be in the order of about Rs. 100 crores annually.

The effects of effluents let out are severe and not known as most of the abattoirs and poultry plants do not have proper facilities for effluent treatment. Many of the abattoirs and poultry dressing plants in India, therefore, need improvements for hygienic production of meat as well as effective recovery and better utilisation of by-products along with efficient disposal of effluents. Thus abattoir is a nucleus place for release of various proteinacious materials, waste products, condemned carcasses and obnoxious odours. These materials contribute to the environmental pollution and public health hazards. Therefore, treatment and processing by-products, condemned carcasses and effluent have become necessary and inevitable to control environmental pollutions to counteract the rising of cost of raw materials, labour and power and ultimately for the betterment of human welfare.

In the light of the above facts, an attempt has been made in this chapter to project the present situation of effluents from abattoirs and poultry processing plants and the possible utilisation of their waste and by-products.

ABATTOIR OPERATION AND EFFLUENTS

Abattoir is the place for production of hygienic meat, clean handling of by-product, rendering of by-products and treatment of wastes and effluents. Important operations in the abattoir systems are:

1. Resting of animals in the lairage prior to slaughter.
2. Adequate facilities for antemortem facilities.

3. Humane method of slaughter.
4. Removal of skin, dressing and evisceration on the overhead rail system.
5. Handling by-products.
6. Post-mortem inspection of carcasses and edible offals.
7. Washing the carcasses.

Abattoir is also provided with the rendering plant and effluent treatment system.

The sources of effluents are faeces and urine from lairage, blood, washings from carcasses, floor and equipment, paunch contents, tissues, trimmings, and fat contaminating the wash water. These materials increase the concentrations of solids resulting in high BOD. The effluent discharge widely varies on the capacity of abattoirs and on the method of handling the by-products.

Lairage

Animals are kept in the lairages till they are slaughtered. Lairage waste consists of excreta (both liquid and solid). The amount and strength of the wastes vary widely, depending on how thoroughly or how often manure is removed. These effluents are known to contain high concentration of solids and BOD.

The abattoir effluents consists of many and varied waste materials (Table 8.1).

Table 8.1. Sources of wastes in an abattoir.

Source	*Waste*
Stockyard/lairage	Dung, urine
Slaughter floor	Blood
Skin	Hair, dirt, soil
Stomachs, intestines, rectum	Ingesta, liquor, dung
Floor washings	Tissues, hair, blood
Carcass dressing	Trimmings, fat, blood
By-products	Fat, offal, tissues
Rendering	Stick liquor or press liquor

If blood is not recovered properly and if it is let out into the drains, it causes major problem in effluents. Blood has the highest BOD of any liquid material emanating from abattoirs and increases the pollution load. Blood consists of solids about 20 per cent and a BOD of 100,000 mg/l. Failure to recover blood increases the BOD of the effluents by 72 per cent.

Paunch contents

The paunch contents are about 5.5 per cent (on live weight basis) in sheep. Paunch waste solids are troublesome primarily because it increases solid concentration in the effluents.

Casing-cleaning operations

These contribute substantially to pollution. Waste from the deslimer increases the BOD level considerably.

The quantity of effluents and the concentration of solids in abattoirs vary to a greater extent and depend on the types and number of animals killed and on the recovery of by-products.

A survey conducted on the characteristics of waste-waters from slaughter of different animals in Saidapet and Perambur slaughter houses and in the Delhi slaughter house indicated the high level of pollutants and BOD level (Table 8.2). The Table shows the levels of waste flow, suspended solids, BOD level and grease in the effluents are high and these effluents are potential pollutants and pose public health hazards.

Table 8.2. Composition of effluents from some of the slaughter houses in India.

Analysis	*Perambur (Chennai)*	*Delhi slaughter-house*
pH	7.5	–
Total solids (mg/l)	26560	9025
Volatile solids (mg/l)	23800	6900
BOD (mg/l)	12600	4290
COD (mg/l)	22600	6320
Nitrogen (mg/l)	1200	680
Phosphorus (mg/l)	450	360
Grease (mg/l)	375	2540

It has been observed that the slaughter capacity of abttoirs in places like Mysore city is around 500 sheep per day. An estimate has been made on the effluent load from the slaughter of 500 sheep per day. An estimate has been made on the effluent load from the slaughter of 500 sheep and presented in Table 8.3.

Table 8.3. Expected effluent load (approximately) from the slaughter of 500 sheep.

Analysis	*Range*	*Average*
Flow (l)	60,000–2,00,000	70,000
BOD (mg/l)	670–900	800
Suspended solids (mg/l)	500–800	670

WASTES AND THEIR DISPOSAL

Urbanisation of industry has necessitated disposal of wastes by conversion into innocuous or usable products. Industrial wastes are unutilised parts of either the raw material or the manufacturing process which have to be disposed off suitably without causing inconvenience to society.

Abattoir wastes being biological in nature can attract undesirable elements like rodents, scavengers and bacteria, and cause hygienic and health hazards.

Conventional methods of separation of solids, and suitable treatment of the clear effluent to acceptable BOD/COD standards by dilution, drying, chemical or thermal treatments, etc. are applicable with suitable modifications to abattoir effluents.

Waste-disposal systems should be designed taking into account the volume of waste, proximity to built-up human habitation, drainage (municipal or natural), topography, wind direction, sub-soil condition and future development.

In modern abattoirs, all the parts removed during conversion of the live animal to meat are utilised. There are not wastes, only by-products.

At Deonar abattoir in Mumbai, designed with international collaboration, proper arrangements for adequate hygiene, inspection, collection machinery, receptacles and skilful collection have led to wide use of by-products. Pharmaceutical outlets covers such products as fresh blood, bile, pancreas, ovaries, thyroid, spinal cord, stomach and kidney. Clotted blood meal is utilised for animal feeding. Commercial outlets have developed for guts (surgical), bone (bone meal), fat, horns and hoofs, and hides and skins (perhaps the oldest outlet). A much wide range of uses could be developed even for these products, and for others such as intestines, trotters, paunch contents, and other glands like the adrenals, lung, pituitary, and placenta.

Pollution Effects

The effects of pollution are many and varied:

1. Depletion of dissolved oxygen: When the effluents are discharged into surface waters, depletion of dissolved oxygen takes place. The effluents from abattoirs contain all essential nutrients for bacterial life, and the temperature and pH of the effluents are ideal to bacterial growth. The rate of decomposition is high resulting in anaerobic conditions followed by bad odours, conditions suitable for fly and mosquito breeding.
2. Pathogens: Abattoir effluents carry a variety of pathogens such as *Salmonella, S. aureus, Bacillus, Clostridium* that affect human health.
3. Off-odours: Abattoir effluents are known to be readily biodegradable and cause unpleasant odours.

Control of Pollutions

Pollution in abattoir can be controlled in three ways:

1. Waste reduction.
2. Processing of by-products responsible for effluent load
3. Treatment and disposal of effluents.

Waste reduction

Waste reduction for abattoir can be achieved by adopting the following steps:

1. Dry clean-up: Pollution loads can be substantially reduced by the use of dry clean-up prior to the wet clean-up. This will also achieve greater recovery of scraps for inedible rendering. Another effect of dry clean-up is the reduction of waste-water volume.
2. Recovery of blood: Blood is recovered from killing operations in drains or collection tanks.
3. Fat recovery: Fat recovery is usually accomplished by means of baffled basins or trays.
4. Paunch contents: Large quantities of undigested materials are released from the stomach. The method by which the contents of the stomach are removed and the ultimate disposal of the contents will affect the waste load of an abattoir. These materials are disposed off as follows:-
   (a) Dry dumping: This method of paunch-handling is one in which the undigested material is dumped into a hopper and conveyed by mechanical means to a truck. Paunch material may be used as landfill or soil conditions. Recent studies indicated that ruminal contents along with blood can be used in poultry rations.
   (b) Wet dumping: This process involves use of water under pressure to flush the entire paunch or the dumping of the paunch contents into a flowing stream of water which will be passed over vibrating or rotating screens. Screening separates the coarse material from the fine. The gross material is then trucked away. The solution containing the suspended fine material and any dissolved matter passes into the sewer.

5. Feathers: Better way of disposing of the feathers is by a conveyor. This helps in reducing the concentration of solids in the effluent.
6. Trimmings and scraps: These materials are normally removed manually. Suitable meshes or containers are to be provided in abattoirs to trap these materials for easy collection and disposal or further use.
7. Condemned carcasses: The carcasses are kept separately and sent for rendering.
8. Use of water. Minimum water required should be used.

Processing of by-products

The important by-products or unfit carcasses responsible for pollution load need special attention for further processing in order to control environmental pollution. By-products for processing include intestines, blood, offals, bones, feathers, heads and hooves. Processing of these materials is briefly discussed as follows:

1. Intestines: Intestines are important by-products from the abattoir. The intestines are used for making animal casings (used in sausages).
2. Blood: Blood is a rich source of protein. Blood is processed into inedible products such as blood meal, which can be used in animal feeding. When blood is collected in hygienic conditions, it can be processed into edible products such as plasma and haemoglobin.
3. Offals: Offals are converted into offal meal by dry rendering process. The defatted and powdered offal meal is used in animal feeds.
4. Condemned carcasses: These are processed by dry rendering into carcass meal. The meal is packed in bags. Carcass meal is rich in protein and used in animals feeds.
5. Feathers: This is one of the main by-products for poultry processing plants. Feathers will be converted into feather meal by dry rendering process. Feather meal is used as a source of protein in poultry rations.
6. Bones: Bones are also important valuable by-products obtained during deboning of carcasses and have several uses, such as bone meal as a source of calcium and phosphorus in livestock rations and also a source of gelating and glue. In light of the above, it is clear that the materials released from the abattoir and poultry processing plants are mainly organic in nature which can cause adverse conditions if discharged into the natural water courses. Therefore it is necessary that the effluents are to be treated to reduce its BOD to the lowest possible level so that it will not cause any harmful effects when discharged into inland waters or sewers and it is in conformity with the requirements of appropriate pollution control board.

Treatment and disposal of effluents

Many different systems exist for the treatment for meat plant effluents. In general, there are two methods of treatment of effluents:

1. Primary treatment where physical methods are employed for removing solids, suspended solids and grease.
2. Secondary treatment where biological methods are adopted for the degradation of organic matter.

Primary treatment

This consists of: (i) screening out solids and removing of fats by mechanical skimming; and (ii) sedimentation.

Screening by rotary wiremesh screen removes coarse materials such as hair, flesh , paunch manure and floating solids. Sedimentation takes place in settling tanks. One to three hours detention period is sufficient in settling tanks. Primary treatment results in about 35 per cent reduction in BOD level.

Secondary treatment systems

The secondary treatment methods commonly used for the treatment of abattoir effluents include: (i) catch basins; (ii) air floatation; (iii) trickling filters; (iv) lagoons (aerobic and anaerobic); (v) activated sludge system and (vi) biological discs. All these treatments are capable of providing complete treatment and can achieve BOD reductions of 70–99 per cent.

Lagoons are presently the most common method of treating abattoir and poultry processing effluents. This is primarily a result of the availability of low cost land. There are two types of lagoons—aerobic and anaerobic lagoons.

Aerobic lagoons

The aerobic pond depends upon algae to supply oxygen for reducing BOD level. Because the algae are dependent upon sunlight to produce oxygen, the pond must be shallow. The depth of the pond is around (1 to 1.5 metre). A BOD reduction of 80 per cent can be achieved by this method.

Anaerobic lagoons

Anaerobic ponds are widely used for treatment of abattoir and poultry processing effluents. These effluents have appreciable amounts of fats and proteins, high concentrations of nurients and an elevated temperature all of which are essential for good anaerobic biological treatment. Reduction of the BOD is performed by bacteria in the absence of oxygen. The ponds are usually much deeper (4 to 5 metres). A BOD reduction of 60–80 per cent can be obtained by anaerobic lagoon systems. Advantages of anaerobic lagoons system are low initial cost, ease of operation, ability to accept stock loads and ability to handle large amounts of grease. Effluents are given in Table 8.4.

Table 8.4. Effluent limits.

Description	*BOD*	*Suspended solids*
Broiler processing g/450 LWK (14 l/bird)	100 (16 mg/l)	280 (48 mg/l)
Abattoir g/450 kg LWK (4.8 L/kg LWK)	7 (3.5 mg/l)	100 (50 mg/l)

Note: LWK (Live Weight Kill).

UTILISATION OF BY-PRODUCTS FROM MEAT INDUSTRY

Animal industry contributes only 10 per cent to our national economy against the 40–50 per cent contributed by agriculture. In the advanced countries, meat industry by-products account for 30 per cent of the total value of the meat produced, with benefit both to producer and consumer.

Dressed meat forms 45 per cent of the live weight of an animal, and the balance of 55 per cent constitutes by-products Skin, and to a smaller extent intestines, are being utilised at present. Other by-products like blood, stomach, pancreas, horn and hoof, fat etc. are not used to any significant extent.

By-products other than skin can be put to good use in a number of ways. Blood meal and composite meal (scrap meat, stomach, horn/hoof etc.) can be utilised as feed material. Trimmed fat can be used both for edible purposes and for soap-making. The internal organs are sources of pharmaceuticals. Bone meal is used as source of minerals in feeds and as a manure. A study of by-product economics shows that for every 500 animals slaughtered, the value of the by-products including skin works out to Rs. 40,000 if fully utilised.

The advantages of proper utilisation of by-product are: (i) additional income to the producers; (ii) higher standards of sanitation in the slaughter-house environment; (iii) recovery of valuable feedstock; and (iv) cheaper meat to the consumer.

POSSIBLE UTILISATION OF SLAUGHTER-HOUSE MATERIAL AS A SOURCE OF FUMARASE

Immobilised proteases and pectinase are in use for clarification of fruit juices, wine and beer, and immobilised glucose oxidase for removal of glucose and oxygen from beer and other beverages. While micro-organisms are at present the main sources of industrial enzymes, slaughter-house wastes could yield trypsin, papain, pancreatin, etc.

Immobilised beef liver mitochondria were examined as a source of fumarase for L-malic acid production. The mitochondria were isolated by centrifuging liver homogenised suspensions in isotonic sucrose solution, discarding the debris, and again centrifuging the supernatant at 6,500 g for 10 minutes, followed by washing/sedimentation. The mitochondria were immobilised in polyacramide gel using gamma-irradiation by BARC procedures, and suspended in phosphate buffer. Maximum retention of ADH enzyme activity was obtained in gels containing 16 per cent acrylamide, 0.8 per cent Bis and a 100 kR dose of gamma rays.

The immobilised fumarase preparation was more stable to heat, radiation and 15 day storage than the untrapped enzyme. The immobilised preparation could stand 12 repeated uses for L-malic acid production in a batch process in which the packed gel (2.5 g) was stirred for 24 hours with 10 ml of 1.0 M fumarate in 0.1 M phosphate buffer of pH 7.5, after which the gel was washed for re-use.

A continuous enzyme reaction system could have several industrial advantages over present batch production of L-malic acid from fumaric acid using *Lactobacillus brevis* broth carrying soluble enzymes. Fumarase has been immobilised in cellulose triacetate for malic acid production. Immobilised whole cells combine the advantages both of immobilised enzymes and fermentation, and avoid tedious enzyme purifications steps. Intact yeast cells have been immobilised, and also whole cells of *Brevibacterium ammoniagenes* for fumarase immobilisation. Cellular organelles rich in particular enzymes, if immobilised, could eliminate inactivation of particle-bound enzymes by cytoplasmic proteinases.

PROCESSING AND UTILISATION OF POULTRY INDUSTRIAL BY-PRODUCTS

Following rapid growth of the poultry industry, as judged by several indices, wastes have multiplied also. Head, feet, feathers, blood and viscera constitute 26 to 29 per cent of bird weight. Collectable by-products per 1000 kg broiler weight, expressed in kg, are: feathers 54, blood 37, offal/viscera 118, feet 30, lungs 9, unwholesome carcasses 10, fat/grease 6 and settlings/sludge 10. Yields of processed by-products meal vary between 23.0 and 27.9 per cent depending on bird type and age.

Feather meal, best obtained by cooking clean feathers at 142 to 153°C (3 to 4 kg/sq. cm) for 30 to 60 minutes, contains, as percentages: protein 85 (mostly derived from the keratin in feathers), ether-extract 6.7, ash 3.9, calcium 0.2 and phosphorus 0.75, besides B-vitamins. Because of deficiency in lysine, methionine, histidine and tryptophan, feather meal can make up only 4 to 5 per cent in a poultry ration to supply a quarter of the protein requirement. Feathers are used in pillows, quilts and sporting goods, or cooked for fertiliser use.

Offal meal or poultry by-product meal, the ground material derived from clean heads, feet, intestine, etc. (but not feathers) by dry rendering or wet autoclaving (one kg/sq.cm. for 30 minutes, followed by fat separation, drying and grinding), has, as percentages: protein 72, ether-extract 3.1, ash 3.0, calcium 1.5 and phosphorus 0.64. It can satisfactorily constitute 12 per cent of a broiler ration, replacing three-quarters of the fish meal component. Blood meal (coagulated, after anticoagulant addition, in a tank using steam) can be had in a yield of 0.7 to 0.8 on bird live weight; its high protein quality is reflected also when feather and blood are processed together, an economical procedure. Mixed feather, by-product and blood meal can constitute upto 5 per cent of chick and broiler rations and 7.5 per cent that of growing pullets, laying hens and breeders. Hatchery wastes, constituting infertile eggs, dead embryos, empty eggs shells, unhatched eggs and unusable chicks, yield a high-quality animal feed by immediate cooking, drying (80°C for 8 hours) and grinding with 0.02 per cent antioxidant (BHA), or alternatively by dry-heat dehydration. Better chick hatchability decreases the protein level, which is of the order of 22 to 32 per cent, fat constituting 10 to 18, calcium 25 to 18, and phosphorus 0.33 to 0.66 per cent. Protein quality equals that of fish meal, and considerably replacements are possible for growers, broilers and layers (at 8 to 15 per cent levels).

Damaged and unsound eggs, constituting 6 to 12 per cent of the eggs produced, can be used to prepare several products. Albumen flake, in demand by the textile and printing industries, is made from the homogenised, desugared and fermented yolk, which is finally dried slowly at 56–58°C to below 7 per cent moisture. Egg shell meal, with 91 per cent ash and 36 per cent calcium, is prepared by grinding shells that have been washed free of albumen, alkali-boiled to dissolve shell membrance, dried and powdered. It is a source of calcium in poultry feeds, and 1,000 eggs, at a processing cost of below one rupee, yield 5 kg of shell powder and 100 g of residual albumen.

PELLETISATION OF POULTRY WASTE

Poultry farming is an important economic activity in India. An estimated population of 100 crores birds are being reared in the country. For every lakh of birds, nearly 3000–3500 MT droppings are generated annually with an average moisture content of 75 per cent. The droppings are very good organic manure in view of high nutrient contents. Though it is used as manure in the poultry belts, its usage is limited in other areas. Bad odour coupled with difficulties associated with its storage and transportation are found to be the limiting factors in extending its usage in non-poultry areas. The above problems can be eliminated by the removal of the odour and converting it into pellet form. In many developed countries, poultry manure pellets are extensively used. India with its vast poultry base and cultivable lands, offers lot of potential for manure pellets.

Poultry manure is an extremely rich source of nitrogen and organic material. It is richer than cow dung in nitrogen, phosphorus and potassium content. The manure may be pure droppings or old built-up litter, which are cleared once in a year or once in two years. The average nutrient content of poultry droppings as compared to other animal refuse is shown in Table 8.5.

Table 8.5. Comparative average of nutrient contents.

Manure	*Nitrogen (%)*	*Phosphoric acid (%)*	*Potash (%)*
Poultry manure	1.0–1.8	1.4–1.8	0.8–0.9
Cattle dung	0.3–0.4	0.1–0.2	0.1–0.3
Sheep dung	0.5–0.7	0.4–0.6	0.3–1.0
Horse dung	0.4–0.5	0.3–0.4	0.3–0.4

Relevance and Importance of Organic Manure

There are several compulsions that render organic manure more relevant to India. These are:

1. The spiraling cost of chemical fertilisers renders their use uneconomical.
2. The exchequer also suffers adversely due to the increase in imports of fertilisers as the requirement has been outstripping the supply by around 35 per cent.
3. The over dosage of chemical fertilisers is increasingly damaging the ecology.
4. The use of chemical fertilisers as represented by micro-elements depletes the soil fertility. A test reveals that chemical fertiliser give high yield initially, but leads to micro-nutrient deficiency and decreased yields later.
5. The bulkiness of organic manure enables moisture retention, aeration, filth and soil temperature.
6. Organic manure improves microbiological regimes augmenting the reserves of energy, nitrogen and other metabolic requirements. Without organic matter the soil will become dead soil.
7. Indian soils, like all tropical soils in general, are deficient in organic matter, nitrogen and phosphorus. Potassium deficiency is encountered in certain pockets.
8. Pests and disease damage of grassland on organic farms is rare according to a recent survey of grassland established practices on organic farms conducted by the Organic Advisory Service, Elm Farm Research Centre, UK.

Poultry Waste—A Promising Organic Manure

Poultry manure accrues to the extent of 10 to 12 kg/day from 100 laying hens. Pathogenic micro-organism and high moisture content must be reduced to enable its use as poultry and ruminant feeds and this is done by initial sun-drying, autoclaving (15 psi for 10 minutes), forced-draft drying and grinding. True protein content of dried poultry manure is 10 to 11 per cent, with another 17 per cent of non-protein nitrogen, mostly uric acid, which poultry cannot use but ruminants can convert to protein. Poultry litter has been successfully used to raise ewes and lambs, dairy cattle and pigs. Beef cattle and calves utilise large quantities of dried poultry manure.

Layers can be used from 12.5 to 25 per cent in their diet, and starter layers, broiler starters and finishers and growing chicks from 10–20 per cent. Poultry manure is a better fertiliser than that from cow or sheep, one tonne of it containing 33 kg of nitrogen, 22 kg of phosphorus and 22 kg of potash, besides magnesium, sodium and calcium, and trace elements.

Biodegradation of raw manure can be brought about by micro-organisms or fly larvae; in 8 days, the latter yield, from the excreta of 100,000 chickens, 250 to 500 kg of pupae which, properly processed, are a good protein source for growing chicks.

Poultry waste has all the above mentioned relevant features that arc common to all organic manures. It scores over all other organic manures with respect to the following aspects:

1. The rising prices of other organic manure such as groundnut oil cake render them uneconomical.
2. The high increase in poultry population as compared to cattle population assures increased and regular availability of poultry manure as compared to cattle manure.
3. Poultry manure has a higher nutrient content than all other animal refuse. Experiments conducted in Andhra Pradesh for example, revealed that 200 kg of groundnut cake along with 300 kg of ammonium sulphate yielded a production of 30 tonnes of sugarcane whereas 4 tonnes of poultry manure yielded 70 tonnes. In other words, the yield was 2.33 times higher with the use of poultry manure.

Constraints and inhibitions in using poultry waste as manure

Poultry manure has been traditionally used in India, however, the usage is predominant in areas where poultry farms abound. The constraints and inhibiting factors that restrict the use of poultry manure in the country are reported to be the following.

Non availability

Poultry manure is not available in bulk as and when the farmers require it. This reasons is that the poultry farms in the country have never taken up marketing poultry manure in a cohesive manner, nor ever attempted to promote or popularise poultry manure as an alternative to the expensive and less nutritious chemical fertilisers. For the poultry farms, the product has always been an irritant as it created disposal problems to the management.

Most farmers are not aware of all the advantages of poultry manure, the dosages recommended and such aspects.

Storage problems

Poultry manure unless cleared from the farm or the farm premises periodically could lead to environmental hazards as the accumulated droppings often attract flies and rodents, spreading diseases. While the poultry farms are ever eager to get rid of the droppings at the earliest, farmers hesitate to buy and store it for fear of messing up their place, inviting bad odour, flies and rodents.

Handling and transportation problems

Unprocessed poultry manure emits a severe odour, unless cleared on time. Even otherwise transporters hesitate to handle and transport the raw manure as it makes their vehicles messy and strinking.

It is against this back drop that processing and palletising of poultry manure assumes significance. The process involves drying, de-odouring, making it germs-free and converting it into pellets for easy and convenient packing.

Advantages of pelleted manure

The pelleted product has the following advantages as compared to both unpelleted poultry manure and chemical fertilisers:

1. Free from disease germs.
2. Free from bad smell.
3. Indefinite storage possible.
4. The plant nutrients are fully preserved.
5. User-friendly.
6. Provides very high yield as compared to unprocessed manure.
7. Cost effective when the higher yield is taken into account.
8. Environmental-friendly.
9. The product can be transported over any distance without messing up the vehicles or polluting the environment.
10. The product can be made available in retail packs.
11. As the product is in pellet form it cannot be adulterated.
12. The product is soil-friendly, as it has not bad side effects unlike chemical fertilisers which depletes the soil fertility.

13. Unlike most chemical fertilisers which provide single or few nutrients, chicken manure pellets supply a range of nutrients.
14. Ease of handling.

Technical aspects

The typical specifications of pelleted poultry manure are given in Table 8.6.

Table 8.6. Typical product composition.

Description	*Quantity (%)*
Primary nutrients	
Nitrogen	3.0–4.0
Phosphorus	1.5–2.0
Potassium	3.0–4.0
Secondary nutrients	
Calcium	4.0–6.0
Magnesium	0.6–1.0
Sulphur	0.10–0.15
Micro nutrients (PPM)	
Iron	Upto 0.10
Zinc	0.03–0.05
Manganese	0.04–0.05
Copper	0.002–0.003
Boron	0.04–0.05
Molybdenum	0.002
Chloride	0.70
Moisture	10.0–12.0
Organic matter	75.0–80.0
Product form	Pellets, 3–4 mm

Manufacturing process

The droppings collected from the farm are to reduce moisture. Then odours are removed by de-odourising material brought to mixing chamber is mixed with binder. After thorough mixing, the material is send to the pelletising machine wherein it is converted to pellet form of 304 mm size. The pellets are packed in 50 kg HDPE packs and taken to the market.

End use segments

The products has potential application in the following end use segments:

1. Horticultural farms.
2. Floriculture.
3. Green houses.
4. Plantation crops.

5. Oilseeds cultivation.
6. Spice cultivation.
7. Mushroom cultivation.
8. Ayurvedic/herbal plantations.
9. Terrace gardens/kitchen gardens.
10. Lawns—hotels/corporates etc.
11. Landscaping.
12. Golf courses.
13. Amusement parks.
14. Municipal parks/botanical gardens.
15. Residential townships.
16. Industrial townships.

The individual crops/plants for which the product can be used for cultivation are mentioned in Table 8.7.

Table 8.7. Crops for which the products can be used for cultivation.

Plantations	*Fruits*	*Vegetables*	*Spices*	*Floriculture*	*Herbal plants*
Tea	Mango	Beans	Pepper	Rose	Tulsi
Coffee	Banana	Brinjal	Ginger	Jasmine	Henna
Cashew	Citrus	Cabbage	Turmeric	Anthurium	Bilva/Bael
Coconut	Lemon	Cauliflower	Clove	Tuberose	Amla/Aavla
Arecanut	Grapes	Chillies	Cardamom	Lily	Neeli
Cocoa	Orange	Onion	Cinnamon	Gerbaria	Hibiscus
Rubber	Guava	Soya	Garlic	Orchids	Brahmi
Sugarcane	Papaya	Tomato	Tamarind	Chrysanthamum	Vetiver
	Pumpkin	Radish		Marigold	Vinca rosea
Oil seeds	Jackfruit	Potato		Champak	Sarpaganta
Sunflower		Green peas		Crossandra	Arugampul
Safflower		Gourds		Carnations	Cenna
Groundnut		Mushrooms		Grass	Satavari
				Creepers	Aswagandha

The growing importance for organic cultivation is expected to boost the market for organic fertiliser. Poultry manure is one of the cost-effective organic manure. Hence, pelleted manure is expected to have good future.

POULTRY PROCESS AND POLLUTANTS

Large and medium size poultry processing units add to the load of environmental pollution by way of effluents.

Characteristics of Waste-Water from a Poultry Plant

The waste-water discharged from a poultry processing plant may range from 14 to 28 litres per bird. Poultry processing waste-water is typically organic in character, high in suspended solids and floating material. The total liquid waste from the poultry dressing process contains varying amounts of blood, feathers, fleshings, fats, washings from evisceration, digested and undigested foods, manure and dirt. The manure from lairages and blood from killing and stricking operations contribute the largest amount of pollution from the process. Table 8.8 shows some characteristics of waste-water from a poultry plant.

Table 8.8. Characteristics of waste-water from a poultry plant.

Character	*Range*	*Average*
pH	6.3–7.4	6.9
BOD (mg/l)	370–620	473
Suspended solids (mg/l)	120–296	196
Total solids (mg/l)	–	650
Volatile solids ppm (mg/l)	–	486
Fixed solids (mg/l)	–	164
Settleable solids (mg/l)	15–20	17.5
Grease (mg/l)	170–230	201

The effluents from the broiler processing plants (Table 8.9) show high BOD and high suspended solids.

Table 8.9. Typical poultry plant effluent (untreated) for broiler processing (1000 birds).

Flow (l)	16,000 liter
BOD (450 mg/l)	13.0 (14 kg)
Suspended solids (345 mg/l)	10.5 kg
BOD – 5 day BOD	

DESIGN CRITERIA FOR CONSTRUCTION OF EFFLUENT TREATMENT PLANTS

If an effluent treatment plant is to be constructed for an abattoir or poultry dressing plant, there are four stages in the entire process.

Design Criteria

Characteristics of waste: (i) average flow; (ii) maximum flow per day; (iii) BOD; (iv) suspended solids; (v) weekly load; and (vi) effluent limitations.

Selection of Treatment Method

Selection of treatment method depends on land availability, proximity to residential or commercial areas, size of poultry plant initial construction cost, operation and maintenance costs, ease of operation, pumping requirements, quality of in-house waste conservation and municipal/pollution control board requirements.

Design of the Treatment System

It is important to collect reliable topographical information on ground elevations, location of existing property lines, building and sewers, sewer invert elevations and a benchmark elevation on which to base proposed construction.

Model Design Suitable to Indian Conditions for Effluent Treatment

Two model design of effluents treatment system (500 sheep slaughter capacity and 1000 broiler processing capacity) incorporating the basic principles of effluent treatment. The treatment system includes the facilities for screening, sedimentation and final treatment by anaerobic lagoons. The screening, sedimentation and final treatment systems are shown in Figs. 8.1 and 8.2 respectively.

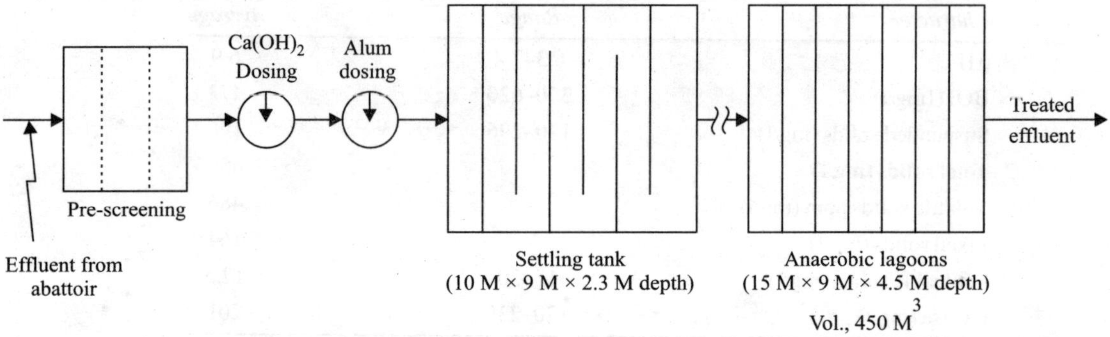

Fig. 8.1. Schematic layout of effluent treatment system for sheep abattoir (cap., 500 sheep/day) 10,000 kg live weight (Approximately).

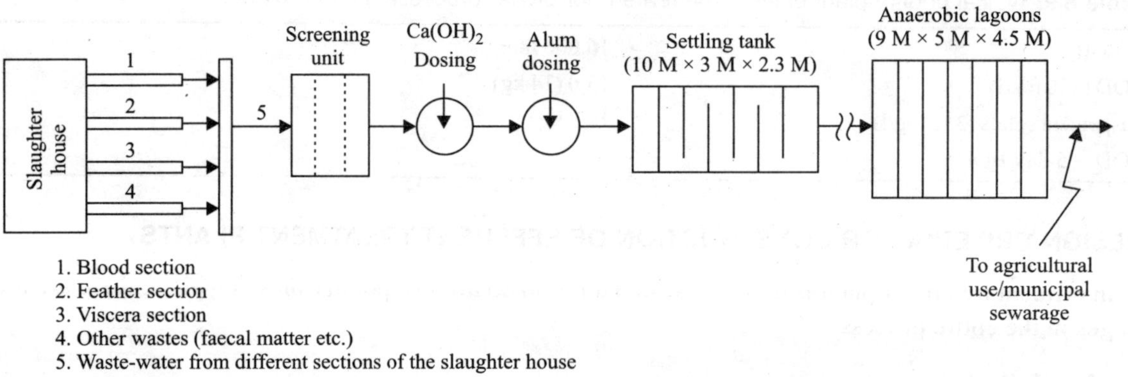

Fig. 8.2. Schematic layout of effluent treatment system for poultry dressing plant (Cap: 1000 birds/day) 1500 kg live weight (Approximately).

Thus, the above discussion justified the need for improvements in the effluent treatments system in Indian abattoirs to protect the environment. Government is making earnest efforts in the modernisation of abattoirs and poultry processing plants in our country. This will in near future definitely result in the pollution control providing thereby a clean environment for betterment of mankind.

9

Coffee and Tea Wastes

INTRODUCTION

Coffee pulp and coffee hulls are the two main by-products that arise from the production of coffee beans from coffee cherries. Both have been widely investigated over many years for agricultural use. More recently, another coffee by-product, coffee grounds, has become available in increasing amounts due to the widening popularity of instant or soluble coffee, and it has been examined as an animal feed. Before describing the work that has been undertaken on the three by-products, it may be beneficial to those not familiar with coffee production and processing to briefly describe the systems by which they are obtained.

Coffee cherries or berries are preferably picked when they are red in colour and fully ripe in order to give the best quality coffee. Normally, each cherry contains two seeds or beans enveloped by four coats. These coats (in the order of outside to inside) are the pericarp (or fruit wall); a sticky, slimy substance which is often called the pulp; a stiffer outer covering of the beans themselves, known as the parchment; and a thin almost transparent inner lining covering the seeds which is called the silver skin or seed testa. All these coats have to be removed to prepare the beans for human use. This is achieved in one of two ways—by the older or dry method, or by the modern or wet method (which is often referred to as the West Indian process).

METHODS OF PROCESSING COFFEE

Dry Method

In the dry method, the cherries are either dried on the trees or picked and spread in a layer a few centimetres deep on stone drying grounds known as barbecues. During the first two days, the cherries are frequently turned so that all are exposed to the sun, after which precautions are taken to avoid wetting. The pulpy layer ferments and slowly dries, and the whole process requires about 15 to 25 days. When thoroughly dry, the cherries can be stored. Next, the beans are separated from the dried, shrunken coats. This has been done in a primitive manner by pounding in a mortar, but a hulling machine is now commonly employed.

In either process, the dried covering is broken and the beans are set free. Later, the beans are separated from the coats leaving behind large amounts of mixed residues. Although this method is simple and requires inexpensive machinery, it is becoming obsolete since it yields a low quality product known in the trade as 'hard coffee' and it is too dependent on the persistence of fine, settled weather over a considerable period of time.

Wet Method

In the wet method, the cherries are pulped as soon as possible after picking; that is, within 24 hours, to avoid fermentation. They are placed in large tanks full of water immediately after harvesting to separate the immature from the ripe cherries. The former are light and float on the surface, and the latter sink to the bottom from where they are drawn through pipes leading to the pulpers. There are various types of pulpers, all designed on much the same principle. The cherries cannot pass through them without being reduced to a pulp by the rasping action of revolving cylinder or other moving surface. The mixture of beans and pulp is passed into a vat full of water, where mechanical stirring causes the heavy beans to settle to the bottom and the lighter pulp to be carried away by an overflow of water. The beans are drawn off in a stream of water and then separated by being passed through a form of sieve. The slimy layer which still adheres to the beans is removed by fermentation with yeasts and bacteria in a special vat. This usually takes 18 to 24 hours, but the process may be hastened by adding enzyme preparations or 2 per cent NaOH solution. Washing of the beans with successive amounts of water together with agitation and stirring leaves them completely free of mucilage. They are then dried until yellow-green in colour and either sold as such—that is, covered by the last two coats—or dehulled and polished before being shipped and sold. For the dehulling or threshing, a variety of machines are employed, all of which are designed to crack the parchment without injuring the beans. This is most readily achieved if the material is thoroughly dry, since the parchment is then brittle and relatively easily broken. Furthermore, the beans that result store well, provided the relative humidity is not high. The light pieces of parchment are separated by winnowing from the heavy beans. Finally, simple rubbing and winnowing removes the silver skin and leaves clean beans as ordinary, unroasted coffee. Both the dry and wet methods are shown in outline in Fig. 9.1.

Instant coffee

Instant coffee is made by treating beans, usually a mixture of different varieties, with boiling water. The solution obtained is freeze-dried to give instant or soluble coffee. However, a considerable amount of residue or grounds remains which contains about 80 per cent water. This is dried by heating to give dried coffee residue.

From 1 tonne of cherries, about 430 kg of coffee pulp are obtained, which in terms of dry matter, represents about 30 per cent of the original cherry. Dehulling at the latter stage yield about 60 kg of coffee hulls, equivalent to approximately 40 kg of dry matter, which is close to 12 per cent of that in the cherry. The yield of beans is about 390 kg, which contains about 190 kg of dry matter (i.e., about 55 per cent of that in the cherry). In the manufacture of instant coffee, about 67 per cent of the dry matter in the beans is left in the residue.

COFFEE PULP

The high water content of coffee pulp causes difficulties in its transportation, handling, preservation and direct use as an animal feed. To overcome this problem, two approaches have been considered. At the Institute of Nutrition of Central America and Panama in Guatemala, *Bressani* have examined ensiling as a method of storing the pulp for feeding to animals. The pulp is allowed to dehydrate to a water content of about 65 per cent by spreading on a clean surface. With sufficient sunshine and air movement, the dehydration is achieved in 8 to 10 hours. It is then ensiled in a trench silo along with 3 to 5 per cent sugarcane molasses. Analysis has shown that there is very little change in nutrivtive value, but there are significant and advantageous decreases in caffeine and total tanning content. On removing from the silo, the pulp silage is initially readily accepted by animals. Unfortunately, exposure to air causes it to darken in colour and become less palatable, and the voluntary intake by animals decreases markedly. The dark colouration is due to oxidation of polyphenols present in coffee pulp, so it is possible that this could be controlled by introducing inhibitors of the activity of polyphenol oxidases. Another possibility to overcome this palatability problem is to produce a mixed silage of pulp with another feed such as grass or green maize. In a laboratory study, *Murillo* ensiled coffee pulp with equal amounts of either Napier grass (*Pennisetum purpureum*) or maize fodder. An appreciable amount of molasses, about 16 per cent, was added to each mixture and the products were considered to have a better quality than coffee pulp silage. There was too little of the mixed silage to carry out animal feeding trials.

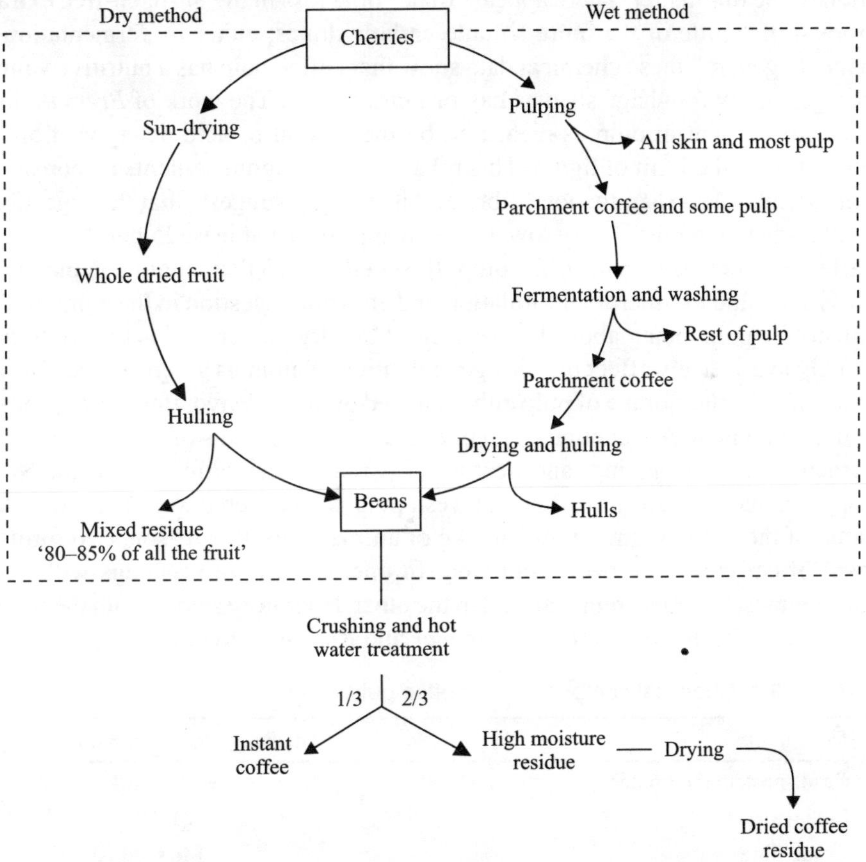

Fig. 9.1. A flow diagram outlining the steps in the processing of coffee.

The other approach used to facilitate handling and preservation of coffee pulp is dehydration either in the sun or by heat. To dehydrate in the sun, the pulp is spread on a clean surface to a thickness of about 5 to 8 centimetres. It is turned three to four times daily and is dried to a moisture content of about 12 per cent in about 24 to 32 hours. During the drying, the colour changes from red to black, but the product appears to store well and lose some of its caffeine. Sun dehydration is a cheap way to dehydrate pulp, but a large area of surface is needed and the change in colour is associated with a decrease in the voluntary intake by the animal. Drying with heated air has been examined and it can be a speedy process—approximately 60 minutes. A modification of this drying method is grinding the wet pulp, followed by continuous centrifugation to remove some of the water, and then dehydrating in a drum drier. The product has a better appearance than the sun-dried material and may be more palatable.

Chemical Composition

The composition of the dry matter in coffee pulp as indicated by a number of chemical analyses is shown in Table 9.1. Fresh coffee pulp invariably contains between 60 per cent and 75 per cent water, and the dehydrated material has a moisture content of about 12 per cent. Dehydration causes some changes in the composition of the dry matter. There appears to be some loss in the nitrogen-free extract, which may be due to removal of a little of the more soluble carbohydrates, either by fermentation or in solution through seepage. In general, these chemical data show that coffee pulp has a nutritive value equivalent to that of an average quality roughage such as hay or grass silage . The work of *Bressani* indicates that of the total fibrous constituent in pulp as measured by the neutral or acid detergent fibre determination approximately half is in the form of lignin. This relatively high lignin contents responsible for the crude fibre being considerably less than the total fibre and it strongly suggests that the digestibility of pulp in ruminants is not high and in pigs is very low. It would appear that at least 25 per cent of the total protein may also be affected by the tannin in coffee pulp. It is well known that tannin combine with protein and render them less available for microbial formation and enzymic digestion in the animal's gut. Caffeine is also process in coffee pulp, being about 1.5 per cent of the dry matter, and is known to have as diuretic effect that is likely to adversely affect the nitrogen retention of animals given appreciable amounts of the by-product. Fortunately, the storage of pulp either ensiled or in the dehydrated form—for several months reduces the caffeine content. A reduction of 70 per cent after 15 months of storage has been recorded by *Bressani*. Fermentation, autoclaving, and heating also reduce the caffeine content. Nevertheless, the presence, of appreciable amounts of caffeine in fresh pulp is considered to be a disadvantage and it may reduce the value of the pulp by limiting the intake of animals. For this reason, two forms of processing for the removal of caffeine have been examined. In one, successive washings with water have been shown to achieve almost complete removal, and in the other, laboratory studies with the fungus *Penicillium crustonsum*, which has been reported by *Schwimmer* and *Kurtzman* to utilise caffeine, could be fruitful.

Table 9.1. Chemical composition of coffee pulp.

Description	*Range (g/kg dry matter)*
Crude protein (N × 6.25)	90.1–128.1
Ether extract	21–29
Crude fibre	146–240
Ash	64–96
Nitrogen-free extract	534–678
Calcium	6.34
Phosphorus	1.33
Magnesium	Trace
Sodium	1.14

(Contd...)

Description	Range (g/kg dry matter)
Potassium	20.2
Zinc	4.6 (mg/kg)
Copper	5.7 (mg/kg)
Manganese	7.2 (mg/kg)

The results of *Bressani* that give the amino acid composition of the protein in coffee pulp indicate that it has a relatively high quality for non-ruminants. It appears to compare favourably with that in soybean meal, but is even more deficient in the sulphur-containing amino acids. Such a limitation may be enhanced if the tannins in pulp make some of the amino acids partly unavailable to the animal.

Coffee Pulp as a Feed for Ruminants and Pigs

To evaluate the material for ruminants, most of the studies have been carried out with growing and fattening cattle given either sun-dehydrated pulp or coffee pulp silage. These studies have shown that in production ratios, the weight gain and food intake of cattle decreased as the level of pulp in the diet increased. These effects were more pronounced when the level was greater than 20 per cent. In this respect, a similar result was obtained in earlier work with lactating cows. The inclusion of coffee pulp at levels below 20 per cent did not affect milk production. There is some evidence, reviewed by *Bressani*, that the depressing effect of coffee pulp on the weight gain of cattle is less marked when the level of dietary protein is high. If this is correct, it would seem that the value of pulp on the weight gain of cattle is less marked when the level of dietary protein is high. If this is correct, it would seem that the value of pulp is limited by poor utilisation of the protein it contains, which is consistent with comments made earlier concerning the characteristic of tannins. Without doubt, pulp for ruminant feeding needs to be classified as an average quality roughage, and avoid discernible reductions in animal performance, its inclusion in production ratios should be 20 per cent or less. At higher levels, *Cabezas*, have shown that young steer previously adapted to dried coffee pulp for 102 days gained weight, while those which had not received the pulp lost weight. Such an adaptation with time may occur, but may be less important in cattle given the lower levels of pulp.

Very few studies have been made to evaluate coffee pulp as a pig feed, probably because the composition indicates that it is less suitable for this purpose than for feeding to ruminants. In Guatemala, the work of *Rosales* with growing pigs has shown that at three different physiological ages, the inclusion of sun-dehydrated and ground coffee pulp at dietary levels of 8.2, 16.4, and 24.6 per cent caused a decrease in growth rate and efficiency of feed utilisation which was greater as the pulp in the diet increased. Similar but more striking results were obtained by *Malynicz* in Papua, New Guinea. When dried coffee pulp was given as 10, 20 or 30 per cent of the diet in place of sorghum, the daily gain of growing pigs was 213, 159 and 113 g, respectively, compared with 495 g for controls. From these results, it may be concluded that coffee pulp is not a suitable feed for growing pigs unless some reduction in growth rate and fat deposition is desirable, as in the production of lean bacon pigs.

Other Uses of Coffee Pulp

Other than its traditional use as an organic fertiliser in coffee plantations, which provides a route for recycling some of the nutrients in coffee cherries, the pulp has been used to make coffee molasses and has been investigated as part of a supplementary feed for fish, *Tilapia aurea*. Coffee molasses is made by drying the sugars from the mucilage and pulp. *Buitrago* have found it has a value a little less that of sugarcane molasses when given in diets based on maize at a level of 30 per cent for growing pigs. The pulp itself gave promising results when given in mixtures containing cotton seed meal with and without urea to *Tilapia* in El Salvador. Further work is needed to specify the amounts to pulp which can be used in relation to other dietary ingredients, including vitamin and mineral supplements.

COFFEE HULLS

Coffee hulls are a dry, stable material that is easy to store and handle. Its decomposition is very slow because it is covered by a waxy film and it has a high lignin content.

Chemical Composition

The composition of the dry matter of coffee hulls, as indicated by a number of chemical analyses, is shown in Table 9.2. Coffee hulls contain between 5 to 10 per cent water. The protein content is very low, similar to the amount found in very poor quality roughage such as wheat straw or mature, topical standing hay, and the fibrousness of the by-product is pronounced. The work of *Bressani* has shown that the total amount of cell wall constituents in hulls, measured as neutral detergent fibre, may be as high as 88 per cent of the dry matter. Of this, half is cellulose, and approximately one-quarter and one-fifth are hemicellulose and lignin, respectively. This high lignin content render the by-product very indigestible and it may be described as a filler in animal feeds. Its inclusion in almost any diet is likely to lower the overall nutritive value. Attempts have been made to improve the value of hulls by chemical treatment causing delignification. The use of either sodium or calcium hydroxide solutions of three different strengths has been shown to cause an appreciable reduction in content of acid detergent fibre, especially at the highest concentration of 10 per cent.

Table 9.2. Chemical composition of coffee hulls.

Description	*Range (g/kg dry matter)*
Crude protein (N × 6.25)	26.3 – 101.7
Ether extract	6.5–20.2
Crude fibre	297–754
Ash	5.4–75.3
Nitrogen-free extract	204–515
Calcium	1.61
Phosphorus	0.30

Coffee Hulls as a Feed for Ruminants

The very high fibre and very low protein content make coffee hulls an unsuitable feed for non-ruminant animals. Several workers have evaluated the by-product as a constituent of diets for cattle. In Cuba, *Rodrigues* and *Zamora* compared hulls with rice husks as the main roughage source in complete diets for dairy cows. Those receiving the hulls gave lower yields than did the cows given rice husks. *Ledger* and *Tillman*, in Kenya, examined the effect of replacing ground maize in a highly concentrated fattening diet with 10, 20 or 30 per cent coffee hulls, given to Boran steer for 89 days. The animals were fed individually to appetite. At the 10 to 20 per cent level of inclusion, feed intake, weight gain, and efficiency of feed conversion were unaffected, but at the 30 per cent level all were reduced. Similar results were obtained by *Jarquin* in Guatemala, using Holstein calves, approximately 3 months old and about 90 kg in weight. They were given diets with no coffee hulls and 15 to 30 per cent. Rate of growth decreased as the hulls in the diet increased. Intake of diet containing 30 per cent hulls was less than that with none. The effect of adding 1.5 per cent urea to the 30 per cent coffee hulls diet and at another time increasing the molasses content from 20 to 28.5 per cent were examined. The performance of the calves eating these modified diets remained below that of those receiving the diet without hulls. From all these trials it may be concluded that hulls may be included in diets for older cattle upto levels of 20 per cent without appreciably affecting their performance. Any decrease that did occur would be difficult to discern in practice, and in the experimental situation would not be statistically significant unless a large number of animals were used. For young cattle, the inclusion of hulls may be more detrimental, and levels less than 20 per cent may give a perceptible decrease in performance.

Other uses of coffee hulls

Coffee hulls have been used as a fuel. On coffee plantations, large amounts have been spread on fields, where it slowly decomposes and acts as a soil conditioner.

DRIED COFFEE RESIDUE

There are several alternative names for this by-product from the manufacture of soluble or instant coffee. It has been referred to as extracted coffee bean meal, coffee residue, coffee grounds, spent coffee grounds, and, in certain commercial circles in the United Kingdom, Cherco. All these names apply to the same product, but there are variations in the composition of this material which may be related to differences in the original beans and in the drying of the residue. When dry, the residue is a stable, brown, powdery material which stores well and is relatively easy to handle.

Chemical Composition

The composition of the dry matter of dried coffee residue, as indicated by a number of chemical analyses, is shown in Table 9.3.

Table 9.3. Chemical composition of dried coffee residue.

Description	*Range (g/kg dry matter)*
Crude protein (N × 6.25)	102–130
Ether extract	225–283
Crude fibre	466–510
Ash	7–8
Nitrogen-free extract	143–168
Calcium	0.55–0.65
Phosphorus	0.50–0.70
Magnesium	0.15–0.25
Sodium	0.40–0.60
Potassium	0.35–0.45
Zinc	9–12 (mg/kg)
Copper	25–29 (mg/kg)
Manganese	29–36 (mg/kg)

The residue contains between 5 to 10 per cent moisture. It has an average protein content, approximately 10 per cent dry matter, but a high fibre and a relatively high fat and gross energy content. Its total mineral or ash content is very low, with none of the major minerals being present in more than small amounts. The studies have shown that the dry matter contains 73 per cent acid detergent fibre, 48 per cent cellulose, 23 per cent lignin, and traces of tannin (0.12 per cent) and caffeine (0.05 per cent). The fat or oil in dried coffee residue is predominantly triglycerides, which are made upto 50 to 60 per cent unsaturated fatty acids, including a high content of linoleic acids. Measurements of *in vitro* digestibility have given very low values, with indicate it may be poorly digested by ruminants. Since the high lignin content may be partly responsible for the low digestibility, the effect of treating with alkali and other delignifying agents has been studied. Considerablc improvements in digestibility measured *in vitro* have been obtained, but the treated material is still mainly indigestible.

Dried Coffee Residue as a Feed for Ruminants

To date, most of the work undertaken to examine the residue as a practical feed has been carried out with ruminants. Conflicting results have been obtained when the digestibility of the by-product was determined using sheep and, to a lesser extent, cattle. In Italy *Fornaroli* and *Perotti* found that adding the residue–which has a relatively high moisture content—to lucerne hay lowered the digestibility in sheep of some chemical constituents of the diet. At a high level of inclusion of the residue, the digestibility of lipids was low, and at both levels of inclusion, the digestibility of fibre in the diet was markedly depressed. Similar results were obtained by *Ali* in the United Kingdom. Dried coffee residue was included in a concentrate diet at 7.5 and 12.5 per cent, and the concentrate only and the mixture were fed with hay in the ratio of 6 to 4. Increasing the amount of coffee residue in the diet caused a decrease in overall digestibility of the diet and a more pronounced decrease in fat digestibility. For these results, the digestibility of the residue was calculated by difference, assuming there were no associative effects between it and the remainder of the diet. The values obtained were low, especially for fat and nitrogen-free extract. Contrary results have been obtained in Canada by *McNiven*. The digestibilities in sheep of two diets, one without coffee residue and the other with 25 per cent coffee residues were compared, and except for protein, found to be higher in the coffee diet. In this work, the coffee residue was calculated to have 76 per cent total digestible nutrients. It should be noted that in all three studies, the protein in dried coffee residue was found to be poorly digested. However, the digestible energy value and, to a large extent, the nutritive value of dried coffee residue depends on its energy digestibility, which is greatly influenced by the extent of digestion of fat. It is possible that in the drying process, if excess heat is used or it is uncontrolled, may adversely effect the nature of the fat. Such an effect may explain the difference between the result of *Ali* and those of *Campbell* with cattle; *Ali* found that inclusion of dried coffee residue depressed the digestibility of dietary fat, while *Campbell* obtained an increase. A few feeding trials have been carried out in which diets with and without dried coffee residue have been compared. *Mather* and *Agpar* used lactating Holstein cows and found that the residue had no significant effect on milk yield, butter fat percentage, or flavour of milk, but the body weight of the cows were significantly reduced, probably due to low intakes. In the same study, feed intake and growth rate of calves were reduced when the residue was included in calf starter ratios. *Ali* included upto about 7 per cent of dried coffee residue in the diet of fattening steer without adversely affecting the overall performance. However, the rate of growth of steer in this trial may not have been high enough to fully test the diets.

In the majority of these studies with sheep and cattle, the dietary inclusion of dried coffee residue in more than very small levels caused a marked diuretic effect. For example, *Campbell* found that 10 per cent or

20 per cent inclusion of the residue more than tripled or quadrupled, respectively, the urinary output of Holstein steers. The diuresis resulted in increased losses of urinary nitrogen and sodium. It is doubtful that this effect is due to caffeine, since the caffeine content appears to be too low to cause pronounced diuresis. Further work is needed to establish the likely cause. Another problem with dried coffee residue is that the material is not readily acceptable or palatable to cattle or sheep. In all the studies considered the authors make a special mention of lowered intakes as the level of inclusion increased or the procedures used to induce the animals to eat the residue or both. In summary, it may be concluded that unless dried coffee residue has a high fat digestibility, it should be classified as a very low quality concentrate. When included in conventional diets at levels of more than a few per cent, it is likely to depress intake and the digestibility of some of the constituents, in particular that of protein. At dietary levels of 10 per cent or more it has given a pronounced diuretic effect. For these reasons, the inclusion of dried coffee residue in ruminant diets should be no more than about 5 per cent or less. At these levels, it will have the advantage of increasing the total energy content of the diet, and possibly the digestible energy value if the fat fraction is well digested.

Other uses of dried coffee residue

Attempts to show that dried coffee residue is a suitable feed for poultry or for other non-ruminant animals by carrying out rat tests have been singularly unsuccessful. *Hammond* and *Carew* showed that ratios containing the residue had a deleterious effect on growing chicks. Similar adverse effects were obtained by *Ali* and *Campbell* when adult and weanling rats, respectively, were given diets containing dried coffee residue. Weight gains or growth rate were retarded and at the higher levels of inclusion, a high proportion of the rats died. These results suggest that dried coffee residue is not suitable for feeding to non-ruminant farm livestock.

With its high calorific value, it has some use as a fuel. Alternatively, it is akin to coffee hulls as a soil conditioner, and if economically worthwhile, it could be a small industrial source of oil and fat.

COCOA WASTE

The disposal of large quantities of waste from cocoa processing is a world wide problem. Some 70 to 75 per cent of pod material is discarded at farm level, while 10 to 15 per cent is wasted during industrial processing as shell, germ etc.

Useful by-product which can be prepared from these wastes are animal feed, fertiliser, furfural, pectin, vinegar, power alcohol, food colour, protein and fatty oil.

The constituents of pod husk suggest its economical use as cattlefeed. A 50 per cent substitution in feed is found as beneficial as is corn meal. Its low content of theobromine (0.2 per cent) when compared to shell (1.3 per cent) makes it less toxic to animals. A high potash value (2.8 to 5.7 per cent) suggests its use as good farm manure for young cocoa plants.

As a source furfural (9 per cent), pod husk is not the equal of corn cobs or oat hulls, but can be compared with groundnut husk (10 per cent). Pectin is in the range 5.3 to 7.0 per cent on a dry basis, but its quality is inferior to citrus pectin. Mucilage material can be a source of alcohol or vinegar. Freshly-collected sweat liquor can be used for making clarified juice or jelly.

Use of cocoa bean shell for the extraction of proteins (13 per cent), tanning (3.3 per cent) or food colour has limited scope due to the complex nature of their occurrence. Extracted fat (3.5 per cent) from germ contains 300 IU of vitamin D and can be used in pharmaceutical preparations. Germ is also a good source for extraction of theobromine (3.0 per cent) and protein (24 per cent).

TEA WASTE

Very little can be said about the wastes from tea. The dust which arises from the manufacture of tea can act as a useful organic fertiliser. It has a composition similar to that of tea leaves and, as such, would supply a useful amount of nitrogen and potassium to subsequent crops, but only small amounts of the other essential major elements. In recent years, the consumption of instant tea has increased, especially in North America, and the spent tea leaves, which contain approximately 50 per cent moisture, have been marketed as a mulch. Recently *Croyle* have examined the potential of this by-product as an animal feed and for composting. Chemical analysis showed that the dry matter in the spent leaves has a relatively high crude protein content (26.4 per cent) but the material is very fibrous (acid detergent fibre, 64 per cent) and much of the fibre is lignified (lignin, 36 per cent). Furthermore, much of the nitrogen appears to be bound to the cell wall constituents, which would make the crude protein fraction less available for fermentation and enzyme digestion. For these reasons, and since the by-product may contain appreciable residues of tannins and caffeine, it is unlikely to be a suitable material for feeding to farm animals, with the possible exception of small inclusions in some diets for ruminants. To make an acceptable compost, *Croyle* have suggested that it should be moistened to 60 per cent water content, inoculated with a cellulolytic organism, and than treated with certain minerals to make it appropriate for nursery or garden crops. It remains to be seen whether such processing is worthwhile.

10

Cereals, Pulses and Oil Seeds

INTRODUCTION

Paddy husk and rice bran are available as by-products of the cereal and grain processing industries. Currently the husk is burnt mainly for its heat value while the rice bran is used for animal feed. Wider and more economic utilisation of these by-products is necessary to improve the economy of the grain-milling industry.

Paddy husk has to be burnt under conditions that produce a fully-combusted white ash if the latter is to be used for producing either cement or silicate. Maximum economy is attained if these two processes are integrated. Methods must be explored of getting a white ash from the commercial boiler ash that accrues when paddy husk is used as fuel.

Rice bran should be considered as a national asset and source of edible oil, and fully mobilised for extraction. Only a third of the available rice bran potential is now extracted to yield a commercial-grade oil. The extracted rice should be compounded into animal feed and returned to farmers in exchange for rice bran collected from them, so that they are not deprived of the feed requirement of their animals. Standardisation of the technologies for rice bran stabilisation and refining of crude rice bran oil would hasten their adoption by industry. Non-solvent methods for recovery of oil from rice bran, as reported to be practised in China, have to be studied.

Edible-grade wheat germ and rice germ should be separated from the million fractions so that, after oil extraction, the germ meals could be used as components of nutritious foods. There are edible possibilities for rice bran fractions rich in protein mixed with brans from the pulse milling industry. Hygienic purity of the raw materials processed, and freedom from stones and grit, must be assured if isolated fractions from the rice bran are to be put to edible use.

The economic feasibility of hardboard and furfural manufacture from husk needs careful examination.

UTILISATION OF BY-PRODUCTS FROM PADDY

Rice and by-products consists of paddy husk, rice bran and straw respectively. Each has diverse possible uses, and some pursued at the Regional Research Laboratory Jorhat (RRLJ) are described.

Paddy husk: Cement, normally made by burning limestone-clay mixtures, carries 18 to 26 per cent silica which could come from paddy husk ash which is 97 per cent silica. Either controlled combustion of rice husk followed by mixing and grinding of the ash with lime or Portland cement is employed, or else the rice husk obtained from boilers in rice mills is mixed and ground with calcined lime sludge.

RRLJ has designed a simple kiln for burning paddy husk in the first process. It has low production cost, flexibility of design and ease in operation are conducive to rural operation. Setting time, compressive strength, water retention and concrete compressive strength are comparable to those of masonry mortar and foundation concrete.

Building bricks, long made in kilns by burning valuable agricultural top soil clay, can be made from paddy husk ash mixed with binder and cured not by firing but partly in the open and partly in water. Fuel is conserved and cost is half that of clay bricks. Compressive strength values meet ISI specifications, and drying-shrinkage and moisture movement are close to them.

The silica level of paddy husk ash also conduces to manufacture of refractory/insulation bricks, which can be prepared at lower temperatures than in conventional processes. The pyrometric cone equivalent is 28, the apparent porosity 66 per cent, the crushing strength 60 kg/sq. cm and the bulk density 0.65 g/ml.

Stack-burnt paddy husk silica has been used by RRLJ in a process to prepare sodium silicates of various Na_2O/SiO_2 ratios at a lower temperature than the 1,400°C required to calcine sand-sodium carbonate mixtures. From such sodium silicates, silica gel and metallic silicates can further be made.

Molecular sieves are crystalline aluminosilicates with large surface area, well-defined pores and ability to incorporate various cations, that are used in the industrial drying and separation of gases and liquids, and as catalysts or catalyst supports. Alkali-washing of paddy husk yields a soluble sodium silicate-soda lignin mixture which is admixed with sodium aluminate and processed by crystallisation to molecular sieves of pore size 4Å and 9Å. These compare well in their water-adsorption capacities, thermal stability and structure with corresponding Linde products. Sieves of other pore sizes have been prepared from these by incorporating ions of various sizes.

A fuller's earth substitute for use in the refining of crude paraffin wax by acid treatment has been prepared by controlled burning of paddy husk, and simple household detergent formulation from paddy husk ash. Good quality boards of medium and high density have been made.

Rice bran, derived in about 6 per cent yield from paddy shelling, is used in animal and poultry feed, and only partly for extraction of oil. Parboiled rice bran has 25 per cent oil, other types from 6 to 20 per cent. Immediate stabilisation of bran to arrest lipolytic action is necessary to extract quality oil for use for edible purposes; high-acidity rice bran oil is used to make washing soap and textiles. Many Indian laboratories (CFTRI Mysore, OTRI Anantapur, CDRI Lucknow) have worked on various aspects of rice bran technology. The deoiled bran is used as cattle and poultry feeds both in India and abroad. There is vast potential for bran oil extraction in the country.

Straw from rice and wheat is now used mostly as a roughage in animal feed and as a roof thatch. Lately its use as pulping raw material for board and paper manufacture has commenced and has a great future since it is a renewable resource. Pulping of composite straws, lemon and citronella grass, waste papers and so on, formation of sheets, and impregnation with fire and water-resistant chemicals yield light but strong sheets that have been approved for use as roofing materials in low-cost housing schemes. Mushroom cultivation on paddy straw and spent citronella grass offers promise for enlarging the availability of this nutritious food material.

Difficulties in developing paddy residue industries arise from several causes. One is consumer resistance to unconventional new materials like paddy husk cement even when it meets IS specifications. Another is the need for region-wise surveys to asses the availability of various wastes, prior to devising systems for their collection, transport and storage. It is also necessary to constitute composite working groups that can secure finance, scale-up technologies and subsidise new products till they gain market acceptance.

RICE HUSK—A RAW MATERIAL FOR INDUSTRIAL CHEMICALS

Rice husk, and by-product of rice milling, is now put to worthwhile use. Chemical conversion of rice husk to organic acids or inorganic chemicals, which in turn could support many industries, would be a profitable method of utilisation.

A scheme is presented for chemical conversion of the hemicellulose of rice husk (20 per cent) to xylose and then to furfural, xylitol, xylonic acid, food yeast, etc. Similarly a conversion scheme is envisaged for selective hydrolysis of cellulose (38 per cent) to glucose, which in turn can be used for industrial fermentation to ethanol, sorbitol, hydroxymethyl furfural, levulinic acid, and so on. Lignin (22 per cent) can be converted to phenol, benzene and vanillin. The conversion processes involve hydrolysis at 100 to 150°C and 1 to 5 atmospheric pressures.

Ash obtained by burning, or preferably by organic conversion employing the above schemes, is a good source for a number of silica-based inorganic chemicals.

Considering the above possibilities, a process for successive conversion of hemicellulose, cellulose and lignin of paddy husk is presented and its economics examined. Thus lot of rice husk would be expected to yield 1370 kg xylitol, 80 kg furfural, 1474 kg industrial alcohol, 1400 kg yeast, 2000 kg lignosulphonic acid and 2000 kg siliceous ash. The effect of several variables was examined and a profit was shown even for wide variations in the price of rice husk and the plant processing capacity or for operating at below the rated capacity through seasonal raw material fluctuations. A change in raw material composition would drastically alter yields, revenue and profitability.

Ethyl alcohol production is attractive only if fermentation by-products are suitably recovered and processed.

Utilisation of rice husk for industrial chemicals production is a good economic proposition. It is well-suited to low-tonnage, small-scale operations at rural level, and could thus raise farm incomes. The raw material is widely dispersed throughout the country, and is a renewable agro-resource not required for human or animal consumption.

Detail Description of Uses of Rice Husk

Paddy has become an important crop of the Punjab state after the green revolution of sixties. Paddy production of 75 lakhs tonnes in 1986 to 2001 has increased to 94 lakh tonnes. This is due to the higher level of mechanisation of paddy production operations and better profitability as compared to other khariff crops. Paddy cannot bc caten, cooked or fed to animals in an unprocessed form. Hence paddy, processing has become an important industry of the Punjab state. Rice industry growth is also coupled with the paddy production. Disposal of rice husk poses a big problem to the rice millers because of its limited use and having low bulk density. Now-a-days several ways are available for utilising the rice husk and still research is going on to bring rice husk under industrial use.

As fuel

Husk can be used in furnaces for producing energy. During burning it produces bad odour and high quantity of ash. Counter-current type burner can completely burn the rice husk. Rice husk has low density. Its transportation is a big problem. Research is going on to compact the rice husk and bring it better use as fuel.

Manufacturing of particle board

Rice husk is hard and rich in silica. The particle board made of rice husk shows better physical properties. Particle board has been manufactured by mixing the rice husk with wood particles and lamination with 0.8 mm thick lauen vaneer on both surfaces of the board.

Conversion of rice husk for animal feed and fertiliser

The nitrogen content of rice husk is about 0.6 per cent. It can be increased to 1.5 to 2.0 per cent after ammoniated conversion. As fertiliser, ammoniated rice husk application increases the yield of *asperagus* growth rate of flower and yield of vegetables. As feed for goat and beef cattles, ammoniated rice husk can be used to reduce 20 per cent of grain, so that the cost of the feed can be minimised. Rice husk can be converted into single cell protein for animal feed by fermentation.

Carbon source for acetic acid production

Rice husk can be hydrolysed with 1.5 per cent dilute sulphuric acid solution at 121°C for one hour for the production of reducing sugar in a concentration higher than 5 per cent. This is the potential carbon source to grow *Clostridium thermoaceticum* for acetic acid production.

Soil improver

The rice husk can be mixed in the soil to increase soil organic matter content and lower soil bulk density. The rice husk applications can improve rice growth, yield and soil texture. Rice husk applied once at a rate of 20 tonnes per hectare has a beneficial effect for more than two crops of rice due to its slow decomposition of the husk.

Raising rice seedling for mechanical transplantation

The ground rice husk can be used as part of seed beds materials instead of 100 per cent soil for raising rice seedling. The advantages of this type of seed beds are: (i) to reduce cost; (ii) easy working; (iii) to tolerate the distant transportation; (iv) to shorten seedling days; and (v) to reduce disease occurrence.

The ratio 1:1 of soil and ground rice husk is most suitable for rice bed material.

Adhesive filler material

The rice husk can be used as filler in the domestic plywood industry. Finally ground rice husk flour (100 mesh size) filler is satisfactory from the viewpoint of bonding strength and no difficulty encountered within the whole plywood manufacturing process.

UTILISATION OF RESULTANT ATTA—A BY-PRODUCT OF ROLLER FLOUR MILLING INDUSTRY

Resultant atta (RA), also known as mill atta in the trade, accounts for 5–15 per cent of the total milled products in a commercial roller flour mill. It has been reported to be nutritionally superior to wholemeal atta (WMA). Samples of RA and of the wheat grists from which they were made were procured from several roller flour mills. The wheat grists were milled in a chakki (disc mill) to obtain the corresponding WMA. RA, in comparison with WMA, was lighter in colour, coarser in particle size, carried 2–3 per cent more moisture, and had lower ash, protein and fat contents.

RA needed 10–12 per cent less water than WMA to obtaine chapati dough of a desirable consistency, perhaps because of higher damage to the starch in WMA during severe grinding in the chakki.

WMA also showed a longer dough development time and less dough stability than RA.

RA chapati dough showed a higher resistance to extension (R) and higher extensibility (E) than that from WMA. The R/E ratio, which indicates stiffness of the dough, was higher for freshly-mixed dough from WMA, but after the doughs had been rested for one hour, the ratio for both doughs was almost the same. RA chapati had a better colour than that from WMA, probably because less bran was present in the former. However, the WMA chapati was softer, quite pliable and had a better flavour, probably through the heat generated during grinding of wheat into WMA.

With regard to eating quality, the RA chapati had a bland taste and was quite leathery, perhaps because of the low bran content. The addition of 5 per cent finely-ground bran to RA improved both textural and eating quality of chapati. A certain minimum amount of bran is necessary in atta to obtain chapati a minimum limit for ash content in the specifications for RA of both the PFA and ISI seems desirable. The reasons for the better flavour of chapati made from WMA calls for systematic studies.

PROCESSING, NUTRITIVE VALUE AND FOOD-USES OF WHEAT GERM—A POTENTIAL BY-PRODUCT OF FLOUR MILLING INDUSTRY

Recovery of the germ present from just one-third of the wheat grain available in India would yield 60,000 tonne/annum.

Commercial wheat germ samples contained, as percentages: protein 19 to 32, ether-extractives 5.2 to 11.6, ash 3.6 to 5.0, crude fibre 2.2 to 6.7, carbohydrates 40 to 53, and sugars 10.5 to 16.8, respectively.

Wheat germ has a very poor shelf-life; it becomes bitter within a few days due to high enzyme activity and a large content of unsaturated fat. This poor stability is a major constraint to utilisation.

Wheat germ was subjected to toasting, steaming, roller-drying and defatting. The two moist heat treatments, viz., steaming and roller-drying, totally inactivated all the deteriorative enzymes. Even dry toasted germ, despite only partial enzyme inactivation, kept well for over 26 weeks.

Dry heat-treatment, either toasting a one-cm thick layer of germ at 150°C for 25 minute in an air-circulation oven, or in a coffee roaster at 130°C (product temperature) for 5 minute, resulted in a highly-acceptable product with a pleasing golden-brown colour and improved flavour and taste. Steaming of a 5 cm thick layer of wheat germ spread on wire-mesh trays at atmospheric pressure for 10 minute was found to be simple and practically-feasible method of stabilising wheat germ. Roller-drying of a slurry containing 33 per cent ground germ solids at 35 psig steam pressure also resulted in a highly acceptable product. All the processed germ samples had a shelf-life of over 26 weeks.

At 10 per cent protein level in the diet, the protein efficiency ratio (PER) of raw, toasted, steamed and roller-dried samples was 2.30, 2.46, 2.65 and 2.38 respectively, against 3.00 for skim milk powder. The PER of a roller-dried blend of wheat germ (80) and edible-grade groundnut cake flour (20) was equal to that of a popular brand of weaning food (2.49). The protein contents of these two samples were 32 and 22 per cent respectively.

Of the several processed samples, toasted wheat germ had the least deleterious effect on loaf volume when incorporated in a bread recipe. Acceptable breads could be prepared with even 15 per cent of various processed germ samples. Addition of 0.6 per cent sodium stearoyl-2 lactylate and 60 ppm of potassium bromate was found to increase appreciably the loaf volume of bread containing wheat germ.

Biscuits containing 20 per cent of steamed germ (12 per cent protein) were comparable in their overall quality to control maida-based biscuits.

Subjecting wheat germ to simple heat treatments increases significantly both nutritive value and keeping quality. Such stabilised germ can be used to fortify bread and biscuits at 15 and 20 per cent levels respectively, without appreciably affecting their overall quality. Roller-dried germ has high acceptability and preparation from it of weaning and other protein foods has possibilities.

UTILISATION OF BY-PRODUCTS FROM OILSEED AND FAT INDUSTRIES

Utilisation of by-products from the oilseed and fat industries, besides minimising disposal problems, would add to the production economy.

Coconut: Shells can be used to manufacture active carbon (process available from the RRL Hyderabad) or as a filler in the manufacture of synthetic resin glues and as an extender in phenolic moulding power.

Cottonseed: Linters (8–10 per cent of the seed) contain 72–85 per cent α-cellulose, which can be purified to chemical cotton (process available from the RRL Hyderabad) for use in the manufacture of cellophane, rayon, paper and carboxymethylcellulose. Hulls (35–40 per cent of the seed) are now used as a filler in cattlefeed. Since hulls consist of 29 per cent pentosans, besides 44 per cent cellulose and 22 per cent lignins, they can be profitably used for manufacture of furfural, and the spent residue utilised for production of active carbons or as fuel.

Gossypol, a polyphenolic compounds present in cottonseed, is a much-maligned substrance, toxic to many species of animals. It can, however, be put to good use, for example as an excellent antioxidant for products such as rubber and petroleum, as a stabiliser against polymerisation of vinyl compounds, and as an anti-tubercular drug. More important is its reversible contraceptive property, which is reported to be extremely effective and without adverse side-effects; if indeed found useful, about 10,000 tonnes of gossypol can be usefully extracted from the seeds by say the liquid cyclone process, yielding in addition a protein-rich flour for edible use.

Corn oil, a rich source of both tocopherols and PUFA, is now only a small by-product of corn starch manufacture. It could expand into a good source of edible oil as corn-growing itself expands.

Mango seed kernel contains 4–13 per cent of fat with usage in confectionery, textile-sizing and soap manufacture. The defatted meal can be used as cattlefeed and the shell in the manufacture of active carbon. The 9 million tonne of mangoes produced in the country are a potential source of 30,000 tonne of fat and 270,000 tonne of defatted meal.

Paddy husk, besides its known use as a fuel, can be used to manufacture furfural, silica and activated carbon. Rice polishings (3.5 million tonne), a traditional cattlefeed, are now being extracted for oil, but for edible use of the latter, both the bran and germ constituents must be stabilised against lipase action. The germ is itself a rich source of tocopherols. Solvent-extracted bran contains 2.5–3.5 per cent of wax, recoverable by chilling and centrifugation, which has FDA approval for wax-coating of ruits, as a chewing gum base, and as an ingredient for coating candy. Deoiled bran is an excellent cattlefeed. Germs, separated by sieving and air-classification, can be used as a protein-rich food.

Sal (*Shorea robusta*, potential seed availability 5.5 million tonne, of which only 2 per cent is now exploited) yields about one-third its weight of wings and shells, a possible energy source. The kernels contain about 15 per cent fat. From the deoiled meal the tannins (14 per cent) can be extracted and used in leather processing. Sal meal can also be used for starch manufacture, and to a limited extent in cattle and poultry feed formulations.

Most crude oils contain 0.1 to 3 per cent of lecithins, recovered by hydration prior to alkali refining. A number of so-called non-edible oils like neem (1.0 to 1.5 per cent of bitter nimbidin), karanja, and malkangini (*Celastrus paniculatus*) contain odoriferous and bitter compounds which have proven pharmacological and insecticidal value. These can be extracted from the oils using alcohol, thereby also improving the colour and odour of the oil. The residual meal is now mostly used as a manure and R & D should be directed at alternative outlets such as in adhesives and detergents.

Palm Oil

The red oil palm, botanically know as *Elaeis guineesis* is quite unique among the commercial oil producing crops. It is the largest yielding oil crop producing 4–6 tonnes of palm oil per hectare per year as compared with less than one tonne per hectare from other oil plants.

Extraction of palm oil

Unlike other oilseeds, extraction of palm oil is done using fresh fruit bunches which requires a large amount of water. A schematic flow diagram of the palm oil extraction process is shown in Fig. 10.1. It consists of 5 main steps:

1. Steam sterilisation of fresh bunches.
2. Fruit stalk separation by stripping.
3. Digestion of stripped fruits.
4. Oil extraction by means of screw presses.
5. Clarification comprising separation of oil from the residue, aqueous liquor and debris.

Production of palm oil mill effluent (POME)

Liquid effluent is generated mainly from the sterilisation and clarification processes in which large amounts of steam and/or hot water is used. Another waste stream originates from the hydrocyclone operation in which the broken shell is separated from the kernels. The amounts of effluent generated from sterilisation, clarification and hydrocyclone are 0.9–1.5 and 10 m^3 per tonne respectively of oil produced. Altogether, generally, it varies between 2 and 3 tonnes of POME per tonne of oil produced.

Properties of POME

POME, when fresh, is a thick brownish slurry. It is acidic and contains very high organic matters with high biological oxygen demand (BOD). It is essentially a colloidal dispersion containing particulates of plant origin, dispersed in liquid medium of dissolved minerals and soluble organic substances. The suspended solids in the slurry are mainly cellulose matter mixed with some residue oil. The effluent is non-toxic as no chemical is added to the oil extraction process. Tables 10.1 and 10.2 give the detailed characterisation of POME.

Table 10.1. Characteristics of palm oil mill effluent.

Particulars	*Value*
Colour	Dirty
pH	3.5–4.5
BOD (mg/l)	25,000
COD (mg/l)	60,000
Total solids (mg/l)	40,000
Suspended solids (mg/l)	19,000
Total nitrogen (mg/l)	770
Ammoniacal nitrogen (mg/l)	35
Oil and grease (mg/l)	8,000
Temperature °C	80–90

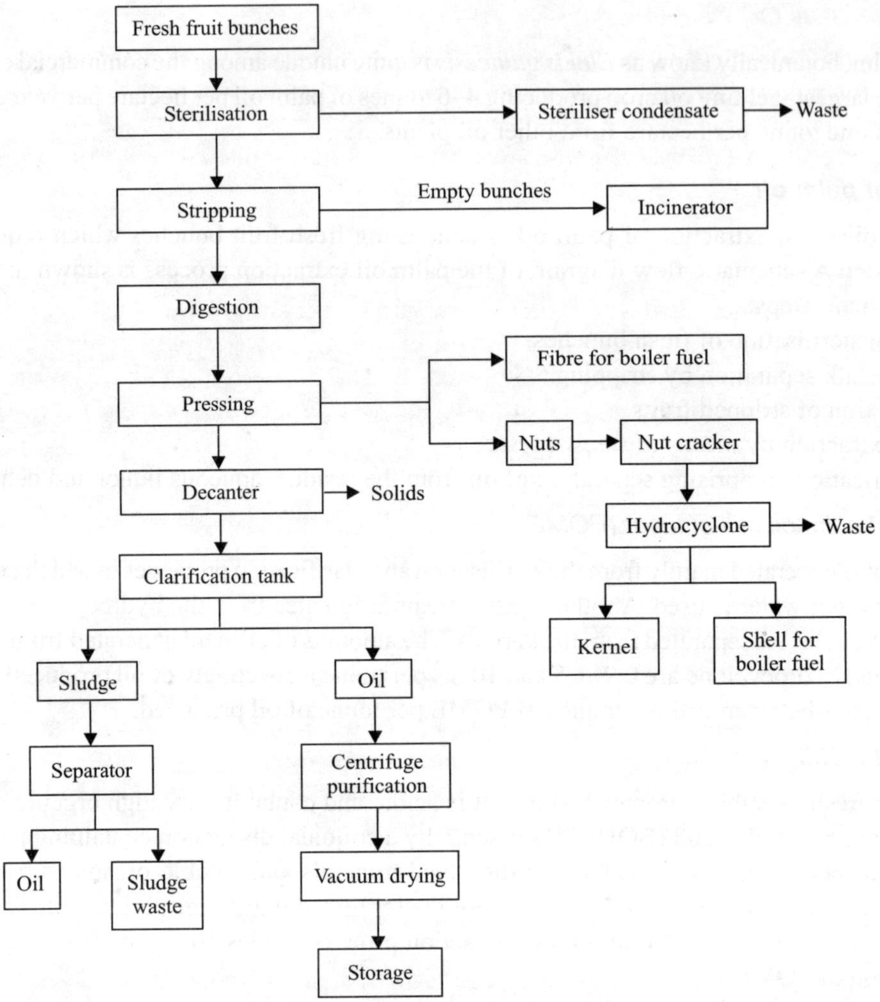

Fig. 10.1. Palm oil extraction process.

Table 10.2. Mineral ions of palm oil mill effluents.

Component	*Range (mg/dm^{-3})*
Phosphorus	110–156
Potassium	1420–2165
Magnesium	215–360
Calcium	230–345
Sodium	26–63
Manganese	1.7–3.8
Iron	62–130
Zinc	1.0–2.1
Copper	1.0–1.5

Treatment of palm oil mill effluent

Traditionally, POME has simply been led to the nearest water course, or where appropriate, out to sea. In view of its high polluting strength and large volumes, there is an urgent need to treat this effluent to prevent the environment from damage. The many and varied methods which have been suggested for dealing with this effluent fall broadly into three categories: (i) simple disposal; (ii) utilisation; and (iii) treatment.

Simple disposal

The effluent is simply run into a nearby stream. Baffles are erected in the stream to trap residual oil in the effluent. However, a dilution factor of 10,000 : 1 would be necessary even if it were acceptable to increase the BOD load of the water course by, for example 2 mg/l, which represent a significant deterioration in water quality. It is thus totally impracticable to adopt this method. High water content of POME would require a very high energy input for incineration. This will be expensive.

Concentration by seepage and/or evaporation on wasteland

Two main methods have been used to produce a solid residue of one-tenth or less of the initial volume of the POME:

1. The effluent after neutralisation with lime is piped into nearby areas and left for seepage and evaporation. Thus, bulk of water is removed leaving a solid layer on the surface. Problems associated are odour and insects, large availability of land and cost. Further, BOD is to be essentially reduced prior to this method.
2. In the forced evaporation method the effluent is sprayed into hollowed out areas while it is hot. Then air is blown across the surface. If possible, hot waste gases are injected in the effluent. The floor of the pond slopes so that the concentrated sludge can be trapped off for further drying in rotary drier. Though by this method less land would be required, problems of odour remain same. Further, initial BOD reduction is again essential.

Utilisation

It is possible to recover the residual oil by direct centrifugation of POME. But this seems a too costly process to be supported by the value of oil recovered. The effluent may be simply run into trenches dug in agriculture land which would then be turned over immediately to bury the POME and so avoid odour and insect problems. The effluent will be absorbed by the soil and broken down biologically, thus providing the land with nitrogen and other nutrients. However, transportation of large volumes of effluent would hardly be practical. Further, pre-treatment to reduce BOD is required.

Crop irrigation

This appears to be an obvious method, as land applications of POME resulted in increased crop growth but there are problems like ability of soil to cope with repeated application and transportation of effluent to irrigation site. Further, BOD is to be brought within stipulated limit.

After partial dehydration, POME can be used as one of the constituents of the animal feed. The pressed fibre can also be used as one of the constituents. However, there are several drawbacks like cost of transporting the raw wet sludge, partial dehydration before use, and storage problem.

Treatment

POME can be treated by: (i) physico-chemical methods; and (ii) biological methods. The former comprises of filtration, settling, floatation and coagulation.

Whiting has reported the use of a series of progressively finer vibrating screens led to a fairly good separation of solids which considerably reduced BOD load of the effluent. However, the particular consistency of the effluent makes the separation of solid mass by filtration impracticable.

Settling for prolonged time or centrifugation fails to give any worthwhile separation.

Floatation appears to be more promising, though it requires the use of various chemicals and compressed air. It depends on attaching small air bubbles to the suspended particles thereby reducing their density below that of surrounding liquid and causing them to float to the surface. A BOD reduction to 10 per cent of raw effluent has been obtained by this method. Secondary aerobic treatment may clean up the remaining organic load. Use of various types of coagulants has been reported to aid liquid-solid separation in POME. The synthetic polymers were found more effective than lime or alumina. Excessively large quantities of latter were required. This might affect the usefulness of the recovered solids in animal feed formation and also lead to a sludge disposal problem. Biological treatment method is the best approach for curtailing the polluting potential of the effluent.

Aerobic method

In aerobic treatment BOD reduction can be achieved by bacteria catalysed oxidation of organic components of POME in open ponds. Forced aeration may also be applied. However, this type of treatment is impracticable with POME due to its very high BOD. Production of single-cell protein (and simultaneously reduction in organic load of the effluent) is also not very attractive idea, a particular strain of micro-organism can only readily assimilate one type of compound.

Hence while processing this way, some of its constituents will inevitably be left unused and run through liquid phase which will then have to be subjected to further processing to remove them. However, mixed culture studies may be an attractive idea. It may have the advantages of: (i) shorter detention times; (ii) use of close tanks avoids smell problems; and (iii) production of protein enriched animal feed.

Anaerobic method

Anaerobic method is the most popular methods for the treatment of POME and presently being followed by most of the palm oil mills. There are various systems, adopted by the industry under the anaerobic digestion. The ponding system essentially consists of a number of ponds constructed in series. A typical ponding system is shown in Fig. 10.2.

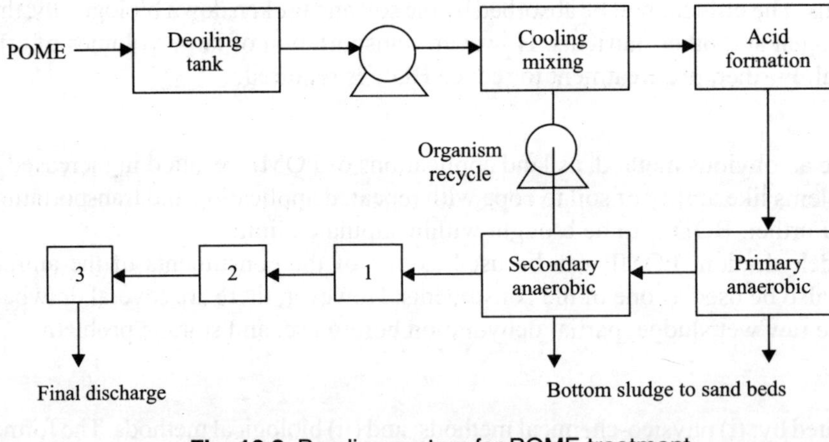

Fig. 10.2. Ponding system for POME treatment.

It involves two-phase anaerobic digestion. The system has proven reliable, stable and capable of producing the final discharge with a BOD of about 100 mg/l. The built-up sludge in the anaerobic ponds is regularly removed. The anaerobic pond is usually 5 to 7 meters deep and the facultative pond is around 1.5 meters deep. The hydraulic retention time for former is 40–100 days and for the latter is 15–30 days. It is operated at low rate, with organic loading of 0.2–0.35 kg BOD/m^3.

The ponding system though cheap to construct, requires a large land area. As it is subjected to environmental changes it is quite difficult to control and monitor. Furthermore, mixing (by biogas) is quite inadequate. It also creates problem of foul smell.

The integrated system to anaerobic digestion of POME in closed tank reactors with biogas recovery and subsequent use of digested effluent for land applications is one of the most promising methods used in POME treatment and is being adopted by several mills. The digesters are operated as a conventional high rate system with loading of 4–5 kg volatile solids (VS) per m^3 day. The hydraulic retention time is about 10 days and operating temperature is 42–50°C. Proper mixing is ensured by recycling the biogas. The system is capable of removing about 90 per cent BOD (Fig. 10.3).

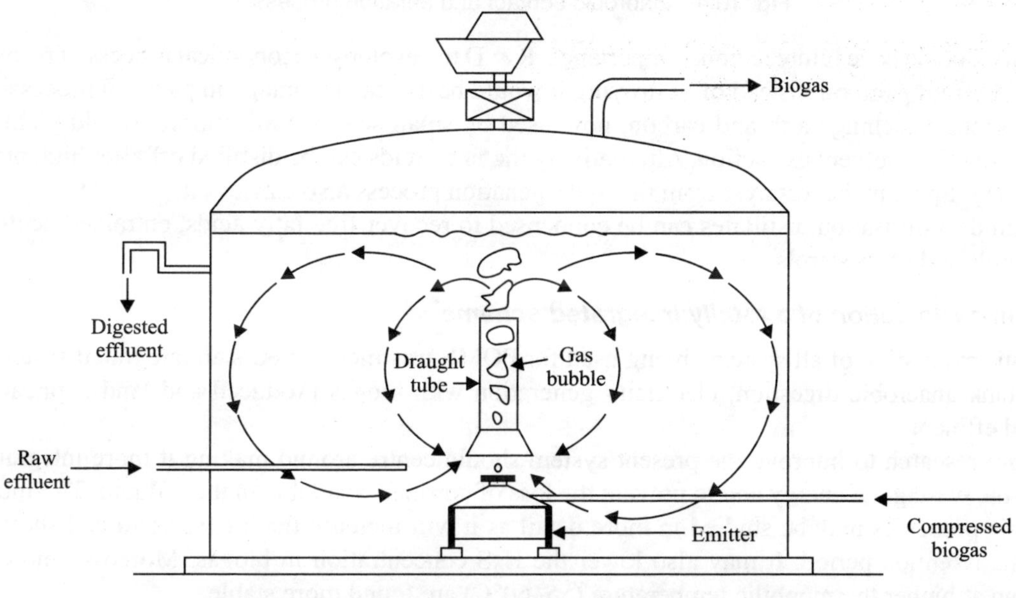

Fig. 10.3. Schematic diagram of anaerobic tank system.

In this process, about 0.59 m^3 biogas per kg VS is produced which can be used for producing heat and electricity. It contains about 65 per cent methane and rest are carbon dioxide. The H_2S level is very low (less than 2000 ppm). Application of digested effluent obtained by this system, to land for irrigation purposes, increases the crop yield and also improves the soil properties.

A typical layout of tank digestion and extended aeration system is shown in Fig. 10.4. The hydraulic retention times of acidification pond, 20 days for tank digester, 1 day for anaerobic settling pond, 20 days for aerobic pond and 2 hour for aerobic settling tank. The organic loading is about 1.0 kg BOD/m^3. An average of about 93 per cent BOD is removed in anaerobic phase while overall BOD removal achieved is about 99 per cent or more. However, the aeration system is very energy intensive and thus requires substantial recurring cost input.

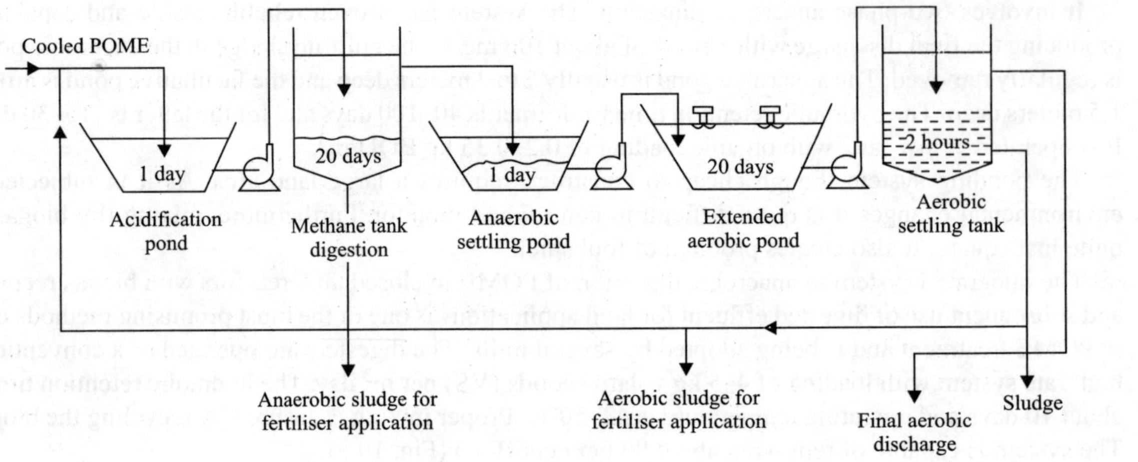

Fig. 10.4. Anaerobic contact and aeration process.

Palm oil could have future national importance. R & D to develop an economical process of recovering β-carotene from palm oil instead of destroying it would be a great advantage in palm oil processing.

The spent bleaching earth and carbon, now used by small soap manufacturers, would yield 30 to 40 per cent oil by solvent extraction. Alternatively the fatty acids can be distilled off after high pressure fat splitting. Spent nickel catalyst from the hydrogenation process also carries fat.

Steam deodourisation distillates can be condensed to recover free fatty acids, entrained neutral oil, tocopherols and other sterols.

Economic evaluation of a totally integrated scheme

Economic evaluation of all systems being used for POME treatment favours an integrated system with closed tank anaerobic digestion, electricity generation with biogas produced and land application of digested effluent.

Future research to improve the present system should centre around making it more integrated, to control the wastage of energy and to prevent the loss of nutrients available in the effluent. Thermophicil anaerobic processes may be studied in more detail as it will increase the biogas yield and shorten the hydraulic retention period. It may also lower the H_2S concentration in biogas. Moreover bioreactors operating at higher thermophilic temperature (55–60°C) are found more stable.

11

Wastes from the Fermentation Industries and Their Use as Feedstuffs and Fertilisers

INTRODUCTION

There are a large number of compounds which are capable of being synthesised by microbial fermentation, but relatively few are manufactured on a large scale by fermentation, as in many cases the synthetic pathway to a particular compound is more economical that a fermentation process. This chapter deals with the solid wastes which remain after the fermentation process is complete and examines their potential as fertilisers and feedstuffs. Of the fermentation wastes considered here, the ones which will be examined in the most detail are those from the antibiotic industry and wine and cider production. Wastes from other industrial fermentations tend to be produced on a smaller scale and companies tend to be very security conscious with respect to the quantities of wastes production. Very few studies have been carried out into the potential uses of these materials.

ANTIBIOTIC FERMENTATION WASTE

Antibiotics are chemical substances that are produced by micro-organisms and have the capacity, in dilute solution, to inhibit the growth or even destroy other micro-organisms. They are produced today on a large scale using the fermentation process. The exception to this is chloramphenicol, which has been produced synthetically.

The wastes with which this section is concerned are the solids remaining after the liquid containing the antibiotic has been extracted from the fermentation broth. This is usually done by filtration. These solids are a heterogeneous mixture of hyphae of the producing organisms, unused media, defoaming oils, and metabolic products of the micro-organisms, the most important of which is the residual antibiotic. Levels of residual antibiotic can vary greatly, and recent studies have indicated that tylosin fermentation waste contains 2000 parts per million residual antibiotic while oxytetracycline waste contain 1500 parts per million residual antibiotic.

Total quantities of waste produced are difficult to obtain, as pharmaceutical companies are naturally reluctant to divulge the quantities of organic waste which they are discharging into the natural environment. Figures for the production of antibiotic fermentation waste in the United States can be estimated from figures obtained by *Howe*, who found that 2.2 kg of antibiotic produced 13,500 to 16,000 litres of waste-water. Based on estimated production figures of antibiotics in the United States, this indicates a total waste volume from antibiotic processes of 140 to 160 × 10^6 litres/day. *Howe* quoted the average solids content of these wastes at about 40,000 parts per million, and this suggests a production of 6350 tonnes of solid wastes per day or 2.3 × 10^6 tonnes/year. Approximately 50 per cent of this will be lost as CO_2 and compounds dissolved in the medium, and therefore the solid waste produced will be approximately 1.18 × 10^6 kg/month, equivalent to 14.2 × 10^3 tonnes/annum from one average plant.

Total world production of antibiotic fermentation waste would have amounted to approximately 50 × 10^6 tonne/annum in the early 2000.

Disposal

Most present day methods of disposal involve biological treatment or the dumping of the waste untreated. Some pharmaceutical companies in the United Kingdom discharge their waste—which can have BOD of upto 32,000 parts per million directly into rivers. In the United States, Pfizer was prevented from dumping its waste into Long Island Sound, Connecticut, by the Environment Protection Agency and now, like many other pharmaceutical companies in the United States. Several pharmaceutical companies have investigated the possibility of biological treatment as a method of reducing the BOD of the fermentation wastes. *Brown* looked at the possibilities of using sewage filters to treat antibiotic wastes, and *Brown* and *Niedercorn* found that the disposal of antibiotic containing fermentation wastes via sewage treatment plants could adversely affect the micro-flora of the filter beds. *Hilgart* appeared to overcome these problems in a treatment plant for penicillin and streptomycin wastes are the Upjohn company, Michigan, USA, when BOD reductions of 97 per cent were achieved. *Barker* examined the possibilities of speeding up the breakdown of antibiotic waste using compressed air, while *Lederman* produced a reduction in BOD using pure oxygen in the lagoons of the treatment plant. Several other pharmaceutical companies in the United States have also set up biological treatment plants, including Merck, Sharpe, and Dohme. Attempts have been made to recycle the waste and use it as base for a fresh batch of medium, but the inclusion of this material into a fresh batch caused a reduction in the yield of antibiotic.

Use as Animal Feedstuffs

In the 1960's and 1970's, much of the waste from antibiotic fermentations was used as animal feedstuffs. Its use as a feedstuff was first suggested by *Stokstad* when he observed a growth promoting effect in chickens fed with crude extracts of *Streptomyces aureofaciens* waste. Similar growth responses were reported when aureomycin was fed to chicks and pigs.

Reviews by *Stockstad*, and *Jukes* indicate that growth responses have been observed in many studies of swine, ruminants, turkeys, and chicks. Optimum levels of antibiotic were found to range from 2 to 50 parts per million, depending on the antibiotic. A report from the Office of Health and Economics indicated that antibiotics stimulated the appetite, increased the efficiency of food conversion, reduced vitamin requirements, and increased the growth rate. These effects appear to be due to the action of the antibiotic on the intestinal micro-flora of the animals. Animals raised in a sterile environment showed no growth response. Antibiotic fermentation waste would therefore appear to be an ideal feedstuff, as not only has it an intrinsic nutritive value, but the residual antibiotic stimulates growth.

There are potential problems with the feeding of feedstuffs containing antibiotics routinely to animals:

1. In the presence of continuously low levels of antibiotic, resistance levels in pathogenic organisms could increase.
2. Antibiotics in animal feeds could be transferred to man via animal products.

Cases of dermatitis have been attributed to the presence of penicillin in milk. But in these cases the presence of penicillin in the milk resulted from the administration of relatively high levels of penicillin to treat mastitis and not from the feeding of sub-therapeutic levels of penicillin as growth promoters. *Stokstad* reported that levels of antibiotics used as growth promoters ranged from 2 to 50 parts per million, 0.2 to 5 per cent of the therapeutic dose needed to maintain effective blood levels. Even at relatively high levels (200 parts per million) of aureomycin administered to chicks, *Stockstad* found that the antibiotic could not be detected in the serum, liver, and muscle. *Jukes* reported that tetracyclines could not be detected in soft tissues and were rapidly excreted in the urine, while erythromycin, tylosin, and bacitracin had not been found in animal tissues after feeding with low levels of antibiotics.

There is still much controversy over the effect of sub-therapeutic levels of antibiotics on the resistance patterns of pathogenic micro-organisms. A review by *Jukes* strongly stated the case for the inclusion of low levels of antibiotics in animal feed and suggested that experiments carried out over a period of 20 years had not demonstrated a significant increase in the resistance of micro-organisms to the antibiotics used. Increasing levels of resistance were attributed to the widespread use of antibiotics in hospitals and general medicine. Therefore, in view of he problems which at present surround the use of antibiotic fermentation waste as an animal feedstuff this section is devoted to an examination of the potential of the material as a fertiliser, together with an examination of its potential effects on the soil eco-system.

Fertiliser Value

Recent analyses of oxytetracycline, tylosin, and penicillin fermentation wastes carried out are presented in Table 11.1. Nitrogen contents ranged from 6.34 per cent in tylosin waste to 2.8 per cent in oxytetracycline waste, on a dry weight basis. *Zvara* reported that a penicillin residue contained 4.59 per cent N; this is less than the figures obtained from United Kingdom penicillin waste and probably reflects the different media constituents in the two processes.

Table 11.1. Composition of antibiotic fermentation wastes.

Description	*Tylosin*	*Oxytetracycline*	*Penicillin*
Waste (%)	83.6	65.7	80.8
pH	6	2.2	–
Organic matter (as % dry weight)	85.5	50	90.8
Organic C (as % of dry weight)	43.7	26.5	40.7
N	6.34	2.80	5.92
P	2.00	0.43	1.36
K	0.59	0.07	0.38
Ca	0.30	0.71	0.72
Na	0.21	0.35	0.25
Microelements (ppm)			
Mg	500	60	550
Fe	275	460	60
Mn	7	11	6
Cu	245	3.5	5
Pb	15	5	5
Cd	1.5	1.5	1
Antibiotic content (ppm)	2000	1500	–

Should the use of antibiotic fermentation wastes as fertiliser become widespread, on the potential problems, apparently ignored in the work of Pfizer, would be the presence of the residual antibiotic. The addition of antibiotics to the soil could affect the micro-organisms in the soil as well as plants growing in the soil. Therefore, it is important to review the potential effect of antibiotics on these aspects of the soil ecosystem, bearing in mind that the levels of residual antibiotics will be low.

BEHAVIOUR OF ANTIBIOTICS IN SOIL

A review by *Pramer* showed that antibiotics can be inactivated in soils, and this inactivation may be as a result of one or more of three processes:

1. The chemical instability of the antibiotic molecule.
2. Adsorption onto clay minerals and organic matter.
3. Microbial degradation.

Jefferys, attributed the inactivation of penicillin, viridin, gliotoxin, frequentin, and albidin to their chemical instability. Using sterile soil, *Jefferys* found that cyclohexamide, gladiolic acid, and penicillin were rapidly inactivated in soils, even under conditions favourable to their stability. The adsorption of antibiotics by clays and organic matter has been studied extensively by *Gottlieb*.

Basic antibiotics—e.g., streptomycin, neomycin, erythromycin, and tylosin—are adsorbed by the clay minerals montmorillonite, kaolinite, bentonite, illite, and vermicullite, although different clays adsorb different amounts of the same antibiotic. Amphoteric antibiotics—e.g., bacitracin, aureomycin, and terramycin—can act either as acidic or basic antibiotics depending on their isoelectric points and the pH of the soil. As the pH of the soil is usually below the isoelectric point of the antibiotic, they behave as bases and are adsorbed onto clay particles. Acidic or neutral antibiotics—e.g., penicillin and chloramphenicol– are adsorbed by clays only in small amounts. Some antibiotics adsorbed by clays can be released by phosphate and citrate buffers.

This release is more prevalent with amphoteric than with basic antibiotics. *Bewick*, found that different cations can produce different rates of release of streptomycin from clays. Microbial degradation can also play an important part in the disappearance of many antibiotics in soil. *Jefferys* found that 72 parts per million penicillin was rapidly degraded in soil, and *Gottlieb* and *Siminoff* showed that of 50 parts per million chloramphenicol added to non-sterile soil, 85 per cent had disappeared after four days. Even with streptomycin, a basic antibiotic rapidly adsorbed by clay minerals, *Pramer* and *Starkey* found that disappearance of 1000 parts per million was more rapid in non-sterile than in sterile soil and that second and third additions disappeared more rapidly than the first addition. This observation was also reported when repeated addition of 20 parts per million griseofulvin were made to soils.

Studies on the release of antibiotics from their fermentation wastes when added to soils have been initiated by *Bewick* using tylosin fermentation wastes. The fermentation waste was added to soil at levels of 3, 8 and 13 tonnes/ha. This resulted in the addition of 3.5, 11.2 and 18.7 parts per million tylosin, respectively, to the soil. These additions were carried out at 4, 10, 15 and 23°C. Increasing temperature was found to have little effect on the uptake of tylosin by the soil, while with the addition of increasing amounts of waste, increasing amounts of tylosin were adsorbed by the soil (Table 11.2). Using the means of tylosin uptake and percentage uptake at the four temperatures examined for each concentration of fermentation waste per hectare, the adsoption capacity of the soil for tylosin had not been reached and the soil had an adsorption constant for tylosin of 0.72 (Fig. 11.1a). Higher amounts of tylosin fermentation waste were added to a limestone waste to examine tylosin adsorption in a mineral soil containing no organic matter. With addition of 13, 25, and 38 tonnes of fermentation waste per hectare, 13.1, 26.2 and

39.3 parts per million tylosin were added, respectively. The adsorption of tylosin at these levels is shown in Fig. 11.1b. The adsorption capacity of the limestone waste was not reached with addition of 39.3 parts per million tylosin and the adsorption constant for tylosin and limestone waste was found to be 0.46. The total amount of tylosin released by the soils in these experiments was determined by leaching soil samples amended with different amounts of tylosin fermentation waste at fortnightly intervals over a 12 week period. Over this period of time, maximum release of tylosin occurred immediately after addition of the fermentation waste to the soils. In both soils, the initial level of tylosin detected ranged from 0.4 parts per million with the addition of 3 tonnes of fermentation waste per hectare to 13.8 parts per million at 38 tonnes of fermentation waste per hectare. Although these levels are above the minimum inhibitory concentration for many sensitive organisms. It must be borne in mind that had these initial concentrations of tylosin been left in the soil, and not removed by leaching, they would have been subject to chemical and microbiological breakdown in the soil, as well as adsoption onto organic matter and clay minerals. The results show the release of tylosin into the soil waste over specific periods of time, i.e., the intervals between leachings, and not the subsequent fate of that tylosin in the soil system. Experiments are at present being carried out on the persistence of tylosin left in contact with soils for different periods of time.

Table 11.2. Uptake and release of tylosin from a modified John Innes compost amended with tylosin fermentation waste.

Concentration of tylosin (ppm)	*Temperature (°C)* 4			10			15			23		
Added	3.5	11.2	18.7	3.5	11.2	18.7	3.5	11.2	18.7	3.5	11.2	18.7
Total release	1.1	3.4	7.2	0.6	2.7	3.9	1.1	3.2	6.1	0.8	3.6	3.3
Uptake	2.4	7.8	11.5	2.9	8.5	14.8	2.4	8.0	12.6	2.7	7.6	14.3
Uptake (%)	68.6	69.6	61.5	82.9	75.9	79.1	68.6	71.4	67.4	77.1	67.9	79.3

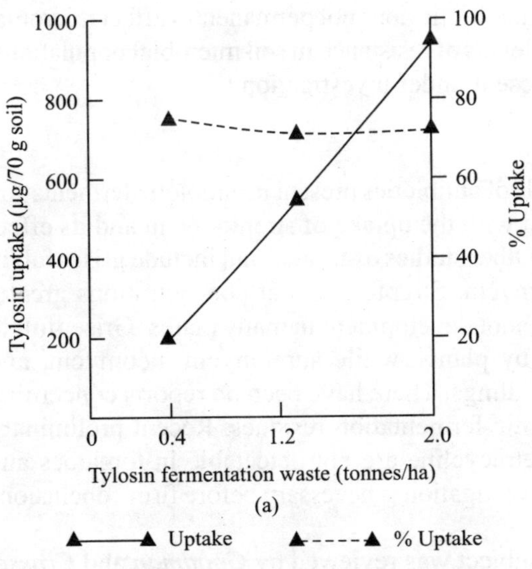

Fig. 11.1a. Uptake and percentage uptake of tylosin from fermentation waste.

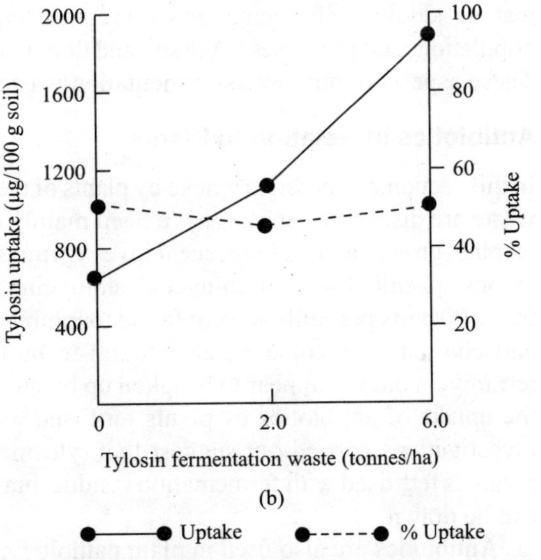

Fig. 11.1b. Uptake and percentage uptake of tylosin from tylosin fermentation waste by limestone waste.

Effects of Antibiotics on Microbial Process in Soil

There have been several studies on the effect of pure antibiotics on microbial populations and processes in the soil. Several of these studies were carried out on pure cultures of soil micro-organisms in laboratory media. However, the activity of an antibiotic in the soil may be very different from that in laboratory media. This is illustrated in a study by *Pramer* and *Starkey* on the influence of streptomycin on nitrification. Nitrification was found to be completely inhibited in solution medium by 1 part per million streptomycin. In sand, a comparable inhibition required 1000 parts per million, while in soil, a concentration of 10,000 parts per million was required. The results of this and other studies indicates that relatively few antibiotics influence the development of micro-organisms in the soil. Clavacin at 50 parts per million and actidione at 5 parts per million were found by *Gottlieb* to inhibit sensitive organism in the soil, while antimycin was found to inhibit *B. subtilis* growth at 6 parts per million in soil. More recent experiments using the antibiotic tylosin have shown that additions of the pure antibiotic at concentrations of 41, 184 and 368 parts per million caused a temporary inhibition in overall microbial activity in soil, as measured by O_2 uptake and CO_2 release, for six weeks after addition. Nitrogen mineralisation in the soil amended with tylosin was found to be inhibited throughout the ten weeks of the experiment.

In contrast to this, experiments carried out using tylosin fermentation waste showed that at all the concentrations of waste used, there was always an increase in microbial activity, as measured by O_2 uptake, CO_2 release, and N mineralisation, compared with the unamended soil. Similarly, the results from research yet to be completed suggest that the addition of penicillin and oxyteracycline fermentation wastes to soil does no cause a decrease in microbial activity. But the results of the effect of pure tylosin on microbial activity suggest that it is possible that the concentrations of tylosin and other antibiotics present in high rates of application of their fermentation wastes to soils could exert a restraining influence on the microbial activity in the soil, and therefore the full potential of the material would not be realised. In the case of tylosin, the amounts of pure tylosin added to soil which resulted in a temporary decrease in microbial activity were those which would be present in addition of 25,125 and 249 tonnes of dried fermentation waste per hectare, equivalent to 152, 762 and 1524 wet tonnes/ha. It is, therefore, important that the addition of fermentation waste containing residual antibiotic does not permanently affect microbial populations and processes in the soil and does not increase levels of resistance in soil microbial populations. These aspects of antibiotics fermentation waste are at present under investigation.

Antibiotics in Relation to Plants

In this section, the possible uptake by plants of the low levels of antibiotics present in antibiotic fermentation waste are discussed. Studies have been mainly concerned with the uptake of streptomycin and its effect on plant growth and include recent investigations by *Bag*. Other studies examined and include griseofulvin chloramphenicol and aureomycin, neomycin, and terramycin. Streptomycin at concentrations greater than 100 parts per million were found to inhibit root and shoot development in many plants. Griseofulvin and chloramphenicol were also found to be taken up by plants, while aureomycin, neomycin, and terramycin did not appear to be taken up by cucumber seedlings. There have been no reports concerning the uptake of antibiotics by plants fertilised with antibiotic fermentation residues. Recent preliminary investigations carried out suggest that tylosin and oxytetracycline are not detectable in tomatoes and potatoes fertilised with fermentation residue, but further investigation is necessary before firm conclusions can be drawn.

Antibiotics are also used in plant pathology, and the subject was reviewed by *Goodman* and *Crosse*. *Berdy* reported that a series of antibiotics active specifically against plant pathogens are manufactured on a large scale, mainly in Japan. These include blasticidin S, kasugamycin, polyoxin, and validamycin.

Berdy reproted the use of fermentation waste from antibiotics produced specifically for use in plant pathology should pose no problems in terms of the uptake of the residual antibiotics by plants, as these plants are already being exposed to much higher concentrations of antibiotic than would ever be present in applications of fermentation waste. It should, however, be borne in mind that although questions have been raised concerning the presence of residual antibiotics in animal products consumed by man, no questions appear to have been put forth concerning residual antibiotics in plant tissues consumed by man, and this could be a problem which will arise in the future. Therefore, it would be advisable for the subject of the uptake of antibiotics by plant tissues to be investigated before questions similar to those asked by concern authorities.

WINERY BY-PRODUCTS

The major waste produced by the wine-making process are stems, pomace and lees. The quantities of stems and pomace produced are large, especially in Europe. *Amerine* and *Joslyn* reproted that 1 tonne of grapes gave rise to approximately 746.7 litres of new wine, 87.6 kg pomace, and 33 kg stems.

Stems

These are the stems and leaves of the vines and represent between 2 and 6 per cent of the total weight of the fruit. There have been few reports on potential uses of this material, although *Schreffler* suggested that the stems could be ground up in water and the resulting liquor distilled for its alcohol content. Potential commercial utilisation of the stems was examined by *Raback* while *Amerine* and *Cruess* reported that in the United States the stems are usually disposed of by scattering on the land, and in Europe, crushers are used to break the stems up.

Grape Pomace

The material is a mixture of the skin, seeds, and pulp of the grapes. When the juice is pressed out for wine production, the solids left represent between 10 and 20 per cent of the weight of the fruit, with a moisture content of 35 to 70 per cent. Reports by *Fuchs* and Amerine suggest that grape pomace contains approximately 11 per cent crude protein, 6 per cent fat, 40 per cent fibre, and 8 per cent ash. The value of grape pomace as a feedstuff was first suggested by *Semichon* and the feeding value was considered to be equal to meadow hay, but experiments by *Folger* suggested that the materials was no better than wheat straw in feeding value. He found that although sheep would eat pomace readily, dairy cows found it unpalatable. *Amerine* and *Cruess* also suggested that the material could be useful as feed supplement. Before feeding it to animals the pomace is dried in rotary, direct-fired drum driers heated by gas or oil to temperatures of approximately 1000°C. The dry pomace is crushed and molasses may be added to improve its palatability. The moisture content must be kept below 6 per cent to prevent spoilage. *Agostine* reported that the feed value of grape pomace could be improved by the addition of lime to raise the pH. The use of grape pomace as an animal feed does not appear to be widespread and recent reports suggest that the material may be of use as a fuel.

The potential of grape pomace as a fertiliser has also been investigated. *Jacob* and *Proebstring* reported that an analysis of grape pomace showed levels of 1.5 to 2.5 per cent N, 0.5 per cent P and 1.5 to 2.5 per cent K on a dry weight basis. They gave its fertiliser value as equivalent to farmyard manure and suggested that it could be used to improve the texture of heavy soils. It was also noted that heavy applications gave rise to phytotoxic effects due to substances formed during the early stages of decomposition. After this initial period, it was found that the toxic effects disappeared and area where waste was dumped supported luxuriant plant growth. *Perrone* advised that the material should be composed before spreading and this would overcome the toxic problems.

Lees

These are the spent yeast cells which remain after fermentation of the wine is complete. Lees contains potassium bitartrate in large amounts, wine-making representing the largest single source of tartrates in the world. Tartrate is also recovered from grape pomace and can often be scraped off the walls of the wine storage tanks as almost pure cream of tartar. In the form in which they are found in wineries, lees are unsuitable for animal feed, as the potassium bitartrate which they contain is a potent laxative. Lees do find a ready market in the distilling industry, where they are mixed with water and the residual alcohol is distilled off. The wastes remaining after this operation are then the concern of the distilling industry.

WASTES FROM CIDER PRODUCTION

The main waste products form cider production are apple pomace, yeast, and filter aid.

Pomace

As with grape pomace, cider pomace consists of the remains of the skins, pulp, core material, and seeds of apples from which most of the juice and soluble constituents have been extracted in the making of cider. Much of the pomace produced is further treated to extract the pectin, which is subsequently sold. The composition of these two types of pomace is given in Table 11.3. Levels of production of pomace are difficult to estimate since there are no recently published figures and not all cider producers are prepared to release figures of the waste which they discharge into the environment.

Table 11.3. Analysis of three types of apple pomace.

Percentage composition	*Dried pomace*	*Dried pomace (After removal of pectin)*	*Wet pomace (After removal of pectin)*
Dry matter	87.5	96.3	21.3
Crude protein	6.2	10.4	10.5
Crude fibre	20.9	36.9	37.8
Ether extract	2.5	4.6	4.7
Total ash	2.0	0.9	1.0
pH	–	–	3.0

In the United Kingdom, much of the pomace, after being depecturised, is dried and sold as feed additive. Many of the reports dealing with cider pomace as a feed date from the first half of this century, and in most of these it is suggested that the wet pomace be ensiled before feeding to prevent decomposition. *Smith* states that the drying of pomace requires a special plant, whereas ensiling can be easily carried out on the farm. In the first half of this century, cider-makers tended to be small organisations and could easily distribute their wet pomace locally, but with the formation of large cider manufacturers with large plants, it became possible to dry the pomace, on site if necessary, and then distribute it as a feedstuff. All cider pomace is readily palatable. In the United Kingdom, the material is marketed as source of a hemi-cellulose-type material, suitable for blending with oils and proteins in cattle cake.

Barker and *Gimingham* suggested that apple pomace could be used as a manure. *Barker* reported that the material had been using in Devon as a manure for grass, after being composted with phosphate.

Barker and *Gimingham* have an analysis of pomace used as manure as 0.2 to 0.6 per cent K_2O, 0.4 to 0.7 per cent H_3PO_4, and 1.6 to 1.7 per cent N. They pointed out that in France, the material was mixed with bone or rock phosphate. The acids solubilised the phosphate, increasing its availability. Today, however, it appears that most of the pomace produced by the cider manufacturers is used as a feedstuff rather than a manure.

Yeast

The yeast produced in the fermentation of cider is usually mixed with the pomace and sold for animal feed in USA and UK.

Filter Aid

Much of the filter aid is used in the production of pectin, and, like the filter aid used in the brewing industry, it is not economic to reclaim or utilise it and it is disposed off by dumping.

OTHER FERMENTATION WASTES

Levels of waste produced by other various industrial fermentations, including enzymes, vitamins, and steroid hormones, are virtually impossible to determine due to the high degree of secrecy in the industry. All that can be discussed here is which compounds are produced on an industrial scale by fermentation; it should be pointed out that the wastes which will be produced could have a market as feedstuffs or fertilisers if the security problems could be overcome.

Of the eight organic acids which are produced by fermentation in Japan, 51 per cent of the production is gluconic acid, while of the sixteen amino acids produced by fermentation, 86 per cent of the production, is L-glutamic acid.

Amylase is the enzyme produced in the largest quantities by fermentation and represents 87 per cent of total enzyme production; together with protease, this represents 95 per cent of total enzyme production. Of the seven nucleic acids produced by fermentation, the most important are 5-IMP and 5-GMP, which together account for 99 per cent of nucleic acid production. Vitamins C and B1 account for 89 per cent of the vitamins produced by fermentation. Very small amount of steriod hormones are produced.

Levels of wastes produced from various fermentations companies are not available, but large quantities of organic wastes from these processes are being produced. But due to the secrecy surrounding many of the fermentation it is unlikely that the materials will be marketed as feedstuffs and fertilisers continue to rise, the companies producing these materials may be induced to process their own wastes and market them. This would depend, of course, on the economics of the process at the time.

12

Dairy Industry

INTRODUCTION

Ideally, waste management should aim at total use of all incoming resources and to recycle or re-use all processing effluents thereby eliminating the need for waste disposal and waste treatment. Proper waste management is closely linked with good resource management. Many industries have, therefore, realised that a waste management system is the logical and profitable alternative for pollution control.

Waste management systems are inherent in nature. These have near perfect and well maintained recycling operations for water, carbon, nitrogen etc. However, the balance of these systems is being disturbed by industrial, agricultural and other activities of man. It has now been realised that this has to be stopped for human survival. A failure to do so leads to long terms irreversible catastrophy. It is in this background that many nations have formulated laws and regulations to protect our environment.

In dairy industry, water has a multipurpose use. Water used for the purpose of processing, cleaning and other general uses should be of potable standard and absolutely free from microbial contaminants. Use of contaminated water in the plant, results in milk and milk products to be unsafe for human consumption.

Several outbreaks of epidemics have been traced to contaminated milk and milk products. Contamination of water invariably arises from the activities of man. It is always desirable to use water of good quality and avoid chances of its contamination either at source or transit.

CONSERVATION OF WATER

The serious concern for prevention of pollution by direct discharge to water bodies has resulted in thinking of possible re-uses of waste-water.

With enormous increase of population and ever increasing demand for water, recycling and re-use by such systems will become necessary for human survival in many areas.

The advantages of water conservation practices are three fold:

1. Water is a commodity and subject to the economic laws of supply and demand. By its conservation, the cost of processing is controlled.
2. Reduction in amount of processing waste-water discharges will means smaller treatment facilities.
3. Higher pollution concentrations per unit volume of waste-water usually means less removal cost for each unit of pollutant.

Recycling and Re-use of Water

Recycling and re-use of water are necessary for in-plant water conservation. Recycling refers to treated water as process water, and also in other applications where the quality requirements are less critical.

In recycling systems, the build-up of organic and mineral compounds, microbial populations and temperature increases or decreases have to be considered. Each successive contact of water with the product contributes to change of quality of water. These are methods to achieve acceptable quality for the recycled application:

1. Recycled water can be treated by screens, filter, cooling towers to re-establish acceptable quality conditions of solid contents and temperature.
2. Microbial populations can be controlled by disinfecting agents.
3. Fresh make up water can be automatically added to recycled water to dilute it to acceptable quality conditions or to maintain desired water volume.

Recovery of By-Products from Dairy Wastes

It is possible to recover certain by-products from dairy waste-waters:

1. Butter milk can be roller-dried and sold as powder or can be incorporated as a source of protein.
2. Whey can be recovered as dry powder by condensing and spray-drying. It can be used as a source of protein in animal feeds.
3. Casein can be recovered from sour milk by treating with acid which can be washed and dehydrated. Casein is used in the manufacture of resins, plastics, paints and varnishes.

HOUSEKEEPING PRACTICES

There are many housekeeping practices which can be used in plants to avoid solid waste from getting into waste-water stream. Practices like dumping or flushing waste materials down gutters have to be avoided. Processing residue which do get into the waste-water stream have to be removed as soon as possible. These residues leach suspended solids and dissolved solids into the waste-water causing increase in treatment costs.

Screening is one method of removing these processing residues. The plant personnel can develop ways to prevent spillage, reduce processing wastes and check wastes from entering the waste-water stream. Housekeeping practices require training and good supervision of employees. A proper attitude towards preventing waste generation must also be developed.

Spillage costs money in four ways: loss of potential product, additional labour to clean up, increased use of water and detergents for cleaning and addition to waste to be treated.

Spillage can be minimised by:

1. Conveyors operating with proper loads and never resorting to overloading.
2. Uniform flow of products in processing lines.
3. Proper maintenance of equipments to prevent breakdowns or breakages in processing line.
4. Arrangements for automatically stopping the flow line in the event of a breakdown.

Besides, non-food residues such as broken glass, paper, metal cans etc., have to be kept out of the waste-water stream, while the food residues in the waste stream are to be recovered and re-used as by-products like animal feed etc.

The system can further be improved by appropriate process modifications.

RENOVATION AND RECLAMATION

Renovation and reclamation of food waste can be dairy industrie's choice to convert waste treatment costs into economic benefits. It is important that dairies strive to conserve water and use more efficient food recovery schemes to reduce effluent treatment costs.

The best way the dairy industry can treat its waste is to avoid producing it. This can be done by improving in-house waste abatement methods which are less expensive than waste treatment. The term 'waste abatement' suggests better in-plant management.

In-Plant Modifications

The in-plant and process modifications are designed to conserve water and prevent pollutants from entering the waste-water stream. These modifications range from simple procedural changes for some plant operations to process modifications involving changes in equipment design and workman's skills. This requires extensive research and development and should be fully explored and approved before putting these modifications into full scale operation.

Modifications In-Plant Clean-Up and Sanitation

Clean-up and sanitation are essential in the operation of a milk plant. It is usual to do clean up by hosing down equipment, walls and floors with vast amounts of water and flush wastes to the floor drain. Water is being used for this without any constraints with the belief that it is cheap and is in unlimited supply and the 'answer to pollution is dilution'. However, many improvisations have been thought of, for clean-up and sanitation of plants. Some of them are:

1. Dry-cleaning of floors and equipments to remove large objects and dirt before washing them with water.
2. Use of high pressure, low volume cleaning equipments.
3. Use of mechanical cleaning devices.
4. Use of new detergents, foams and gels that provide faster cleaning.
5. Improving design of equipments for easy and fast cleaning.
6. Adoption of CIP systems, in which various detergents and sanitary solutions can be recirculated several times before the final re-use cycle.
7. Improvisation in design and construction of plants to have easily cleaned floors, walls and ceilings etc.

Process Modifications

Process modifications have to aim at conservation of wastes and keep solid wastes away from entering the waste stream. This is more easily said than done. It involves re-designing of equipment or even conceiving new concepts of processing. It necessarily requires research and development of concepts and pilot plant studies.

End of Pipe Waste Treatment

Subsequent to statutory Acts coming into force, many plants are planning to have effluent treatment plants for waste-water treatment before disposal.

Several treatment methods are available for treatment of dairy wastes. The choice depends upon the local factors and costs.

A few of common methods being adopted are:

1. Dilution: This can be adopted if dilution water is available.
2. Land irrigation: This can be useful for small plants and generally in isolated places.
3. Septic tanks: For chilling centres and very small plants.
4. Biological treatment.

Trickling filters or biofilters should have fixed support media like graded stones, slag or plastic media to maintain active organisms for absorbing organic matter from the wastes and oxidising the same. Oxygen is normally available by natural ventilation provided in the filters. Contact time of the waste in trickling filters is normally short depending upon the loading or application rate of filters. This is usually increased by recirculation of treated waste-water back through the filters. The filter is provided with under drains for collecting the treated effluent.

ANALYSIS/CHARACTERISATION OF DAIRY WASTES

The most significant phase in the treatment process of dairy wastes is its quantitative and qualitative analysis. This would help in knowing the quantity of waste-water to be treated which in turn becomes a crucial design parameter.

Analysis of waste parameters like physical, chemical and biological give the complete picture of the waste constituents. These would be very much helpful in the selection of treatment method, treatment unit to be designed and the magnitude of treatment to be given. The prominent parameters to be constantly monitored/analysed are biological oxygen demand (BOD), chemical oxygen demand (COD), total dissolved solid (TDS), hydrogen ion concentration (pH), dissolved oxygen (DO) and nutrients like phosphorus and nitrogen.

EFFECTS ON ENVIRONMENT

Dairy waste is basically organic in nature and slightly alkaline when fresh. When these wastes are allowed to go into the stream without any treatments, a rapid depletion of dissolved oxygen content of the stream occurs, along with the growth of sewage fungi covering the entire bottom of the stream.

Though alkaline in fresh condition, the milk waste becomes acidic due to the decomposition of lactose into lactic acid under anaerobic conditions particularly after complete oxygen depletion of the stream. The resulting condition precipitates casein from the wastes, which decomposes further into a highly odorous black sludge. At certain dilutions, the dairy waste is found to be toxic to fish. (Lactic acid is toxic to fish at about 600 mg/l).

The presence of phosphorus, nitrogen etc., serve as nutrients for micro-organisms and thus they are responsible for eutrophication of lakes and other water bodies. Because of highly biodegradable nature, the dairy waste also causes considerable air pollution by emitting foul and irritating odour.

TREATMENT TECHNIQUES FOR DAIRY WASTES

In the modern set-up, proper management of waste-water is a necessity. Public health consequences (negative) of poor waste-water management and treatments will be very high. The practice of collecting and treating waste-water prior to disposal is a relatively recent undertaking. Thus, the treatment and disposal of dairy effluents are assuming great importance on account of the rapid development of the industry. Treatment techniques are categorised in Table 12.1.

Table 12.1. Treatment techniques.

Treatment	*Techniques*
Preliminary treatment	Screening, grit removal, neutralisation, equalisation
Primary treatment	Primary clarification
Secondary treatment	Activated sludge process, trickling filter, biotowers, rotating biological contactors, aerobic and anaerobic digestion
Tertiary treatments (Advanced treatment)	Suspended solids removal, nutrients removal, disinfection etc.

Other techniques available for high strength organic wastes treatment are oxidation pond, aerated lagoon and oxidation ditch in addition to usual conventional treatment methods employing suspended and attached growth systems. The comparison of various treatment options for dairy wastes is presented in Table 12.2.

Table 12.2. Comparison of treatment methods for dairy wastes to achieve a consistent effluent quality of 30 mg/l.

Treatment	*Effluent quality BOD and suspended solids*	*Capital cost*	*Operating cost*	*Land required*
Activated sludge	3	H	H	L
Oxidation ditch	3	H	H	A
Aerated lagoon	3	A	A	H
Stabilisation pond	1	L	L	H
Trickling filter	2	H	H	A
Biological rotating disc	3	H	L	L
Anaerobic process	1	L	L	L
Irrigation	3	A	A	H

Note: 3 = Excellent; 2 = good; 1 = fair; 0 = poor; H = high; A = average; and L = low.

The effluent quality after the treatment must conform to the effluent discharge limits suggested by Bureau of Indian Standards.

Standards to be met by the industry people for discharging the treated effluents on to be met by the industry people for discharging the treated effluents on to the land, surface water body or sewer.

DISPOSAL AND RE-USE OF WASTE-WATER

An insignificant volume of the influent waste-water accompanies sludges and other materials disposed of during waste-water treatment processes. The bulk of the waste-water remains to be disposed off after the treatment processes have been completed. Ultimate receptors of treated waste-waters include surface water and ground water bodies, land surfaces and in some instances, the atmosphere. Recognition of the value of waste-water resource has resulted in an increase in the re-use of treated effluents, particularly in water scarce regions.

Disposal sites of re-use facilities must be found within a reasonable distance of the waste-water treatment plant because of the cost of transporting the effluents over long distances. The disposal methods include: (i) ocean disposal; (ii) land applications; (iii) irrigation; and (iv) rapid infiltration.

The re-use methods include: (i) recreational facilities; (ii) industrial water supply; (iii) ground water re-charge; and (iv) re-use in potable water eystencs.

BY-PRODUCT RECOVERY

This plays an important role in the reduction of organic load in the waste generated from the dairy. Recovery of usable by-products such as buttermilk, whey, casein, lactose and butter oil is beneficial to the industry and as well to public health authorities as an important step towards pollution control and economy.

UTILISATION OF BY-PRODUCTS FROM DAIRY INDUSTRY

Milk is an excellent and balanced food, essential for all young ones. It is processed into many products like market milk (pasteurised, single-toned and double toned milk), cream, butter, cheese, ice-cream, condensed milk, sweetened condensed milk, milk powder, ghee etc.

Skim milk, buttermilk and whey are the principal by-products from the dairy industry. Skim milk contains protein 3.6, fat 0.1 and ash 0.7 (as percentages). Condensed skim milk, sweetened condensed skim milk, skim milk powder, lactic beverages, cheese, bakery products, casein, co-precipitates of calcium, magnesium and potassium caseinate, and meat and egg substitutes can profitably be prepared from skim milk. Skim milk powder is used in bakery and confectionery products to improve their taste, flavour and nutritive value.

Casein is used for the manufacture of sausages, baked cereal goods, simulated milk and whipping cream, coffee whiteners, imitation milk, high protein foods, spreads and beverages, and egg and meat substitutes. It is also used as a wine clarifier, adhesive for plywood, paint stabiliser, and reinforcing agent for rubber, and in the paper-coating, plastics, cosmetics, textile and surface-coating industries.

There are many small industries manufacturing casein, channa, cheese, paneer and other products, from which about 133,000 tonnes of whey is the main by-product. Whey contains lactose 5.1, protein 0.9, fat 0.3 and ash 0.6 per cent respectively. It can be used as such for irrigation, and as cattle feed and binder, or after processing into condensed whey, sweetened condensed whey, whey powder, whey cream and protein concentrate, lactose, alcohol, riboflavin, lactic acid, beverages and cheese. Whey or whey solids are used for the preparation of infant food, dried soups, dried culture media, bakery products, confectionery products, desserts, dry mixes, cheese, and alcoholic and non-alcoholic beverages.

Whey protein is obtained either by cold precipitation or heat coagulation. In cold precipitation techniques, treatment with polyelectrolytes (ferric polyphosphate, ferric chloride, sodium hexametaphosphate, carboxymethyl cellulose or polyacrylic acid), reverse osmosis, ultrafiltration, and gel filtration yield whey protein with all the desirable attributes intact. If heat coagulation is used, the foaming, whipping and other properties of the whey protein are severely affected.

Lactose constitutes about 70 per cent of whey solids. It is used as a flavour and flavour enhancer, controlled browner and crystalliser in bakery and confectionery, and dispersing and emulsifying agent. It is also used in silvering mirrors and controlling the rate of burning in fireworks, and as a polishing agent in toothpastes and a filler and inner carrier for capsules and tables. Galactose, dulcitol, lacitol, sorbitol, lactose monoesters and lactobionic acid can be prepared from lactose either by hydrolysis, oxidation or reduction processes.

Buttermilk can be utilised as animal feed, or processed into cold drinks, lactic beverage, soft cheese, ice-cream, casein or buttermilk powder. A quarter of the serum solids of ice-cream can be substituted by buttermilk solids without any change in product characteristics. Soft cheese prepared from buttermilk can be consumed as such after adding 15 per cent sugar. Also, 10 per cent of it may be incorporated into processed cheese without affecting quality.

Ghee residue contains 21 to 55 per cent fat, 7.5 to 17.2 per cent protein and 2.3 to 2.8 per cent mineral matter. It can be processed into chocolate burfi by mixing 22 parts of ghee residue with 10 parts sugar, 10 parts khoa and 1.5 parts chocolate powder.

Whey and buttermilk are the main by-products generated when manufacturing cheese, casein, channa and butter. Both are similar in composition except that whey contains only 20 per cent of the original milk protein whereas buttermilk contains all of it.

Lactose is the main by-product prepared from the whey derived from cheese or casein manufacture. Whey is vacuum-concentrated to about 50–60 per cent solids, and cooled to room temperature with constant stirring to facilitate crystallisation, after which lactose is separated by centrifuging, and purified.

Whey powder and demineralised whey powder are prepared from whey. The latter may be used for infant food. Whey protein concentrate is made by heat precipitation or ultrafiltration, or by using polyelectrolyte like carboxymethyl cellulose, ferric chloride, ferric polyphosphates, polyacrylic acid or sodium hexametaphosphate.

Rictotta cheese is prepared by acid precipitation of hot whey, after which the precipitate is mixed with salt and pressed. It is consumed as such, or may be mixed with other cheeses for preparation of processed cheese. From 50 to 75 per cent of the protein in whey is recovered by preparing whey cheese.

Infant food and weaning food have been prepared using whey and soyabean as the main raw materials, and adding vegetable fat, citrate and methionine.

Fermented ammoniated condensed whey is obtained by fermenting whey to lactic acid, and adding ammonia to produce ammonium lactate. The fermented ammoniated whey is evaporated to about 60 per cent solids, of which the crude protein content constitutes about 45 per cent. This is used as animal feed along with forage or concentrates.

Alcoholic beverages are prepared by fermenting the lactose present in the whey. Attempts have also been made to prepare single cell protein from whey.

Buttermilk can be conserved by condensing and drying. It can also be processed to lactic beverages or alcoholic beverages using suitable cultures.

MILK LIKE BEVERAGE FROM CHEESE AND GROUNDNUT PROTEIN

Fluid whey is a dairy by-product that contains nearly half the nutrients of milk. Recovery of whey solids is uneconomical because of high energy costs. Whey proteins are of good nutritional quality and have valuable functional attributes such as foaming, emulsifying and acid-solubility.

Use of whey in high protein beverages is attractive to dairy processors since beverages have the potential for population acceptance.

Toning of milk with cheaper proteins is being used to extend the limited availability of milk in this country. Once such product, 'miltone' is made using groundnut protein isolate and 50 per cent milk, which, however, constitute 80 per cent of the beverage cost. Use of a waste product like whey in place of milk will greatly reduce cost.

Groundnut protein isolate was prepared by extraction of edible groundnut flour with dilute sodium hydroxide, removal of the insoluble carbohydrate residue by centrifugation and precipitation of the extracted proteins by addition of dilute HCl to an isoelectric pH of 4.8. The proteins were separated and the residual acid was neutralised.

The acid whey from cheese was neutralised to pH 6.8 with lime water and heated to 40°C. Groundnut protein isolate, sugar and hydrogenated fat were added at levels of 3, 2 and 3 per cent respectively, and the mass homogenised and finally pasteurised. Sodium dehydrogen phosphate at 0.01 per cent level was used to improve the dispersibility of the protein and stability of the emulsion. The beverage had a final composition of protein 3.8, fat 3, carbohydrates 6.8 and mineral 0.51 per cent respectively. Whey proteins are rich in lysine, sulphur amino acids and threonine, and thus enhance the nutritional quality of groundnut protein. The beverage has good stability. Organoleptic evaluation by a trained panel in terms of appearance, flavour, taste and mouthfeel, on a hedonic scale of 1 to 8 ranging from unaceptable to highly acceptable, yielded a high ranking of 6 to 7.

The beverage can be prepared with equipment available in a modern dairy, and could well be a milk substitute at much lower cost.

Planning and Management

Planning in field of waste management is the process by which community needs regarding the management are measured and evaluated and workable alternatives are developed for presentation to decision makers. Planning is accomplished by applying the engineering principles to the need capabilities and goals of community. Thus, planning is both exciting and challenging, because most of the technical, environmental, economic, social and political factors and inter-relationships that involved are now only partially understood. Figs. 12.1 and 12.2 present the flow change of planning and management system for a waste management.

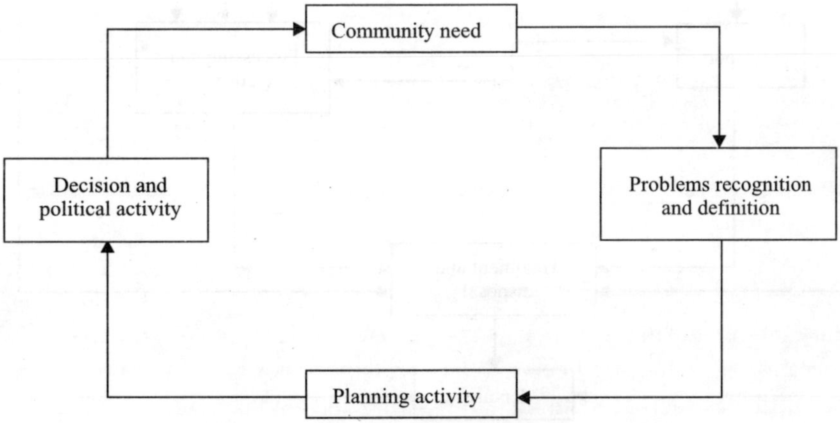

Fig. 12.1. Planning and management activity.

UTILISATION OF WHEY

Whey is the largest by-product of the dairy world. It is obtained during the manufacture of paneer, chhana, cheese, casein and co-precipitates. Maximum whey is produced in those countries where large quantities of cheese is manufactured. Each kilo of cheese gives rise to about 9 litres of whey. Cheese manufacture is one of the best methods of preserving surplus milk solids. Most of the developed countries have enhanced the level of milk production by application of modern technology, to such an extent that they are finding it very difficult to utilise the entire milk production profitably. There is still tremendous potential to accelerate the milk production further by judicious application of biotechnology. The same can be expected to happen in this country where, the dairy industry changing very fast, there is big scope for enhanced milk production by application of modern breeding, feeding and management practices.

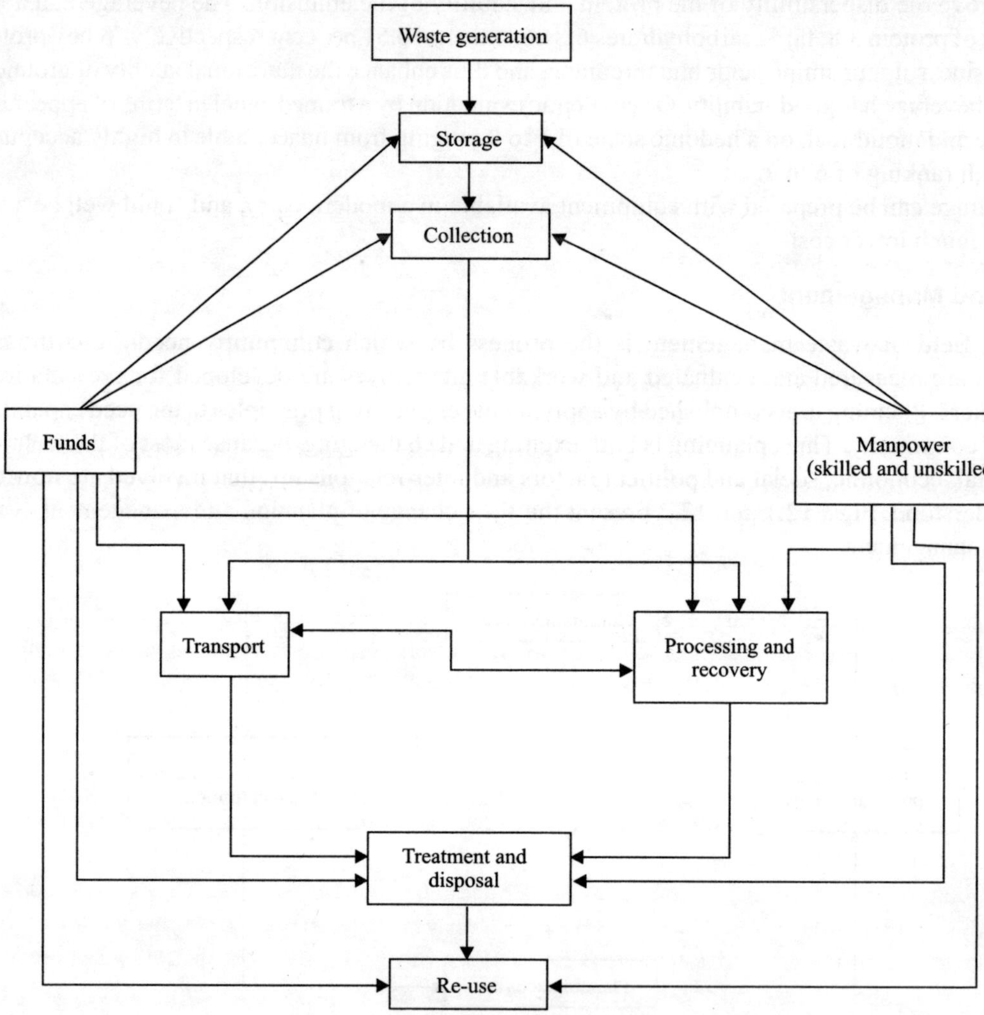

Fig. 12.2. Inter-relationship of functional elements in a waste management system.

In India, milk production is seasonal. There is a lot of milk production during the flush season and considerably less during the lean season. Dairy plants have begun to face problems in utilising flush season surpluses in a useful way. Therefore, the most pressing need to the Indian dairy industry today is to convert surplus milk into such products so that the milk solids could be preserved for long period and extended into the lean season. This could be accomplished by only two of the existing technologies in the dairy field, viz., drying of milk and manufacture of cheese. As far as drying is concerned our milk plant capacities are already fully saturated. Thus the only alternative left is to convert surplus milk into such products so that the milk solids could be preserved for long periods and extended into the lean seasons. This could be accomplished by only two of the existing technologies in the dairy field, viz, drying of milk and manufacture of cheese. As far as drying is concerned our milk plant capacities are already fully saturated. Thus the only alternative left is to convert surplus milk into cheese. This will provide a highly nutritious and palatable and value-added food.

Luckily cheese production and consumption is increasing quite fast in our country and it is likely to increase in future as the standard of living goes up. Also, if our milk production really becomes surplus, it could be converted into cheese which can be exported to middle east countries. Maximum cheese import is made by these countries. India could capitalise on proximity to these countries both geographically and politically. Thus there is tremendous potential and possibilities for generation of large quantities of whey in addition to what is being produced at present.

Need for Whey Utilisation

Utilisation of whey has been a burning problem of the dairy industry all over the world. This has been mainly because of two reasons: Firstly, to salvage precious nutrients and secondly, to check environmental pollution. Its biological oxygen demand is from 38,000 to 46,000 ppm as against 200 ppm for domestic sewage. Nutritionally whey is a very important food item. It has been estimated that nearly 42–44 per cent of the total solids in milk are to be found in whey. Whey contains about 6 per cent total solids of that protein and lactose constitute the major proportion, i.e., 12 and 70 per cent respectively. The whey protein has greater biological value (104) than the major milk protein casein (73). Also whey protein consists of about 61.4 per cent B-lactoglobulin. 20.5 per cent—lactalbumin, 6.0 per cent blood serum albumin and 12.2 per cent immunoglobulins.

Whey proteins is one of remarkable nutritional quality and contain higher total quantity of essential amino-acids than other proteins thus if whey proteins could be salvaged in good condition, they could help fight wide spread protein under-nutrition in our country. Salvaging of whey protein is particularly important when a hectic search is on to extract proteins from non-conventional sources, like bacteria, fungi, algae, leaf and various oil cakes.

The proteins from all these sources are of relatively poor biological value and are difficult to extract and convert into palatable forms. On the other hand, whey is so useful that it has been consumed by people since ancient times.

It was found to be very palatable and thirst quenching to be ancient Greeks and has been so, down the centuries to the present day. Whey has been used to treat a wide variety of ailments, such as arthritis, anemia, and liver complaints. Hence, there is no justification in draining off the whey, wherever it is produced. Efforts should be made to judiciously conserve and process each component of whey for edible purposes.

Mode of Whey Utilisation

The potential use for whey solids are numerous. However, it will have to be critically evaluated as to which mode of utilisation would be most economical and convenient under prevailing Indian conditions. It would be prudent if we took a cursory view of various modes in which whey solids are being used in developed countries (Tables 12.3 and 12.4).

Table 12.3. Whey solids utilisation in USA.

Description	*Human food*	*Animal food*	*Total*
Concentrated whey	7.1	3.8	10.8
Dried whey	31.5	34.1	65.5
Reduced lactose whey and reduced lactose minerals whey	3.2		3.2
Whey protein concentrate	3.0	3.6	6.6
'Lactose	4.9	–	4.9
Dried whey product	–	7.2	7.2
Whey solids in blends	–	0.8	0.8
Undesignated food	–	1.0	1.0
Total	49.70	50.5	100.0

Table 12.4. Whey solids used in human foods.

Market category	*Per cent*
Dairy	26.6
Bakery	22.3
Prepared dry mixes	22.4
Blends	19.1
Confectionary	1.2
Soups	1.1
Margarine	1.0
Chemical/Pharmaceuticals	0.4
Meat processors	0.3
Institutions	0.3
Soft drinks special dietary foods	0.1
All others	5.2
Total	100.0

Under Indian conditions any technology which could use the entire milk solids as a whole would be most economical. Thus the probable mode of whey utilisation in India could be listed as follows: (i) liquid whey; (ii) concentrated whey; (iii) dried whey (iv) fractionated whey; and (v) fermentation products.

Liquid whey

Liquid whey without further treatment is used as animal feed. This is a major outlet and in many circumstances, the best economic use for whey. However, since under Indian conditions whey productions is scattered and in small scales, the best use of whey can be made by converting it into refreshing drinks. These drinks or beverages may be fermented or non-fermented.

Fermented whey drinks

Conversion of whey into a beverage form through lactic acid fermentation is one of the attractive avenues for utilisation of whey for human consumption by employing *Lactobacillus acidophilus* fermentation. The nutritional properties are enhanced and certain therapeutic attributes are added to the beverage. Recently an attempt has been made at National Dairy Research Institute, Karnal (Haryana) institute to convert whey into a palatable, refreshing and low cost beverage named 'acido whey'.

Preparation

Whey from paneer, cheese, chhana, or casein is separated to remove fat and traces of curd particles. It is then heated to 95–100°C for 10–20 minutes Whey is then cooled to 40°C inoculated with an active culture of *L. acidophilus* and incubated at 39 ± 1°C for 20 to 24 hours.

After incubation fermented whey is clarified to remove precipitated cellulose mass, sugar is then added to the product at the rate of 13–14 per cent in the form of 50 per cent sugar syrup which has been earlier pasteurised. Desirable amount of flavour is added. The beverage is chilled to 4°C and packaged in pouches.

Non-fermented whey drinks from deproteinised whey

A patent issued to co-operative Agricol du Val-Dl' described a process for obtaining a beverage in which whey was filtered, centrifuged and pasteurised at 85°C. The product was cooled, and filtered to remove precipitated protein. A flavoured sugar solution containing tartaric and citric acid was added, the product was bottled and pasteurised or sterilised. The sterilised product was said to keep for 3–6 months. This process can be tried under Indian conditions and can be used as a genuine thirst quencher.

Incorporation of whey in bread

Bread is an ideal nutrient enrichment vehicle due to its use as a staple food in various, forms throughout the world, its convenience of use, stability and high acceptance in low income diet.

Addition of 10–12 per cent whey decreases gas production. Bread structure elasticity taste and aroma are also improved.

Concentrated whey

Whey is a dilute and highly perishable product so it is often condensed (as plain or sweetened) to reduce its water content and improve its keeping quality plain and condensed. Whey may be used in human or animal feed products.

Dried whey

It is apparent from Table 12.3 that maximum utilisation of whey solids (65.7) is in dried form alone in USA. This would be very appropriate technology for our country. The whey produced at scattered points may be collected in cheese manufacturing plants and subsequently concentrated and dried. All our dairy plants are well equipped with drying facilities. If need be, additional plant may be installed exclusively for whey drying. The processing cost and product quality may be improved by application of reserve osmosis on the initial whey followed by vacuum concentration, spray drying and finally fluidised bed drying. This process results in excellent whey powder which can be used in variety of foods viz., bakery products, processed cheese spread and processed cheese food, soups etc.

Fractionated whey

The past decade has been the beginning of a major change in dairy processing. The adoption of membrane processing, both ultrafiltration (UF) for whey treatment, and reverse osmosis (RO) as an alternative method of concentration of milk and whey. It is now a major means of whey processing in most of the dairy countries of the world.

The technique of UF involve separation on a molecular basis of the components in the whey. Whey is pumped at high speeds past a semi-permeable membrane molecular weight and size can pass through the membrane whereas those above that size are retained within the system. Thus is principle it in possible to separate the high molecular weight proteins from the lower molecular weight proteins from the lower molecular weight lactose. UF by its very nature, is capable of producing a range of products with a widely varying gross composition.

Commercially, products with protein content of approximately 35 per cent, 50 per cent and 80 per cent are produced throughout the world. It is necessary to employ diafiltration that is the addition of water during UF to obtain 80 per cent protein content in whey protein concentrate. By and large ultrafiltration combined with reverse osmosis in a two-stage process could most efficiently and economically be used for producing of WPC and lactose from whey. WPC has side application in various foods for several reasons.

Versatality of whey protein

The versatality of whey proteins is listed below:

1. Whey protein concentrate has a high and balanced amino acid profile which means that it does not require supplementation with amino acids in the same way that the soy protein (deficient in sulphur containing amino acids) does.
2. It has low lactose content, and the carbohydrate is lactose as opposed to sucrose, which is currently out of vogue in America and starting to become so in Europe.
3. It has low calorific value which is an attraction to the health food market.
4. It has good emulsifying capacity which has proved useful to emulsify the oils present in many food market.
5. It can be used as a supplement to boost other proteins that are low in specific amino acids of nutritional value.
6. It is blend in flavour and has no bitter after-taste, which makes it widely acceptable.
7. It is truly a natural protein which is complete and does not require supplementation. Its origin from milk gives it a wider consumer appeal.
8. It is also low in fat and sodium content and 'in all it is quite a unique protein'.

Uses of WPC

Bakery products

WPC is a profitable substitute for whole egg or egg white in cake-making. The 60–70 per cent protein grade will replace upto 75 per cent of the whole egg used in making cakes of the sponge type. The lower grades WPC containing 55–60 per cent protein can be used as egg extender for making fruit cake, replacing 50 per cent of the whole egg. WPC will also replace egg white for making meringue, provided the whey is clarified first.

Meat product

Whey proteins form gel on heating and therefore find a use in luncheon meat and meat pastes. They also serve as binding agents, and unlike soy protein binding agents they do not contribute off-flavour.

Soups and dessert products

These products make use of the gelling properties of whey proteins. The gel have a smooth taste which makes them very acceptable for this application.

Soft drinks

Soft drinks can be fortified with whey protein concentrate upto 1 per cent protein content without changes in flavour or appearance, provided the proteins are not denatured. The coca-cola company hold a patent for the fortification of soft drinks with WPC with a high protein/ash ratio. It is also necessary to clarify the whey to avoid the development of turbidity and off flavours during storage.

Fermentation products from lactose rich permeate

With the prospects of more cheese made via ultrafiltration or alternatively more whey being ultrafiltered for better use of protein, there will be increased quantities of permeate. This is mainly a lactose solution with a high BOD calling for use in some way. This lactose solution might be subjected to a very wide variety of fermentation processes for the production of biomass, methane, ethanol, lactic acid, ammonium lactate, food gums, amino acids, vitamin and antibiotics.

Thus whey should be utilised because of its nutritional qualities. It helps in minimising the pollution, and above all:

1. As a beverage it is a genuine thirst quencher unlike most soft drinks.
2. Whey drinks are light and refreshing but less acidic than fruit juice.
3. There are health and nutritional arguments.
4. There are good potential profit margins.

CONCENTRATION AND EXTRACTION OF A VALUE-ADDED PRODUCT CASEINO-GLYCOMACROPEPTIDE FROM DAIRY WASTE BY PRECIPITATION

Sweet cheese whey, a renewable effluent of dairy industry contains lactose, whey proteins (α-*lactalbumin* and β-*lactoglobulin*) and caseino-glycomacropeptide (C-GMP) as essential components. Here, whey treatment for the recovery of a value-added product C-GMP is discussed. In the first step whey proteins were precipitated to enrich the concentration of the C-GMP. Heat coagulation, isoelectric precipitation, selective precipitation and ammonium sulphate fractionation for the removal of whey proteins were compared. Heat coagulation was found to be the best method which could precipitate 70.2 per cent of the whey proteins and thereby could enrich the C-GMP content from 12.7 per cent to 37.89 per cent. HCW was treated with ethanol for lactose precipitation. 48 per cent of the lactose from the HCW was precipitated by ethanol and the C-GMP content was further enriched to 70 per cent.

Cheese whey is the largest effluent of the dairy industry. Each kilo of cheese gives rise to about 9 litres of whey. It has been estimated that about one million tonne of whey is annually produced by the dairy industry around the world accounting for the strongest waste of any kind. One hundred kg of whey containing approximately 3.5 kg BOD and 6.8 kg COD, has the polluting strength equivalent to the sewage produced by 45 people.

Cheese whey is of two types. Sweet whey and acid whey. Major components of sweet cheese whey include, lactose, whey proteins (α-*lactalbumin* and β-*lactoglobulin*) and a value-added bioactive peptide called caseino-glycomacropeptide (C-GMP).

C-GMP, serves as a growth factor for bifidobacteria, which play a major role in human health especially in enhancing infants immunity. In addition C-GMP serves as an antithrombolytic factor and antiobesity factor. Presently, C-GMP is commercially produced by the enzymatic degradation of C-GMP, and studies has been carried on the concentration and extraction of C-GMP from cheese whey by precipitation.

Precipitation is a very valuable and relatively simple technique for recovering many biological products. Precipitation is the separation technique most often used during the early stages of downstream processing to achieve partial purification of the product as well as reduction in volume. Since the product is most often found in the precipitant, both enrichment and concentration are accomplished in one step. Concentration reduces cost because smaller volumes go on to further processing.

As both enrichment and concentration of the desired protein is accomplished in one step, the present study was carried out in order to concentrate and enrich C-GMP content of cheese whey by precipitation.

Materials and Methods

Cheese whey

Cheese whey was procured from Ravileela dairy, Jaganguda, Hyderabad. It was filtered immediately through a muslin cloth, autoclaved and stored at 4°C until use.

C-GMP purification from the cheese whey by precipitation—GMP purification from the cheese whey by precipitation.

In the first step, whey proteins α-*lactalbumin* and β-*lactoglobulin* were precipitated using four different precipitation methods such as heat coagulation, thermal isoelectric precipitation, selective precipitation and salting out for the enrichment of C-GMP. In the second step C-GMP was concentrated further by the ethanol precipitation of lactose. Cheese whey thus treated for concentrating C-GMP serves as a base for the further purification of C-GMP.

Cheese whey treatment by various precipitation methods for the removal of whey proteins and enrichment of C-GMP

1. Heat coagulation: Heat coagulation was performed by heating cheese whey at different temperature and pH conditions for one hour. Coagulated whey proteins were centrifuged at 4300 rpm, 23°C for 30 minutes.
2. Isoelectric precipitation: This treatment involves the isoelectric precipitation of α-*lactalbumin* followed by the isoelectric precipitation of β-*lactoglobulin*. α-*lactalbumin* was precipitated from cheese whey at pH 4.1, temperature 60°C it was centrifugally removed and from the supernatant β-*lactoglobulin* was precipitated at pH 5.2, temperature 37°C and was centrifuged at 4300 rpm, 23°C for 30 minutes.
3. Selective precipitation: This treatment is a combination of thermal isoelectric precipitation of α-*lactalbumin* at pH 4.1, temperature 60°C followed by precipitation of β-*lactoglobulin* using 1 M $MgSO_4$ at pH 7.0. Precipitated whey proteins were centrifuged at 4,300 rpm, 23°C for 30 minutes.
4. Ammonium sulphate fractionation: Whey proteins were precipitated at different concentrations of ammonium sulphate added to the cheese whey at 27°C, 100 rpm for one hour and were centrifuged at 4300 rpm, 23°C for 30 minutes.

Supernatants of the centrifuged solutions obtained from the above four treatments were analysed for their protein content and C-GMP content.

Analytical methods

C-GMP standard was prepared by the chymosin digestion of k-casein. As C-GMP is the only N-acetyl hexosamine containing component in whey, we have estimated the C-GMP content of the sample by estimating its N-acetyl hexosamine content and comparing it with the standard C-GMP hexosamine content. N-acetylhexosamine content was estimated spectrophotometrically by the method of *Reissig*. Protein content was estimated by Folin-phenol method. Carbohydrate content was estimated by using phenol-sulphuric acid assay. Precipitated aggregate size was measured microscopically using ocular micrometer.

Chymosin, k-casein, N-acetyl glucosamine, Bovine serum albumin were from Sigma Chemical company USA. Other chemicals were from a local company.

Results and Discussion

Heat coagulation

Effect of pH on the heat coagulation of whey proteins

Cheese whey was heated at temperature of 100°C for one hour under various pH conditions (2.0 to 7.0) and the results are depicted in Fig. 12.3. Whey proteins were coagulated by heating at various pH conditions. Whey protein coagulation was found to be the highest at pH 6.0. At pH 6.0, 70.2 per cent of the whey proteins were found to be heat coagulated.

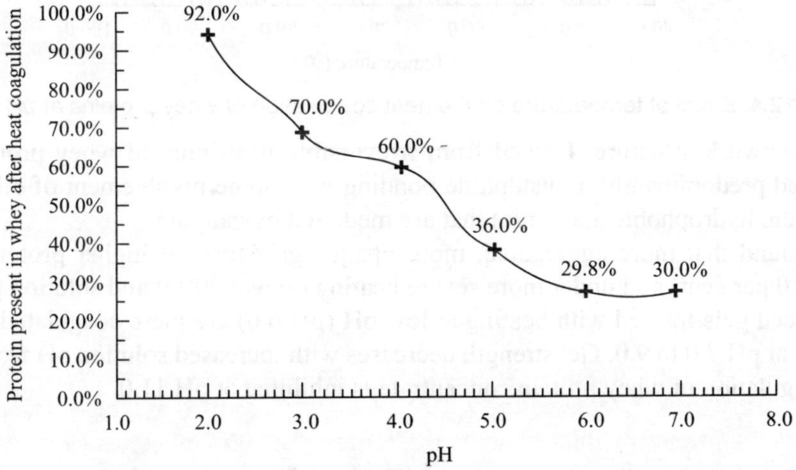

Fig. 12.3. Effect of pH on the heat coagulation of whey proteins at 100°C for one hour.

Effect of temperature on the heat coagulation of whey proteins

Cheese whey was heated at the optimised pH 6.0, for one hour at different temperatures ranging from 0°C to 100°C and the results are depicted in Fig. 12.4. Coagulation of whey proteins increased with increase in the temperature and was maximum at 100°C. By heat coagulation 70.2 per cent of total whey proteins were precipitated and the C-GMP content was enriched from, 12.7 per cent to 37.89 per cent.

Whey proteins in their native soluble or undenatured form have the ability to form heat induced irreversible gels at appropriate protein concentration, pH and ionic strength. The coagulation or gelation mechanism involves an initial step involving unfolding or dissociation of protein molecules followed by an aggregation step in which association or aggregation reactions occur which results in gel formation under appropriate conditions.

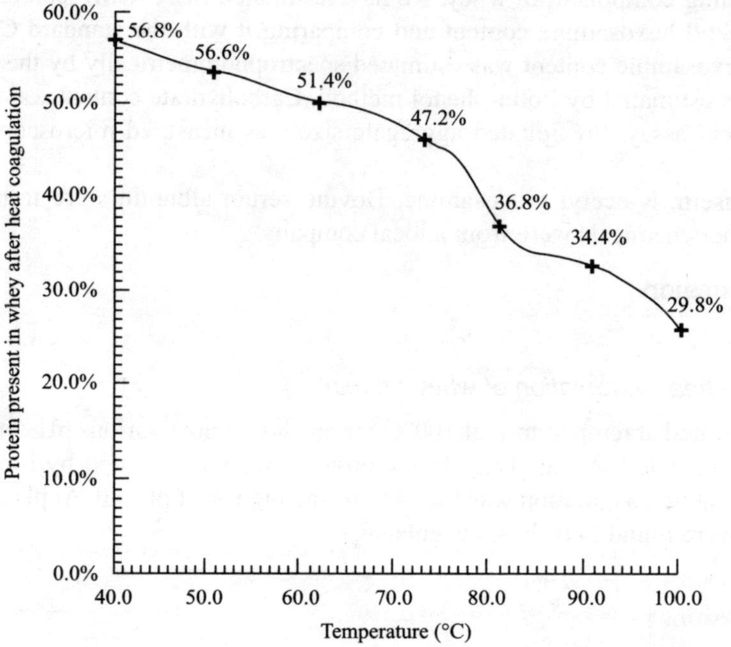

Fig. 12.4. Effect of temperature on the heat coagulation of whey proteins at pH 6.0.

The general network structure, formed from irreversible heat-induced whey protein coagulation reaction, is related predominantly to disulphide bonding with some involvement of other non-specific bonding (hydrogen, hydrophobic and ionic) that are mediated by calcium.

It has been found that more aggregated, more opaque gels form at higher protein concentration (approximately 10 per cent) and under more severe heating (above 90°C) and solution pH dramatically effects heat induced gels formed with heating at low pH (pH 6.0) are mere coagulated and less elastic than gels formed at pH 7.0 to 9.0. Gel strength decreases with increased solution pH from 7.0–10.0 and heat induced coagulation of whey protein concentrate is inhibited at pH 11.0.

Salting out

Effect of ammonium sulphate concentration on whey protein precipitation

Ammonium sulphate was added at different concentrations to the cheese whey (pH 5.5) for the precipitation of whey proteins and the results are depicted in Fig. 12.5. At 40 per cent wt/vol of ammonium sulphate, around 70.2 per cent of whey proteins were precipitated and the C-GMP, along with other whey proteins at this concentration.

It is stated that if salt is added progressively to an exhaustively (> 90 per cent) demineralised whey at pH values between 4.5 and 5.5, two macroscopic effects are observed. Initially protein solubility increases (salting in) and subsequently, after passing through a maximum it starts to decrease (salting out). The salting in process is usually considered as a net increase of protein solubility on the basis of non-specific electrostatic interactions between charged protein molecules and its environment.

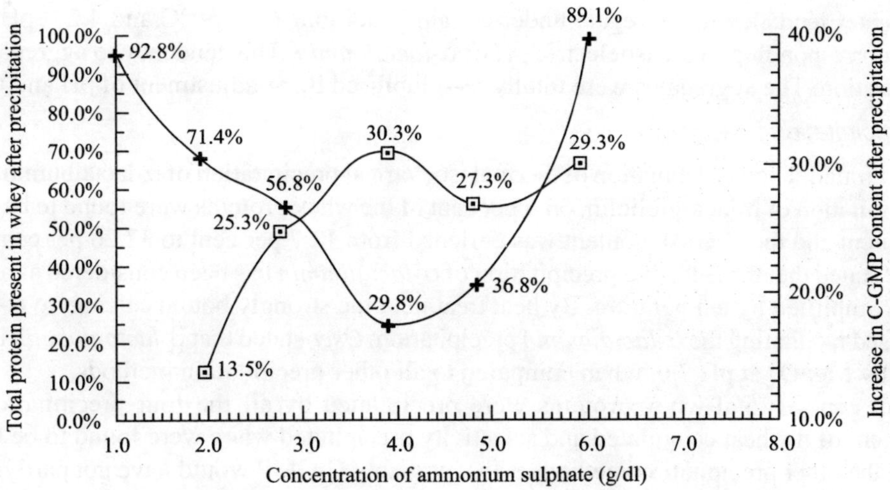

Fig. 12.5. Effect of ammonium sulphate concentration on precipitation of whey proteins from cheese whey at pH-6.

Isoelectric precipitation of whey proteins

Whey proteins were isoelectrically precipitated as per the methodology given and the results are depicted in Fig. 12.6. 64.4 per cent of the whey proteins were precipitated by this treatment and the C-GMP content was increased from 12.7 to 27.5 per cent.

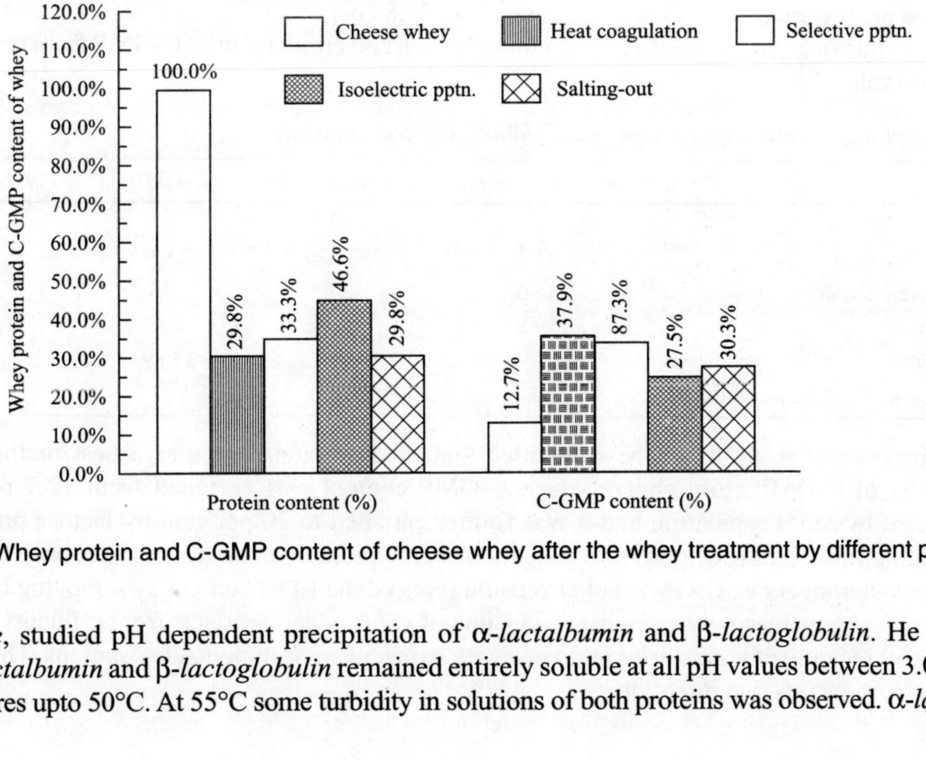

Fig. 12.6. Whey protein and C-GMP content of cheese whey after the whey treatment by different precipitation methods.

Pearce, studied pH dependent precipitation of α-*lactalbumin* and β-*lactoglobulin*. He stated that both α-*lactalbumin* and β-*lactoglobulin* remained entirely soluble at all pH values between 3.0 and 9.0 at temperatures upto 50°C. At 55°C some turbidity in solutions of both proteins was observed. α-*lactalbumin* showed a greater tendency to aggregate under certain conditions ($T > 50°C$ and $4.2 < pH < 4.6$), this range of pH corresponding to the isoelectric pH of α-*lactalbumin*. This tendency to aggregate increased with concentration. The aggregates were totally re-solubilised by re-adjustment of pH and temperature.

Selective precipitation

Selective precipitation is a combination of thermal isoelctrical precipitation of α-lactalbumin followed by $MgSO_4$ precipitation of β-lactoglobulin. 66.6 per cent of the whey proteins were found to be precipitated by this treatment and the C-GMP content was enriched from 12.7 per cent to 37.26 per cent.

Bramaud stated that the selective precipitation of α-*lactalbumin* has been considered as an isoelectric precipitation amplified by temperature. By heat treatment the strongly bound calcium to α-*lactalbumin* may be released facilitating the α-*lactalbumin* precipitation. *Oser* stated that β-*lactoglobulin* is selectively precipitated by $MgSO_4$ at pH 7.0, when compared to all other precipitation methods.

64–70 per cent of total whey proteins were precipitated by all the four precipitation methods. C-GMP content of the heat coagulated and selectivity precipitated whey were found to be higher when compared to the other precipitation methods. Most probably C-GMP would have got partly precipitated by the ammonium sulphate and hydrochloric acid used during precipitation. As heat coagulation is an easier processing step when compared to the selective precipitation, it was selected for the further purification of C-GMP by precipitation.

Effect of lactose precipitation on the C-GMP enrichment of lactose precipitation on the C-GMP enrichment

50 per cent vol/vol cold enthanol was added to the heat coagulated whey to precipitate the lactose from the heat coagulated whey. 48 per cent of the lactose could be precipitated by this treatment. C-GMP content of the heat coagulated whey was increased from 37.89 per cent to 70 per cent by ethanol treatment. A summary of purification steps used for the concentration and enrichment of C-GMP from cheese whey are given in Table 12.5.

Table 12.5. Purification steps and recovery of C-GMP from cheese whey.

Purification	*Total proteins (%)*	*C-GMP (%)*	*C-GMP*	*Purification fold*
Cheese whey	100	12.7	1	–
Selective precipitation	33.3	37.26	2.93	–
Isoelectric precipitation	46.6	27.5	2.16	–
Salting out	29.8	30.26	2.3	–
Heat coagulation	–	29.8	37.89	2.98
Heat coagulation + lactose precipitation	28	70.0	5.51	–

From the present study it can be concluded that precipitation serves as a best method for the concentration of C-GMP from cheese whey. C-GMP content was enriched from 12.7 per cent to 37.89 per cent by heat coagulation and it was further enriched to 70 per cent by lactose precipitation thus facilitating further purification. Above mentioned precipitation methods could not only concentrate the value added product i.e., C-GMP, but also could reduced the BOD content by removing lactose and proteins. It is being increasingly seen that production of value added products from effluents which can be an attractive incentive for industries burdened with stringent environmental legislations. This research is an endeavour towards that direction.

SECTION III

Waste from Chemical Processing and Allied Industries

13

Petroleum Refinery and Petrochemicals

INTRODUCTION

The exploration and production sector of the petroleum business explores, develops and extracts crude oil and gas resources from natural formations beneath the earth. Exploration involves mapping geological features and drilling test wells in areas where there is a high probability of finding reserves. Proven oil fields are then developed through a well drilling programme, which leads to a producing field. The extraction techniques used depend on the geological characteristics of the area, the type and formation of the oil and/or gas deposits, and the design of the well field.

Crude oil refining operations involve extracting useful petroleum products (e.g. gasoline). Crude oil contains fractions of naphthas, jet fuel, gasoline, diesel fuel, gas oils, lubrication oils and asphalt. Refineries extract these fractions thermally or catalytically.

SOURCES AND TYPES OF WASTE

Various wastes are generated during petroleum exploration, production and refining. Large quantities of production wastes–low-toxicity solid waste and waste-water–are produced during exploration and production. The waste stream from refineries is smaller but is considerably more toxic, containing listed and characteristic hazardous wastes. In addition, poor housekeeping practices during exploration, production and refining can produce excessive waste. A general flow sheet indicating sources and natura of wastes generated in refinery in shown in Fig. 13.1.

Waste from Exploration and Production

Production wastes can be grouped broadly into two classes: wastes related to drilling and well completion, including drilling mud, drill cuttings and chemical additives, and wastes related to oil production, primarily produced water. The volume of produced water far exceeds the volume of drilling wastes.

Treatment of disposal

Treatment and disposal of oil drilling wastes take place either on or off the drilling site. Wastes that are regulated by CPCB must be tested for hazardous waste characteristics before disposal. Wastes can be treated and disposed of on the site in reserve pits, by land spreading, or through annular disposal. Liquid wastes can be treated in closed systems or treated and discharged to surface water.

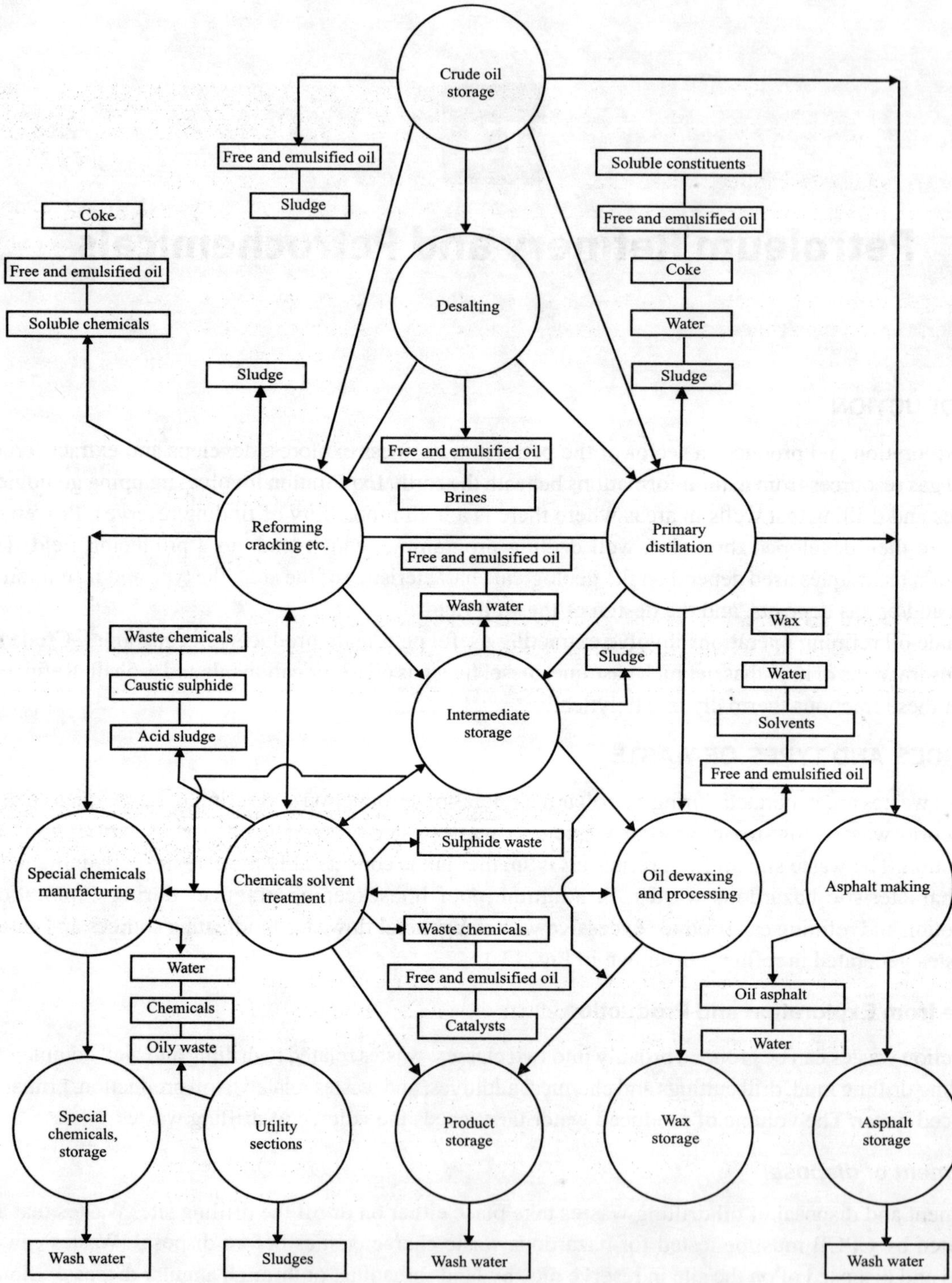

Fig. 13.1. Typical waste produced in petroleum refining.

Offsite treatment and disposal methods include disposal in centralised pits, reinjection and application on commercial landfarms. Drilling muds can be reconditioned. The majority of produced water is disposed of underground through injection wells.

Strategies for preventing pollution

The petroleum industry has no control over the composition of the crude oil they extract. Most waste generated during production is the result of this basic lack of control over the composition of the crude oil. Lower-quality crude oil requires more processing and generates more waste than high-quality crude.

Although source reduction is the preferred strategy in overall waste management, the petroleum industry has limited opportunities in this regard. Therefore, the key elements of pollution prevention in petroleum production and exploration are recycling and careful handling of waste. Pollution prevention techniques used during production include:

1. Collecting used solvents, hydraulic fluids and motor oils that are generated during drilling operations and shipping them to a petroleum recycler.
2. Using smaller reserve pits for disposal of drilling muds and produced water.
3. Segregating drilling muds for disposal as non-hazardous waste before they enter a hydrocarbon zone.
4. Recycling drilling muds in a closed-loop system.
5. Solidifying drilling wastes in reserve pits.

Although implementing pollution prevention practices during production is not mandated by regulation, several factors provide some motivation to do so. These factors include the high cost of commercial disposal, the potential for future liability for environmental damages and the difficulty of obtaining disposal permits.

Solid Waste from Refineries

The hazardous wastes are reduced through recycling, recovery, detoxification and volume reduction. To reduce waste management costs, the petroleum industry is developing waste reduction programmes that limit the volume and toxicity of its wastes. Because of the often 'tight' specifications placed on petroleum products and the limited process flexibility, waste reduction through recycle and reuse is preferred over the often-expensive process modifications.

Listed hazardous wastes

The refining sector generates both listed and characteristic hazardous wastes. The CPCB has identified five waste streams as listed hazardous wastes. These streams are dissolved-air floatation (DAF), slop-oil-emulsion solids, heat-exchanger-bundle cleaning solids, separator sludge, and leaded-gasoline-tank bottoms. Metals and organic constituents form the basis of the hazardous designation for these wastes.

Non-listed hazardous wastes

A major portion of refining sector hazardous wastes are not listed. These wastes include primary treatment plant sludge, secondary treatment plant sludge, biological treatment sludge, cooling tower sludge, ion exchange regenerant, fluidised catalytic cracking (FCC) catalyst and hydrocracking catalyst. These wastes are not listed; however, testing has shown that all of them have one or more characteristics of hazardous waste (that is, toxicity, ignitability, reactivity and corrosivity).

Additional wastes classified hazardous as a result of TCLP

Toxicity characteristic leaching procedure (TCLP) is a modification of the existing extraction procedure, designed to include wastes containing leachable organic compounds. The TCLP procedure measures the potential of toxic constituents to leach out of waste and contaminate groundwater.

Sources of Waste-water

The major sources of waste-water in a petroleum refinery are cooling-tower blowdown, boiler blowdown, oily wastes and other process wastes. Cooling-tower blowdown, which constitutes a large component of the total waste-water flow, contains relatively low concentrations of conventional contaminants such as organics, suspended solids and heavy metals, but contains high levels of TDS. The number of cycles of concentration for the cooling tower may be limited by TDS, alkalinity, magnesium, silica hardness, aluminium, iron, or some other parameter.

Boiler blowdown is dependent on boiler feeder quality, which is dictated by the boiler specifications. Boiler blowdown is high in TDS and minerals, and relatively free of organic compounds and heavy metals. Oily waste-water is generated from processes in which water contacts crude oil or refined products. Typical sources in a refinery include crude desalter bottoms, storage tank bottoms, foul-water stripper (FWS) bottoms and storm water. Other process waste-waters include chemical plant effluent, ammonia recovery unit (ARU) bottoms and sulphide-bearing wastes.

A large amount of water is used in the refinery process and a big fraction of it comes out as waste after getting polluted by oil and other toxic substances.

A major amount of the water used in a refinery is for cooling. Like any other industry, the extent of its recirculation and reuse will determine the volume of the effluent discharge.

Second to cooling, the major use of water is for boiler feed. The steam obtained is utilised for different operations like desalting of crude oil, distillation and steam-stripping in the topping process, and in catalytic cracking. Steam is also used for the stripping of spent catalyst before it is sent for regeneration. Water is also required to wash products like gasoline to remove traces of the chemical reagents used in earlier operation. The process waste-water includes the condensates and wash waters which come into contact with petroleum products at some stage of processing. So waste-water contains free or emulsified oil, spent caustic and acid solutions, impurities of petroleum products like hydrogen sulphides, ammonia, mercaptans, phenols, and spent catalysts. Again where semi-continuous delayed coking is adopted and hydraulic method is used for decoking the chamber, water in the form of water jets is used for cutting and breaking down the cokes from the chamber walls. The next operation of dewatering and drying of coke produces a waste containing a large amount of coke fines.

Oil enters into the waste-water from leaks, spills etc., at various stages. It is not possible to generalise the characteristics of waste-water from petroleum refineries, as these vary widely, with the size of the refinery, type of crude oil used, complexity of processing, water economy adopted, etc. The following table gives the general characteristics of composite waste from a typical Indian Oil refinery:

Free oil	2000–3000 mg/l
Emulsified oil	90–120 mg/l
H_2S and RSH	10–220 mg/l
Phenolic compounds	12–30 mg/l
5 day 20°C BOD	100–300 mg/l
Suspended solids	200–400 mg/l

Effect of wastes on receiving streams/sewers

Even a negligible amount of petroleum in the order of 0.5 mg/l can impart a bad taste to water. Naphthalein also imparts bad odour and becomes toxic at a concentration of 5 mg/l. Mercaptans produces a very strong and unpleasant odour at a very low concentration. The odorous compounds of the waste impair a bad taste of the fish living in that stream. Again as the waste contains various organics, it exerts pressure, an oxygen demand and also the oil film on the surface of the water prevents its reaeration. Cattle drinking such a polluted water, become prone to lethal diseases.

Treatment of Refinery Waste-water

In general, the treatment of refinery wastes is carried out in three steps:

1. Physical separation of free oil.
2. Chemical coagulation of emulsified oil.
3. Biological treatment for the removal of BOD, phenol and other toxic materials.

The physical separation of oil is accomplished in 'gravity separators'. These are essentially rectangular tanks, with arrangements for oil skimming and sludge scraping. The design of the gravity separator is based on fundamental principles of sedimentation and it is assumed that all oil globules of 0.015 cm in diameter and above will be removed in it. A safety factor of 1.2 is applied to the surface area to allow for short circuiting. Whatever may be the theoretical efficiency of this type of gravity separator, in particular, its efficiency declines due to the fact that certain portion of a free oil forms agglomerates with the suspended solids present in the waste, and are not separated in the tank.

Dissolved air floatation, with or without flocculation, has also been used to remove the oil content of refinery wastes and has been found satisfactory in some special cases.

Effluent of the gravity separator is given a chemical treatment, consisting of coagulation, flocculation and sedimentation, to remove emulsified oil and certain other chemicals, and also to condition the waste for final treatment in biological reactors. Various coagulating agents like ferrous sulphate, ferric sulphate, chlorinated copperas, calcium chloride, calcium carbonate and hydrated lime have been used to demulsify an oil-separator effluent. The most suitable type of coagulant, or the combination of coagulants, and their dosage are determined by laboratory tests. The flocs formed in the flocculator adsorbs the oil and becomes considerably buoyant. Therefore, the following sedimentation tank must be designed with low surface settling rate and large surface area. It has been observed that the size of the sedimentation tanks can be considerably reduced by adding coagulant aids like Fuller's earth during flocculation, which almost doubles the settling velocity of the flocs. The chemical treatment not only produces effluent of low oil content, but reduces the sulphide content considerably. It has been shown that a much improved quality of the effluent can be obtained, if the waste is also aerated along with the chemical treatment.

The 'waste decoking water' containing large amounts of coke fines, is subjected to segregation and is treated separately. Coke fines are not easily wetted by water; this is usually accomplished by air blowing, subsequent to which these are settled in a settling tank. An alternative device of treatment of coke containing waste is its treatment with a coagulant aid (2 mg/1 of coagulant aid CA-15 developed by NEERI with flocculation (1 minute) and sedimentation (15 minutes). This can produce an effluent of very high quality. By this method, not only suspended solids, but a certain percentage of oil content can also be removed.

Biological treatment methods include waste stabilisation pond, aerated lagoon, trickling filter, and activated sludge process. Waste stabilisation ponds require large area, but are capable of reducing BOD by 43–96 per cent and phenol by 61–99 per cent with a retention period of 60 days. In an aerated lagoon, with mechanical surface aerator, a retention period of six days may be provided but the microbial mass produced there settles so slowly that good effluent quality cannot be obtained unless a number of ponds following the lagoon in series are provided.

Many refinery engineers favour trickling filter as a biological treatment method. Trickling filters can absorb shock loads of toxic matter and do well without any retardation even with an influent oil concentration of up to 100 mg/l. Deep bed high rate trickling filters have been found to be very efficient in the reduction of phenols. All sulphides are removed in trickling filters. A flow diagram for treatment of a typical refinery employing trickling filter is shown in Fig. 13.2.

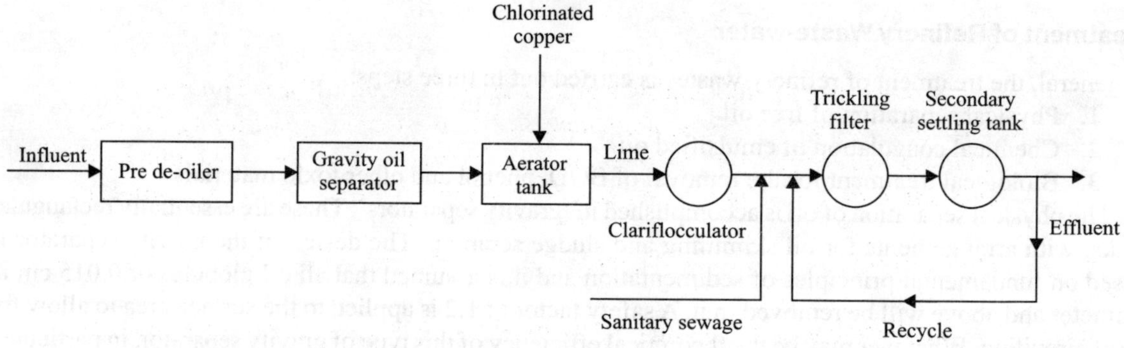

Fig. 13.2. Flow diagram for the treatment of a typical petroleum refinery's waste.

Treatment of refinery waste by the activated sludge process is affected when the oil concentration exceeds about 30 mg/l. It has been also observed that the activated sludge treatment of chemically treated refinery wastes results in very poor removal of BOD and phenol.

To tackle this situation, a modified method designated as 'Ferro-biological' process has been developed. In this method, both microbial mass and ferric hydroxide are used simultaneously in the aeration tank to flocculate BOD, phenol and oil.

In view of the toxic nature of refinery waste, these aeration tanks must be designed as complete mix type. As the biological flocs formed in the above process are absorbed with some quantity of oil, the final settling tank following the aeration tank must be designed with a low surface settling rate and large surface area.

The selection of the most suitable methods and the sequences of the treatment processes depends not only on the raw waste characteristics, but also on the desired effluent quality specified by the local authority. The raw waste characteristics also vary from one refinery to another. All these aspects are to be considered in the final design of the treatment units of refinery waste.

Opportunities for Pollution Prevention

Pollution prevention opportunities in the refining industry include recycling and reusing waste, segregating waste-water, modifying process equipment and improving general housekeeping. The goals of pollution prevention include reduction of waste toxicity or volume.

Recycling and reusing waste

Recycling and reuse consist of recovering materials that will otherwise be wasted and putting them to beneficial use. Treatment can play a significant role in product reuse.

The following waste recycling and reuse opportunities are feasible for the refining industry:

1. Recycling FWS bottoms as crude desalter water.
2. Recycle boiler blowdown for cooling tower make-up water.
3. Soften cooling tower side-stream to increase cycles of concentration.
4. Recycle waste-water treatment plant (WWTP) effluent as cooling tower make-up water.
5. Regenerate spent catalyst.

Most of the water used in refining is for process heating or cooling. The easy availability of water and its high heat capacity have made it the favoured heat-transfer medium in industries. However, recent water shortages and environmental awareness have resulted in an increased interest in conserving water used for heating and cooling. In the past, many systems used water once and returned it to the receiving streams. Currently, water is reused to a much greater extent and as a result, refineries spend a huge amount on treatment chemicals to prevent system corrosion, deposition and fouling. This section presents information on several inplant techniques for recycling and reusing water.

Recycling foul-water-stripper bottoms as crude desalter water

Ammonia and hydrogen sulphide are present in varying concentrations in crude oil stocks. NH_3 and H_2S gases are generated during cracking operations. When steam is used to strip the product streams, these contaminants are transferred to the steam and they dissolve in the water produced when the steam condenses. Steam condensate water is collected at various points and treated in an FWS. When NH_3 and H_2S are removed from the condensate, the NH_3 can be thermally destroyed or can be recovered for the production of fertilisers, and H_2S can be converted to elemental sulphur for resale.

The FWS effluent can then be used as wash water in the crude oil desalter. The FWS bottoms may be high in phenol, and using the water for this purpose has the additional benefit of phenol removal. The amount of FWS effluent that may be used in this manner is limited. Because crude oil is mixed with 5 per cent (of crude volume) water before it is processed through the desalter, the maximum amount of recycled water used for this purpose should be limited to 3 per cent. For a 100,000-barrel-per-day refinery, the maximum water that can be used for this purpose would be 90 gal/min.

Segregating waste-water

Hazardous wastes and waste-water requiring unique treatment should be separated from other waste-waters that require relatively limited treatment. The waste-water collection system and the WWTP are the logical places to begin segregating waste-water because much of the hazardous waste generated by refineries is generated there. Ideally, a refinery storm-water and waste-water collection system should include four sewers: an offsite storm water sewer, a clean storm-water sewer, a non-point-source sewer, and a point-source (oily) sewer.

Offsite storm water sewer

An offsite storm-water collection system collects all offsite storm water, including clean storm water, from areas such as access roads around the process areas. Typically, this storm water should be discharged without treatment.

Clean storm water sewer

Storm water from clean areas (i.e. roofed areas or areas with no process units) should be segregated from storm water originating from dirty areas. Clean storm water requires no treatment before discharge or reuse. The rainwater falling on access roads around and inside the process areas should be directed to a clean storm water sewer. Run-on from adjacent areas also should be intercepted and sent to this sewer. Roofs should be installed over process units that can contribute pollutants to storm water run-off. Roof drains should be connected to the clean storm water sewer.

Non-point-source sewer

A non-point-source sewer collects and conveys washdown water and storm water that falls on the process areas. Effective segregation of point-source contaminants can result in relatively clean water being collected by a non-point-source sewer system. This water should need only minimal treatment for oil removal prior to discharge.

The chemical constituents in the water and the potential for variation must be considered in determining appropriate treatment. Provisions for directing this water to the WWTP during a spill must be included in the design and operation of this system.

Other pre-treated waste-water streams can be directed to this sewer for final discharge. Examples of such streams include neutralised water softener regenerant waste and treated groundwater.

Point-source sewer

Process waste-water containing oil requires a significant level of treatment before discharge. Oily waste-water consists of process waste that has come into contact with process fluids. A point-source sewer collects all point-source discharges from inside the process areas of the units. Waste-water from sampling stations, pump seal water and other contaminated waste-water also should be directed to this sewer.

Opportunities for reducing waste within the actual process units include optimising the process and minimising the volume of washdown water used to clean the process units. Every effort should be made to contain potentially polluting equipment and prevent spreading contaminants to the surrounding and general process areas. Areas around pumps, compressors and other potentially polluting equipment should be curbed and roofed. Waste-water generated inside the curbed areas should be directed to a point-source sewer.

Modifying Process Equipment

Several process equipment modifications that can improve operating efficiency and reduce waste-water generation are:

1. When sampling, use continuous recycling loops on sampling ports to eliminate the volume of water necessary for purging.
2. Substitute pressure wands for fire hoses during unit washing operations to reduce the volume of washdown water.
3. Use an automatic shutoff, rather than a manual shutoff or a holding tank, when removing water from crude, intermediate and product storage tanks to avoid excess oil discharges to the oily sewer and to return oil to the storage tank.

Improving general housekeeping

Excessive waste production often results from poor housekeeping practices. Leaking tanks, valves, or pumps can cause oil spills, requiring clean-up and disposal of the oil, contaminated soil and clean-up material. Key advantages of housekeeping changes are that they can usually be implemented quickly; they require little, if any, capital investment; and they are likely substantially to reduce waste generation.

Spill prevention and control

Spill prevention and control programmes should be developed to enforce the management of waste generated from activities that occur in process areas, including routine sampling, equipment decontamination, equipment cleaning and maintenance, and spilling.

Contained work areas

All areas that have potential for spills should be paved rather than covered with gravel or other granular media. All containment areas should be adequately sloped towards the appropriate collection system.

Designated maintenance areas should be paved with concrete and used for equipment overhaul, cleaning, and maintenance. These areas should be curbed and drained to the oily sewer. High-pressure hot-water cleaners should be provided in the maintenance area so that the paved area can be cleaned after maintenance activities are over. After clean-up, rainwater should be diverted to the clean sewer.

Concrete-paved pads should be installed for areas behind heat-exchanger banks. This will provide adequate safeguards against clean sewer contamination when heat-exchanger bundles are pulled for cleaning or maintenance.

Effluent Recycle Options

The flow schematics for the effluent recycle options are shown in Figs. 13.3, 13.4 and 13.5.

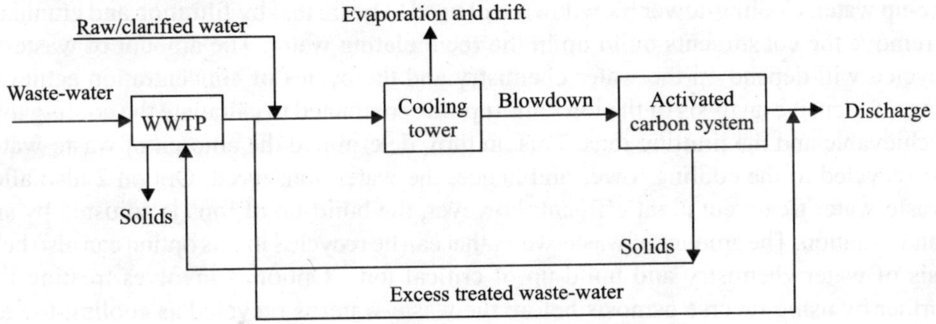

Fig. 13.3. Flow schematic for effluent recycling: Option One.

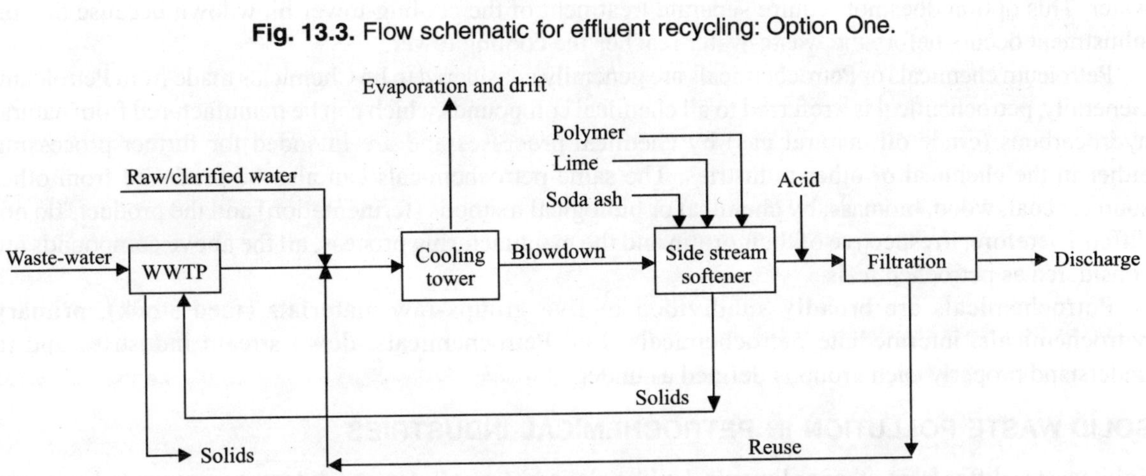

Fig. 13.4. Flow schematic for effluent recycling: Option Two.

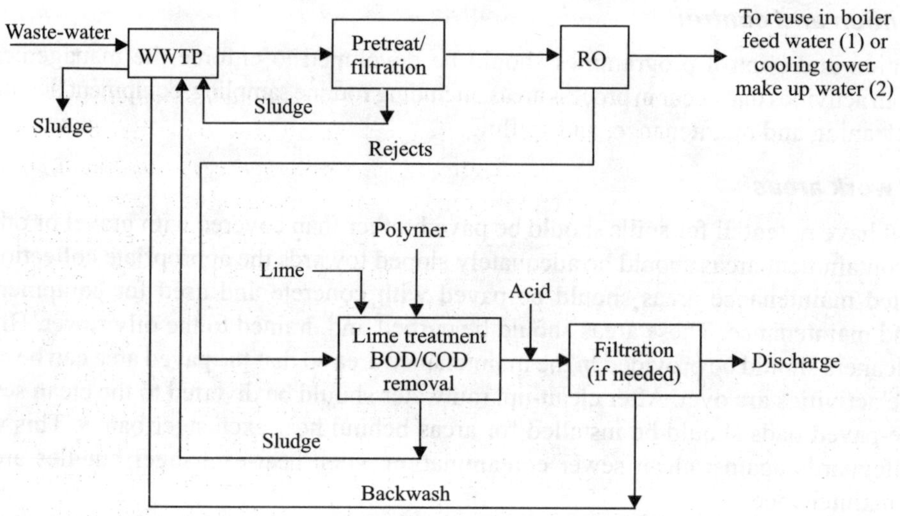

Fig. 13.5. Flow schematic for effluent recycling: Option Three.

In Option 1 (Fig. 13.3), the WWTP effluent can be reused without additional treatment as cooling tower make-up water. Cooling-tower blowdown will need to be treated by filtration and granular activated carbon to remove the constituents build up in the recirculating water. The amount of waste-water than can be recycled will depend on the water chemistry and the cycles of concentration achievable in the cooling tower. Extensive analysis of the ion build-up was conducted to calculate the cooling tower cycles has been achievable and the limiting ions. This, in turn, determined the amount of waste-water effluent that can be recycled to the cooling tower and hence, the water conserved. Option 2 also allows direct reuse of waste-water treatment plant effluent; however, the build-up of ions is adjusted by side-stream softening and filtration. The amount of waste-water that can be recycled in this option can also be calculated on the basis of water chemistry and build-up of critical ions. Option 3 involves treating the WWTP effluent further by using reverse osmosis before the waste-water is recycled as cooling-tower make-up water. This option does not require separate treatment of the cooling-tower blowdown because the ion adjustment occurs before the waste-water reaches the cooling tower.

Petroleum chemicals or Petrochemicals are generally considered to be Chemicals made from Petroleum. Generally, petrochemical is a referred to all chemical compounds which can be manufactured from natural hydrocarbons (crude oil, natural gas) by chemical processes and are intended for further processing either in the chemical or other industries. The same petrochemicals can also be produced from other sources; coal, wood, biomass, by chemical or biological methods (fermentation) and the products do not differ. Therefore, irrespective of their origin and the manufacturing process, all the above compounds are considered as petrochemicals.

Petrochemicals are broadly subdivided in five groups-raw materials (feed stock), primary petrochemicals, intermediate petrochemicals, End Petrochemicals, down stream industries and to understand properly each group is defined as under.

SOLID WASTE POLLUTION IN PETROCHEMICAL INDUSTRIES

Solid wastes differ from other pollutants. Unlike air or water pollutants, solid wastes are yet to have seen any unpredicted growth of public concern. There has been dramatic enforcement of federal legislations in air and water effluents. However, more and more concern on solid wastes too is anticipated in the coming years. Solid wastes differ from air and water pollutants in another sense too., viz. the solid wastes remain at their origin points till such a time a decision is taken by those concerned to collect and dispose them off. Petroleum refining and petrochemical industries are feeding the blood into the veins of today's industrialisation. Considerable expenditure has been made in the arena of treatment of air and water pollutants. But, solid waste treatment is yet to catch up to the required level.

State-of-the-art manufacturing processes, new refining processes etc., have helped to bring down or treat the harmful effects of effluents in general. In the order of priorities among the three kinds of effluents, water pollutants are first followed by air pollutants. Solid wastes fall in the last category. While a review of our past accomplishments alone will not do good to us and there is more room for improvement even in the treatment of air and water pollutants, it is undeniably true to add that we have a long way to go in solid waste management in petrochemical industries. Solid wastes in petrochemical industries can be in different forms:

1. Toxic solids.
2. Non-toxic solids.
3. Actual solids.
4. Semi solids.
5. Combustible solids.
6. Non-combustible solids.
7. Degradable solids.
8. Non-degradable solids.
9. Bio-degradable solids.
10. Organic solids.
11. Inert solids.
12. Dry solids.
13. Aqueous solids.
14. Non-aqueous solids etc.

COLLECTION AND DISPOSAL

The means of collection and methods of disposal of these differ from one to the other. Few are listed below:

1. Recycling.
2. Chemical conversion.
3. Incineration.
4. Pyrolysis.
5. Landfill.
6. Ocean dumping.
7. Lagooning.
8. Salvaging.
9. Biological treatment.
10. Reclamation.
11. Open dump burning.
12. Landforming.

The methods are applicable either alone or in combinations. Before reviewing the basic factors that are required to be considered in depth before attempting to design the relevant method of storing, collecting and disposing the solid wastes, a look at the various types of such wastes will be of great interest. Some of them are:

1. Lunchroom wastes.
2. Catalysts.
3. Organic chemicals.
4. Waste-water treatment sludges.
5. Filter cakes.
6. Viscous solids.
7. Ashes.
8. Fly-ashes.
9. Incinerator residue.
10. Plastics.
11. Ferrous metals.
12. Non-ferrous metals.
13. Water treatment sludges.
14. Inorganic chemicals.
15. Polymers - pellets, dust, powder.
16. Off-specification products.
17. Contaminated products etc.

Considering the wide varieties of types, forms etc., as stated above, the major factors considered while designing the facilities are narrated here.

Physical Characteristics

1. Volume.
2. Weight.
3. Density.
4. Handling problems.

Chemical Characteristics

1. Combustibility.
2. Degradability.
3. Toxicity.
4. Potential value of materials salvaged for recycling.
5. Availability of land for disposal.
6. Anticipated future effects on land due to the disposal.
7. Effect of method on environment.

Cost of process

Wasteless processing probabilities

Almost every petrochemical industry creates some or other type of solid waste that can be categorised in any of the above listed types etc. For their processes, many petrochemical units need water treatment facilities. The chemicals that are required for satisfactory operation of these treatment units are of various types and nature. Flocculants, the precipitates they form etc. too add to the complexity of the solid wastes that are generated. 'Plant trash' is the generic term coined to define miscellaneous wastes that may be found in the manufacturing facility. Trash is considered harmless if they consist only of non-combustibles. Striking examples of such a category are glass, metal scraps, waste wood, floor sweepings etc. Trash can originate from boilers, incinerators etc., too. The ash or residues left of fuel (solid) is trash. The incinerated maternal (such as sludge) may also leave trash. Power generation facility also is capable of inducing trash. The commonly identifiable scrap in any petrochemical unit is the metal chips, parts or portions of discorded, corroded or damaged vessels, tanks, pumps etc. and are mostly ferrous in nature. This ferrous scrap contains within it the capability of getting contaminated by the exposure to toxic substances generated elsewhere in the plant. Recent day technologies play around catalysts and they have become a must for any petrochemical unit. By definition, catalysts do not take part in the reactions they catalyse and hence is a commodity with that highly beneficial and economic factor of recoverability.

Over a period of usage, they too get 'Spent'. Spent catalysts may be solid, semi-solid etc. Catalysts are generally possessing complexity in their configuration, nature and chemical characteristics. Spent catalysts of solid/semi solid configuration impart challenges in selection of treatment/disposal methods. Production processes that encompass combinations of multiple complexities end up with equally complex solid pollutants. A wide variety of other organic/inorganic chemicals have every chance of entering the waste streams as a result of all these complexities. The family of plastics, viz. polypropylene, polyester, polystyrene, polyvinyl chloride, polyurethane etc., is present to a high degree in the solid wastegroup of petrochemical units. These polymers can be identified in any of the following forms: (i) pellets; (ii) chops; (iii) cleanouts; (iv) shreds; (v) dust; (vi) powder; and (vii) fines; (viii) granules; (ix) emergency dumps; (x) off-specification products; and (xi) contaminated products.

Not only physical but also chemical processes are made use of to treat waste water streams emanating from any petrochemical unit. Unit operations are available to effect the process via. biological routes. Methods like usage of trickling filters, aeration, stabilisation ponds and activation of sludge do work well but not without their own solid waste derivatives. Variance of production techniques or combinations of treatment processes alter the end waste characteristics by a huge variance.

DISPOSAL/TREATMENT

Chemical Treatment

This is a factor often neglected or overlooked as a method of useful waste management in petrochemical industries. Processes that help in chemical conversion of wastes may be enumerated as: (i) stripping; (ii) distillation; (iii) leaching; and (iv) extraction etc. Some physical characteristics of materials also are made use of in conversion processes of wastes. They are: (i) separation by magnets; (ii) Flotation; (iii) hand picking; (iv) jigging; and (v) electrical conductivity etc.

Saleable products are possible outcomes of adoption of the right combination of selective methods in the correct sequence during waste treatment. Even if such a cleverly worked out strategy fails to derive fruitful products, they can render innocuous materials that are easier to dispose off or are not detrimental to the environment. Some possibilities in this direction, are enumerated below:

1. Organics are hydrolysable to product fuels.
2. Waste cellulose is acetylated to give cellulose acetate.

Recycling

This is a way to set up a wasteless closed circuit flow scheme. Here, a stream is returned back into the process. Recycling helps in more vigorous and effective conversions with savings in materials, energy and safety from objectionable emissions to surroundings. Usage of catalysts and solvents are acquired to the maximum extent possible in this process, when-performed at optimum conditions. Reversible reactions such as the ones represented by:

$$A+B \leftrightarrow C+D$$

are well effected via recycling. This is more true and proves cheap especially when the cost of the reactants are high or when treatment of residual reactants before disposal etc. is costly and/or hazardous. The cheaper reagent is used in excess and the costlier one is circulated repeatedly to get more yields of the products. Here, reactor design plays a vital role, however.

Incineration

This is a combustion process, conducted in controlled conditions, for burning solid wastes (in addition to the liquid and gas phase wastes), and hence is a valuable route for solid waste disposal. The common products of incineration are carbon dioxide, water, ash etc. These are non-combustibles. The wastes are converted to a less bulky, toxic or noxious material. However, this process has its limitations also, in the sense that, compounds containing the following elements must not be incinerated as they will product oxides: (i) sulphur; (ii) nitrogen; (iii) chlorine; (iv) fluorine; (v) bromine; (vi) iodine; (vii) mercury; (viii) arsenic; (ix) selenium; (x) lead; and (xi) cadmium.

Incineration is considered useful where availability of land is scarce and methods other than land disposals are to be used. Few variables, however, have a great effect on completion of proper incineration. They are:

1. Residence time of waste in incinerator.
2. Flammability limits of wastes.
3. Temperature inside the incinerator.
4. Ignition temperature of the wastes.
5. Turbulence in the reaction zone.
6. Combustibility of wastes being incinerated.

Reduction of operating costs of incineration can be envisaged by heat recovery through steam generation. Incineration helps not only reduce one's investment in pollution control equipments but also gives out products which are a very small fraction of the original weight and volume of the input wastes for landfill. More of organic wastes find incineration as their funeral, compared to inorganics. This also constitutes a way out for harmful bacterial, viral and toxic constituents. Necessity of support fuels may be considered as the negative aspect of this process, in addition to the limitations of items that can be handled in this process. Several processes are in use for incineration of petrochemical solid wastes, such as:

1. Fluidised bed reactors.
2. Dual chamber incinerators.
3. Rotary kilns.
4. Open chamber incinerators.
5. Liquid burner furnaces.
6. Hearths.
7. Multiple hearth furnaces.

8. Catalytic chambers.
9. Molten salt incinerators.
10. Pyrolysis units.

Factors that tend to limit the selection of incineration as the favoured method of treating solid wastes are as below:

Physical state of wastes

1. Solid.
2. Semi-solid.
3. Others.
4. Ease of handling.
5. Reacting capacity
   (a) Within themselves at higher temperatures.
   (b) With the material of construction of the incinerator.
6. Combustion temperature.
7. Moisture content.
8. Combustion products.
9. Ash content.
10. Combustibility.
11. Possibility of disintegration at elevated temperatures etc.

The incombustibles are the ones that require a support fuel for sustaining combustion and thus incineration. These are generally identified to be due to the following factors: (i) moisture content; (ii) inorganic salts; (iii) halogens; (iv) nitrogen; (v) sulphur; and (vi) phosphorus.

Combustibles are born for easy incineration. They do not need any support fuels and hence their incineration is cheaper, compared to the incineration process of non-combustibles. Functioning of combustibles, during incineration too depends on the following factors: (i) mixing with air; (ii) physical state; (a) solid and (b) others; (iii) spraying (for liquids only) etc.

Dual chamber incinerators have been designed to combust solid wastes at the rate of 5700 kg/day from petrochemical units. Here one of the two chambers is for vapourisation and the other for combustion of the vapour. The latter is operated at three times higher a temperature of the first, which operates at 400°C or so. Mixtures of organics with inorganics pose the toughest task to incineration. This is easy to understand by considering the fact that inorganics may contain water and that inorganic salts form oxides, carbonates etc. unlike the organics which tend to decompose. Usage of equipments like scrubbers are necessitated.

Hazardous wastes that are given the incineration treatment as a method of disposal include organohalogen compounds, and chlorine, fluorine and bromine derivatives. Presence of halogens depresses the combustion temperatures. The organohalogen wastes are first hydrolysed to form respective acid gas that are water soluble and thus treatable through easy and well known routes like water absorption or alkaline absorption methods. Compounds that may pose explosion hazards (like organo nitrogen compounds) are treated using long residence, low temperature incineration methods. Incinerators can also handle organo phosphorus compounds and sulphur bearing compounds also, of course followed by few secondary treatment methods.

Pyrolysis

While normal incineration demands 50–100 per cent excess air over stoichiometric value, pyrolysis is a Zero-Air process, as this functions on indirect heating principles only. Temperatures as high as 1650°C are envisaged in an air-free or air starved chamber in pyrolysis. Basically, this resembles processes like distillation or cracking. Pyrolysis wastes are distilled to combustible forms that gets discharged from the furnace. pyrolysis poses comparatively less air pollution than incineration. This process is capable of handling large volumes. Burnable materials are degraded marginally in pyrolysis. The greatest plus point of this technique is the possibility of recovering chemicals or synthesis gas. Partial combustion of pyrolysis gas within the furnace provides the process heat. This results in using the unoxidised portions of the combustible gases as fuel elsewhere and recovering that heat. Only wastes having high calorific value are processed in this route. Problems of clinkering or fusion of ash generally encountered with the incineration process are absent here, mainly because of the application of abnormally high temperatures in pyrolysis. Waste polymers are conveniently converted to fuel oils by pyrolysis. The pyrolysis unit can be a tubular, plug flow reactor and the process adapted is thermo cracking. It is proposed in many experiments that if operating conditions are maintained properly, even other polymers can be converted into various other useful end products. Few of these are:

1. Additives for gasolene
2. Fuel oils
3. Gaseous fuels
4. Stain removers
5. Solvents of inferior strengths
6. Lighter fractions
7. Feedstocks for few other petrochemical processes etc.

Waste sludges of non-autogenous character are often combined with high calorific value materials to facilitate releasing of high energy levels required for pyrolysis. It is quite interesting to note here that few wastes such as spent filter media, shredded wood, torn paper wastes, unusable oils etc., comprise the family of such 'high calorific value materials'.

Open dumping

Open dumping or open dump burning is an unacceptable method of solid waste disposal and is illegal in many countries. This demands availability of huge area of land which will become ultimately unusable. Contamination of ground water or even surface water, is also likely.

Landfill

This simple technique involves spreading and compacting solid wastes into cells and covering them. This method of 'sanitary landfill' requires low level labour and is low in cost in addition to requiring low degree of expertise and ease of operation. Wastes that are allowed to be dumped must be inert or capable of degradation of microbial attack to harmless products. Production of inflammable gases during this degradation is possible in some cases.

Land forming

Land forming is the spreading of wastes over a vast area wastes in a thin layer and is applicable for semi-solids or solids mixed with liquids. On spreading, the liquid either evaporates or seeps down the earth, depending upon the type of liquid, thus leaving the solids for degradation. Predominantly paraffin wastes are given this treatment. To some extent, even aromatic sludges are land-formed. Heavy metal concentrates pose a limitation for this method. Materials having no contamination of the following metals or their salts can be disposed off using this process: (i) lead; (ii) nickel; (iii) manganese; (iv) copper; (v) chromium; and (vi) arsenic.

Lagooning

A method generally found suitable for solids, semi-solids and liquid wastes generated by petrochemical industries, lagooning comprises dumping wastes into ponds or pits. Inspite of its advantages (such as easy operation, low labour requirement, low construction cost and low degree of expertise required), this process is inherent of a bad characteristic viz. if being odouriferrous and unsightly. As degradation of wastes is limited in lagoons, this method is necessarily to be followed by a secondary treatment of concentrates the lagoon will produce.

Hazards can be faced in either of the following four forms: (i) corrosive; (ii) ignitive; (iii) reactive; and (iv) toxic. At least in case of these wastes, waste management extends further to:

1. Minimise production of such wastes.
2. Investigate possibilities of using these in industries elsewhere as a desirable material.
3. Reprocessing to minimise end wastes.
4. Separate these from other common and non-hazardous wastes.
5. Treat to give a non-hazardous waste (however high that treatment may cost).
6. Dispose with care, security and caution in landfills.

Various products that produce hazardous waste streams are listed below in their descending order of quantity generated in major petrochemical units: (i) atrazines; (ii) lead alkyls; (iii) furfural; (iv) explosives; (v) perchloroethylenes; (vi) methyl methacrylates; (vii) epichlorohydrin; (viii) parathion; (ix) malathion; (x) chlorobenzene; (xi) vinyl chloride; (xii) trifluoroaniline; (xiii) ethanolamines; (xiv) toluene di isocyanate; (xv) maleic anhydride; (xvi) chloromethane; (xvii) acrylonitrile; (xviii) nitrobenzene; (xix) fluorocarbon; (xx) chlorotoluene etc.

It may be hard to believe that many wastes do contain valuable substances that can be extracted from concentrated wastes economically. As pollution control facilities aim at the production of the least environmental effect processes within the constraints of cost, it is mandatory to look for indirect generation of pollutants while selecting such methods that may seem suitable and cost-effective at first sight. Means of avoiding the inadvertent creation of 'indirect pollutants' must also be derived for implementation.

Some of industries and pollutants generated are given in Table 13.1. Technologies do exist for tactful and efficient management of any waste. But they may be unaffordable. It is much more simple to assign monetary values to the long-term damage to humanity and mother earth or environment that may result from a poor management of such wastes, than to set aside the process of coining a suitable treatment and disposal method of neglecting the total issue as 'highly costly'.

Regulations that are not available from governments or agencies may be another possible reason of not caring for a proper waste management system. But then, there will be no government or agency tomorrow to pose any regulation whatsoever, as we ourselves have not cared to pose self-regulation in spite of our being aware of the fact that tomorrow's world is to be made worthy of living (if not beautified) for our off-springs. After all, what is shown today is to be reaped tomorrow and nothing is costlier than an eroded environment.

Table 13.1. Some industries and pollutants generated.

Name of industry	*Pollution generated*
ABS	Acrylonitrile
	Aromatics
Acrylic	Cyanide
Latex	Acrylonitrile
	Acrolein
Alkyd	Acrolein
	Aromatics
	Poly aromatics
Epoxy	Aromatics
	C_3 chlorinated compounds
	Phenol
Phenolics	Aromatics
	Phenol
HDPE	Aromatics
PP	Aromatics
PVC	C_2 chlorinated paraffins
SAN	Aromatics
	Acrylonitrile
SBR	Aromatics
UPE	Phenol
	Aromatics
Fibres	
Acrylics	Acrylonitrile
Others	
Cellulose accetate	Isophones
Polycarbonates	Phenol
	Chloroaromatics
	Halomethanes
Polyester	Aromatics
	Phenol
Polystryene	Aromatics

Treatment of Polymers

Polymers are generally in the form of fine powder. It is a valuable product. Because of economical consideration it has been always endeavour of enterpreneur to recover maximum product escaping alongwith effluent. Moreover most of polymers manufactured in the petrochemical industries has specific gravity very close to oils (hydrocarbons). Hence polymers when comes in contact with oil, emulsified scum is formed which interferes with deoiling process as well as biological treatment. Polymers are removed by under given methods.

Vibro-screen

Screen having very fine mesh size and vibrating mechanisms are used to recover fine polymer powder. Frequent chocking of screen is common phenomenon. Screen requires quite a bit of maintenance.

Side hill screen

This type of screens are used in the paper industry for recovery of fine fibre. Generally screens are inclined at 30° to 45° and normally mesh size of screen is kept at 60, effluent containing polymers is screened through for recovery of polymers. Good recovery is possible and quite economical.

Gravity separators

Polymers being light floats on the surface of water, hence slot pipe arrangement is used to recover polymers from the surface of water and water under flows for further treatment.

HYDROCARBON RESINS—WEALTH FROM WASTE

Hydrocarbon resins present an opportunity to convert waste C9 aromatic and C5 aliphatic streams at reasonable investments into value-added products. According to estimates made by the US-based petrochemical consultants, Dewitt and Co., the market for these resins has been growing at an impressive 6 per cent annum since 1997. Main applications include formulation of adhesives, inks, toners, paints and coatings and in the compounding of rubber products.

Hydrocarbon resins can be made from a variety of feed streams. The shortest and most economical route is the polymerisation of aromatic components from cracker waste streams with a boiling point above 180°C. The next alternative is using the bottoms of a C5 stream rich in dicyclopentadiene (DCPD), and the third (and most expensive) route involves separating piperylene to make aliphatic resins. The first two methods are the ones commonly used and will form the basis for the ensuing discussion.

Hydrocarbon resins have a considerable history of use. The first resins were made in Europe before the First World War, and production in the US began around 1920 with the introduction of coumarone-indene or coal tar resins. Shortages of these resins during the Second World War spurred the search for alternatives that led to the development of processes from crude oil cracking. While the initial resins were unstable, dark, soft and tacky materials, improved technology (better catalysts, lower temperature, etc.) led to the development of stable, hard and light coloured materials. Commercial production, which began in the 1940s near Pittsburgh, was initially based on C9 aromatic resins, the emergence of world-scale naphtha crackers led to the production of considerable quantities of raw materials that spurred downstream development. Today a variety of aromatic resins are available ranging from aliphatic through naphthenic to aromatic and mixtures of one with another.

While the original aromatic resins were made using sulphuric acid catalysed polymerisation, practically all the C9 resins produced today use aluminium chloride or boron trifluoride and/or its phenol complexes as catalysts and produce resins with differing character. Manufacture can be either a simple batch operation or a computer-controlled continuous one, depending on the scale of operation. Basic C9 resins so produced can be modified with a number of other chemicals (phenol, phenol-formaldehyde, rosin, maleic anhydride, sulphur trioxide, DCPD) to produce derivatives that have niche markets. The minimum size C9 resins plant is a batch plant with a capacity of 10,000 TPA and construction costs are estimated at below US$10 mn. At a price of US$1000.per MT of dry resin, operating costs need to be pegged below US$500 per MT with raw material costs at 70 per cent (US$350 per MT), for the economics to be attractive.

The second product line—C5 aliphatic hydrocarbon resins or DCPD resins—are the lowest priced in the hydrocarbon resins family due their poor storage stability and are available in a wide range—from viscous oligomers to high softening point resins. They are prepared by thermal polymerisation of DCPD with or without a catalyst, and while less versatile than the aromatic resins, find application as co-vulcanisate in rubber compounding and in vegetable oil refining.

In countries where there is no production of hydrocarbon resin, rosin resins are commonly used in applications such as adhesives, inks, paints, rubber compounding and plastics processing and/or imports of hydrocarbon resins meet the requirement. This is the case in India, as well, where small production by IPCL is insufficient to meet local needs. Elsewhere in the world the markets are large enough for cracker operators to consider the additional investments in recovering the C9 and C5 streams for processing into value-added derivatives. Dewitt estimates global production to have grown from about 530 kT in 1997 to 923 kT in 2001, which not surprisingly was a disastrous year for the industry: demand grew just about 1 per cent and prices fell sharply impacting bottom-line. While the products are considered specialities, they are not immune from the boom-bust cyclicalities that characterise commodities. Global consumption of hydrocarbon resins is given in Table 13.2.

Table 13.2. Global consumption of hydrocarbon resins (MT).

Name of the country	*Production*	*2000 Imports*	*Exports*	*Consumption 2000*	*2001*	*Difference*
Americans						
North	417,000	49,697	90,530	315,550	376,200	19%
Central	0	2,662	0	4,160	2,700	–35%
South	10,000	15,529	134	23,650	25,400	7%
Sub-total	427,000	67,888	90,664	343,360	404,200	18%
Europe						
Western	273,000	45,318	11,282	309,070	307,000	–1%
Eastern	8,000	222	2,432	6,760	5,800	–14%
Middle east	0	6,776	0	11,860	6,800	–43%
Africa	0	3,300	0	7,930	3,300	–58%
Sub-total	281,000	55,616	13,714	335,620	322,900	–4%
Asia	233,000	94,539	113,665	259,090	213,900	–17%
Global total	941,000	218,044	218,043	938,070	941,000	0%

Not surprisingly the US is the largest producer and consumer of hydrocarbon resins in the world, followed by Canada, Brazil, Mexico and Argentina in the Americas. The primary resin consumed in the US is the aliphatic C5 resin made from piperylene used in hot-melt pressure sensitive adhesives and packaging tapes, followed by water-white resins used in non-woven hot-melt formulations. In Europe, France is the leading consumer (~145 kT), followed by the Netherlands, Germany, Italy, Spain and Great Britain. Total consumption in W. Europe is about ~307 kT, while E. Europe accounts for about 6 kT. In Asia, Japan is the largest market (~109 kT), followed by China (21 kT), S. Korea (21 kT), Thailand (17 kT), Indonesia (8 kT) and Malaysia (8 kT). The Indian market is estimated at about 2 kT and nearly the entire requirement is met by imports. Hydrocarbon resins do not seem to be on the radar of petrochemical majors here in India, due certainly to the small size of the local market. But petrochemical players will do well to recognise that once a local plant makes material available locally, dormant demand can be plodded to great benefit. This has been seen in other countries, as well as in India in a number of other chemicals, and there is no reason to believe this case is different. Elsewhere in the region, China has made tremendous progress in setting up local capacity and it will come as no surprise if investments are made in a petrochemical centre like Singapore to meet the demands of Asia.

14

Sugar and Distillery

INTRODUCTION

Sugar industry is the second largest agro based industry in India next to textile. A brief insight into the process of converting stalks of sugarcane into glittering white sugar crystals with purity level of 99.8 per cent shall convince any technocrat, the relevance of chemical engineering in sugar industry. The process of sugar manufacture encompasses predominantly, most of the unit operations covered under chemical engineering, viz. solid liquid separation. Heating, precipitation-coagulation, decantation (under clarification), evaporation and concentration, crystallisation through vacuum boiling and cooling, centrifuging, drying and so on.

These operations require special skill from machinery, and also the men behind the machines for the two reasons:

1. The raw material, i.e., cane, has highly variable ingredients and does not follow any fixed chemical formulation, thus requiring continuous watch on the process by the process engineer.
2. The juice extracted is highly thermal and pH sensitive and undergoes marked change in characteristics, viz. brix, pH, consistency, viscosity, solid; liquid ratio etc.

For the above reasons, the role of chemical engineering becomes vital right from selection of material for the vessels and pipes to the design of equipment, thermodynamics of the process, methods of conveying the material from one point to other and optimum application of energy (heat as well as power) for extracting sugar from juice. With all-round development and technological upgradations on one hand and sustained efforts for augmenting the productivity while conserving every form of energy, the relevance of chemical engineering in sugar industry under the present scenario is far more pronounced than it was in the yester years. The guiding torch therefore lies on the shoulders of the chemical engineer who has to innovate newer technologies, which while being adopt able to Indian conditions, cater to the following district goals the sugar industry is striving to achieve:

1. Production of export-worthy, white crystal sugar with minimum residual sulphur content.
2. Reduction of sugar losses in the process of conversion, thus attaining higher recovery and lower cost of production.
3. Optimisation of the total energy (heat + power) for the process.
4. Making the industry more environment friendly by reducing the liquid effluent as well as air pollution through stack gases.

SUGAR MANUFACTURING PROCESS

Sugarcane is normally harvested manually in India, which eliminates the carriage of soil and trashes to the factory. Sugarcane is cut into pieces and crushed in a series of rollers to extract the juice in the 'mill house'. The milk of lime is then added to the juice and heated, when all the colloidal and suspended impurities are coagulated; much of the colour is also removed during this lime treatment. The coagulated juice is then clarified to remove the sludge. The clarifier sludge is further filtered through filter presses and then disposed of as solid waste.

The filtrate is recycled to the process, and the entire quantity of clarified juice is treated by passing sulphur dioxide gas through it. The process is known as 'sulphitation process'; colour of the juice is completely bleached out due to this process.

The clarified juice is then preheated and concentrated in evaporators and vacuum pans. The partially crystallised syrup from the vacuum pan, known as 'massecuite' is then transferred to the crystallisers, where complete crystallisation of sugar occurs. The massecuite is then centrifuged, to separate the sugar crystals from the mother liquor.

The spent liquor is discarded as 'black strap molasses'. The sugar is then dried and bagged for transport. The fibrous residue of the mill house, known as 'bagasses' may be burnt in the boilers, or may be used as raw-material for the production of paper products. The black strap molasses may be used in the distilleries. A flow diagram of the process of manufacturing in a typical sugar mill is given in the Fig. 14.1.

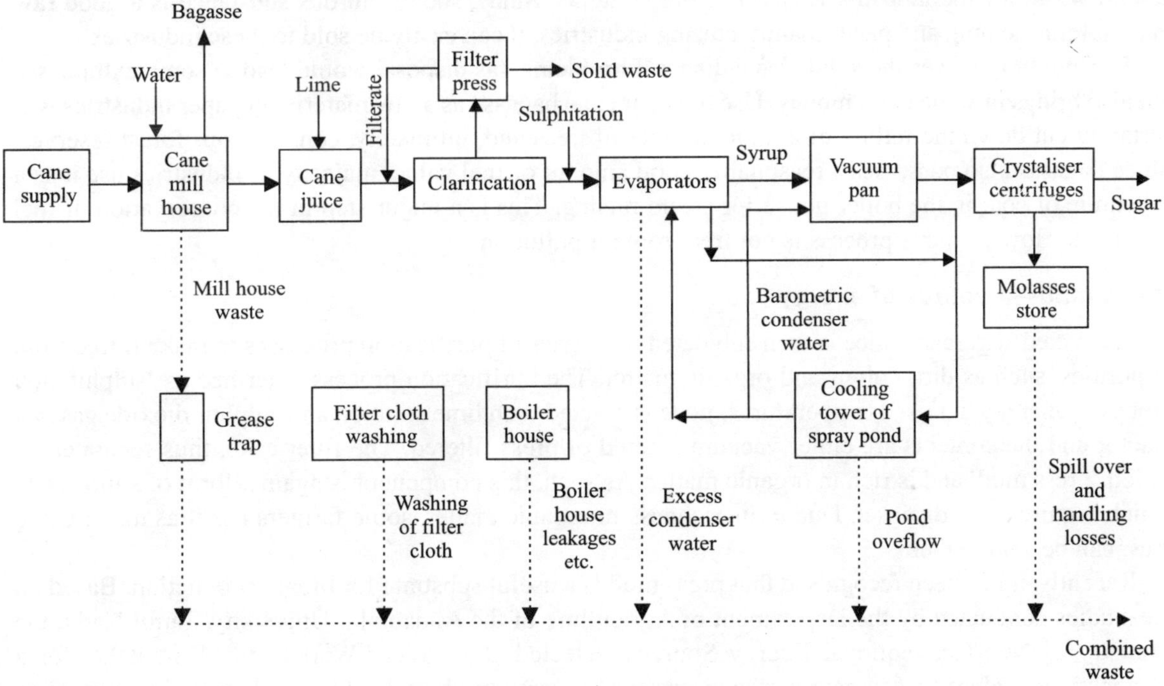

Fig. 14.1. Flow diagram for the sugar manufacturing process.

SUGAR AND DISTILLERY

Sugar

The effluent from sugar industries contains a large amount of dissolved organic matter. This organic matter is readily decomposed by biological action. Consequently, discharge of these liquors to surface water causes serious damage to aquatic life.

In *sugar mills,* the waste generated includes water used as splashes to extract the maximum amount of juice, and those used to cool roller bearings. As such, mill house waste contains high BOD due to the presence of sugar and oil from the machineries.

The filter cloths, used for filtering the juice, need occasional cleaning. The wash water thus produced, though small in volume, contains high BOD and suspended solids.

Additional waste originates due to the leakages and spillages of juice, syrup and molasses in different sections, and also due to the handling of molasses. The periodical washings of the floor also contributes a lot to the pollution load. Though these wastes are small in volume and are discharged intermittently, they have a very high BOD.

Bagasse utilisation

It is well known that sugarcane (Fig. 14.2) is the basic raw material for the extraction of sugar in India. After crushing of sugarcane, the juice is squeezed and the left-over cellulosic bagasse, which otherwise is a solid waste for the industry, is rich in fibre contents. Since, such a fibrous substance is a good raw material for the pulp and paper manufacturing industries, it can easily be sold to these industries.

This not only solves the solid waste disposal problems, (as disposal would lead to some extra cost), but also brings in some extra money. Use of sugar cane bagasse as a raw material in paper industries will certainly cut down the felling of a large number of trees and, ultimately, conserve our forest reserves. Since bagasse also possesses a reasonably good amount of fuel value, majority of industries use it as a substitute of coal in the boiler house for steam raising. This is a major step in the conservation of fuel resources. However, this process is not free from air pollution.

Press mud—a source of energy

The extracted sugarcane juice is then subjected to a series of purification processes to make it free from impurities, such as dirt, colour and organic matter. The purification process is termed as 'sulphitation process' whereby at raised temperatures, juice is reacted with lime solution and sulphur dioxide gas in a reactor and the contents are either vacuum filtered or press filtered. The filter cake, thus separated, is called 'press mud' and is rich in organic matter. As such, this component is again a form of solid waste which requires safe disposal. Due to its richness in organic matter, some farmers use it as manure and thus, can be sold to them.

Recently, it has been recognised that press mud is a useful substrate for biogas production. Based on the studies conducted by the Department of Agriculture of the Annamalai University, Tamil Nadu, the Ministry of Non-Conventional Energy Sources, selected Ugar Sugar Works Ltd., Karnataka, for a demonstration plant to generate methane gas out of press mud, and a biogas distribution system to 120 families residing in the factory premises. The results indicated that as high as 80 Nm^3 biogas/tonne of fresh press mud could be produced. The effluent slurry is also useful as a nutrient rich bio-fertiliser. This clearly reflects wealth recovery out of waste by-products.

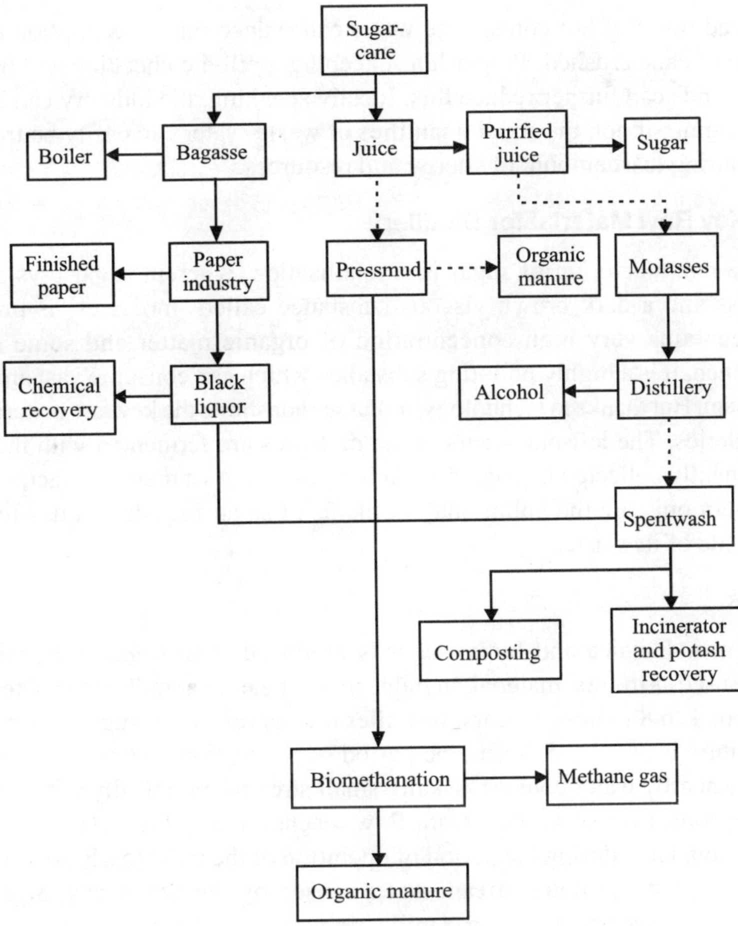

Fig. 14.2. Sugar industry—an example of waste utilisation.

Hot condensate

After the purification of the juice, it is transferred to evaporator pans for concentration prior to crystallisation of sugar. In the evaporator pans, the juice is vacuum evaporated so that its water content is reduced. These water vapours are, in fact, one of the purest forms of water when condensed, without any carry over of sugar, and can be used as boiler-feed water. In this regard, the 'poly-baffle entrainment catcher' in pan evaporation for sugar recovery provides an alternative over the conventional pan evaporation technique. Such water will not only reduce the scaling in the boiler tubes, but also conserve large quantities of fresh water to be used. In addition, this hot condensate needs far lesser amount of energy (when compared with the fresh water) to be transformed into steam for obvious reasons, resulting in money saving to the industry. More recently, it is emerging that if hot condensate is used to extract the last traces of sugar from the crushed sugarcane, not only does this leads to increased recovery, but it also saves large quantities of fresh water that is used otherwise. It also requires less heat energy for the sulphitation process. According to an estimate by the Central Pollution Control Board, in general, approximately 1500 to 2000 litres of water is used per tonne of cane crushed.

The recycling and reuse of hot condensate water can reduce this consumption to as low as 100 to 200 litres per tonne of cane crushed. Proper housekeeping, periodic checking and maintenance of pipe joints, valves and glands can further reduce this. Ideally speaking, the industry can be brought down to zero effluent discharging; if not, the small quantities of waste-water can easily be treated in the end-of-pipe treatment, requiring least amount of energy and resources.

Molasses—The Key Raw Material for Distillery

The last step in the manufacturing of sugar is crystallisation, wherein sugar crystals are removed by centrifuging process and a dark brown viscous substance called 'molasses' is produced as a waste by-product. This contains very high concentration of organic matter and some sugars that escape crystallisation. As such, it is a highly polluting substance which can consume vast amount of energy and resources for treatment. But thanks to technology, molasses constitute the key raw material for the extraction of alcohol by distilleries. The left-out sugars in the molasses are fermented with the help of yeast into alcohol and subsequently collected through distillation process. As a matter of fact, once the molasses is sold to distilleries, not only are the pollutional problems of sugar industry eliminated, but it also earns revenues from the sale of its wastes.

Sugar Mill Wastes

In countries like Cuba, Jamaica and India, sugar is produced from sugarcanes, while in many other places, beetroots are used as the raw material. In India, most of the sugar mills are situated in the countryside and operate for about 4 to 8 months a years, just after the harvesting of sugar canes. A large volume of waste of organic nature is produced during the period of production and normally they are discharged onto land or into the nearby water courses, usually small streams, practically without pre-treatment.

The condition becomes worse as the stream flow reaches a very low level and eventually enough dilution water is not available during the period of operation of the mills (early November to end of May or June). Putrefaction of the polluted stream water caused by the heavy discharge of organic waste, resulting in the odour nuisance near the sugar mills is a very common phenomenon.

In fact, all the concerned bodies, both sugar industry and pollution control agencies, are aware of these problems and are trying to find an economical means to check the nuisance created by sugar mill effluents.

Sources of waste-water and the characteristics of the wastes

Wastes from the mill house include water used as splashes to extract maximum amount of juice, and that used to cool the roller bearings. As such, mill house waste contains high BOD due to the presence of sugar, and oil from the machineries.

The filter cloth, used for filtering the juice, needs occasional cleaning. The wash water thus produced, though small in volume, contains high BOD and suspended solids.

A large volume of water is required in the barometric condensers of the multiple effect evaporators and vacuum pans. The water is usually partially or fully recirculated, after cooling through a spray pond. This cooling water gets polluted as it picks up some organic substances from the vapour of the boiling syrup in the evaporators and vacuum pans. The water from spray pond when it overflows, becomes a part of the waste-water, and is usually of low BOD in a properly operating sugar mill. But because of poor maintenance and bad operating conditions, a substantial amount of sugar may entrain in the condenser water; this polluted water, instead of being recirculated, is discarded. These discharges contribute substantially to the waste volume and moderately to BOD in many sugar mills.

Additional waste originates due to the leakages and spillages of juice, syrup and molasses in different sections, and also due to the handling of molasses. The periodical washings of the floor also contribute a lot to the pollution load. Though these wastes are small in volume and are discharged intermittently, they have a very high BOD. The periodic blow off of the boilers produce another intermittent waste discharge. This waste is high in suspended solids, low in BOD and is usually alkaline.

Characteristics of wastes from the different sections of common sugar mills and that of the combined waste are given in Table 14.1. Composition of the waste and its volume varies widely from mill to mill, depending upon the availability of water.

Table 14.1. Characteristics of sugar mill wastes.

Characteristics	*Mill-house waste*	*Filter cloth washings*	*Condenser water*	*Boiler house and floor washings*	*Combined waste (excluding condenser water)*
Rate of flow litres per tonne of cane crushed	730	360	1640	230	–
pH	6.7	9.5	–	7.2	4.6–7.1
COD, mg/l	–	–	–	–	600–4380
BOD, mg/l (5 day 20°C)	210	1765	–	5150	300–2000
Total solids, mg/l	1760	6970	–	5130	870–3500
Total volatile solids, mg/l	–	–	–	–	400–2200
Total suspended solids, mg/l	910	4000	–	120	220–800
Total nitrogen mg/l	–	–	–	–	10–40
COD/BOD ratio	–	–	–	–	1.3–2.0

Effects of the waste on receiving water

The fresh effluent from a sugar-mill decomposes rapidly after few hours of stagnation. It has been found to cause considerable difficulties when their effluent gets an access to water courses, particularly the small and non-perennial streams in the rural areas. The rapid depletion of oxygen due to biological oxidation followed by anaerobic stabilisation of the waste causes a secondary pollution of offensive odour, black colour, and leads to fish mortality. No question of the discharge of this waste into the seweres arises, as most sugar mills are situated in the unsewered rural areas.

Treatment of the wastes

Like any other industry, the pollution load in sugar mills can be reduced with a better water and material economy practised in the plant. Judicious use of water in various plant practices and its recycling, wherever practicable, will reduce the volume of waste to a great extent. Volume of mill house waste can be reduced by recycling the water used for splashing. Dry cleaning of floors or floor washings using controlled quantity of water will also reduce the volume of waste to certain extent.

The organic load of the waste can only be reduced by a proper control of the operations. Overloading of the evaporators and the vacuum pans, and the extensive boiling of the syrup lead to a loss of sugar through condenser water; this, in turn, increases both volume and strength of the waste effluent. Disposal of the effluent on land as irrigation water is practised in many sugar mills, but it is associated with odour problems.

The reasonable COD/BOD ratio of mill effluents indicate that the waste is amenable to biological treatment. However, generally it is found that the aerobic treatment, with conventional activated sludge process and trickling filters, is not too efficient, even at a low organic loading rate. A maximum BOD reduction of 51 per cent was observed in a pilot plant study at Kanpur, where both trickling filter and activated sludge process were tried. In view of the high cost of installation and supervision of the treatment units, and the seasonal nature of the operation of this industry, it is generally observed that the conventional aerobic treatment will not be economical in India. Anaerobic treatment of the effluent, using both digesters and lagoons, have been found to be more effective and economical. A BOD reduction of about 70 per cent was observed in a pilot plant study with an anaerobic digester, where BOD loading was 0.65 kg/m^3/day with a detention time of 2.4 days at a controlled temperature of 37°C. In the same study, the BOD reduction was found to be 60 per cent in an anaerobic lagoon, where a BOD loading of 0.23 kg/m^3/day and detention time of 7 days was provided. In another laboratory study, a BOD reduction of 88.9 per cent was observed in an anaerobic digester, with a detention time of 2 days and controlled temperature of 36 ± 1°C, at a higher BOD loading of 1.79 kg/m^3/day. The effluents of the anaerobic treatment units are found to contain sufficient nutrients (nitrogen and phosphorus). As such, further reduction of BOD can be accomplished in aerobic waste stabilisation ponds.

Where sufficient land is available, two-stage biological treatment, with anaerobic lagoons followed by aerobic waste stabilisation ponds, is recommended for Indian conditions. The mill effluent, however, is to be pre-treated primarily in bar screens and grease trap. The flow sheet for the complete treatment of the sugar mill waste is shown in Fig. 14.3. With the process parameters as indicated in Fig. 14.2, it is expected that the BOD reduction in the anaerobic process will be in the order of 60 per cent, while overall BOD reduction may be in the order of 90 per cent.

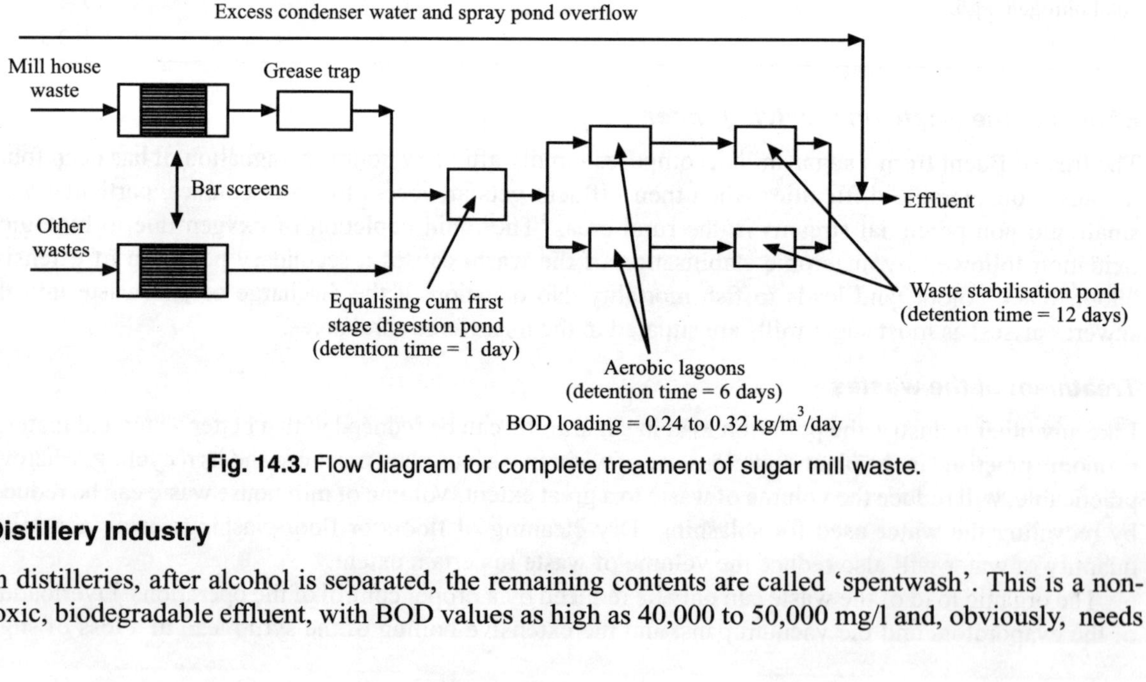

Fig. 14.3. Flow diagram for complete treatment of sugar mill waste.

Distillery Industry

In distilleries, after alcohol is separated, the remaining contents are called 'spentwash'. This is a non-toxic, biodegradable effluent, with BOD values as high as 40,000 to 50,000 mg/l and, obviously, needs huge resources for treatment and disposal. The cleaner technology has also provided a number of ways to recover wealth from the spentwash. The first and the foremost way is 'biomethanation' of spentwash.

Biomethanation

Biomethanation is a process involving decomposition of organic matter (and to some extent, inorganic matter) in the absence of molecular oxygen, resulting in methane-rich gas that has good fuel value. The generation of biogas can meet more than 60 per cent of the fuel requirement of the distillery.

During such bioconversion, approximately 80 to 90 per cent of BOD is also removed from the spentwash. Over the years, a number of distilleries in the country have adopted bio-methanation for the treatment of spentwash. Although this is a cost-intensive technology, yet the investment made in this system is usually recovered within 4 to 5 years.

For a 30 kld alcohol producing industry, generally the spentwash generated is of the magnitude of 450 kld and, on an average, the total biogas production is about 12,000 Nm^3/day. Based on the net calorific value, the coal saving is as high as 12 tpd, which amounts to a net saving of Rs 72 lakh a year.

Composting

In yet another process, the spentwash is mixed with filler materials, such as press mud, rice husk, wood chips, bagasse and pith. The mixture is seeded with cowdung or specially developed micro-organisms to hasten the process. The process is aerobic and it takes 12 to 14 weeks to be completed, resulting in black compost, which can be used as manure.

Incineration and potash recovery

The third possible alternative is based on the principle that the spentwash, when concentrated to 60 per cent w/w, develops sufficient calorific value and can burn by itself without any external input of energy. This process results in potash rich ash which can again be used as a fertiliser.

Utilisation of distillery effluents

Shortage of raw materials for industry and inputs for agriculture is the major problem facing the country today. Energy crisis has aggravated the situation. Rapid industrialisation programmes have resulted in the generation of liquid, solid and gaseous wastes in such huge quantities that a serious threat is likely to be posed to the 'quality' of life in the years to come. All these problems can be solved only through a change in our philosophy and attitude to consider wastes not a nuisance or a disposal problem but as new resource materials for recovery of by-products.

Manufacture of alcohol and generation of effluents

Alcohol is manufactured by fermentation of molasses in the presence of yeast. The fermented wort is fed to the distillation (analyser and rectifier) columns. The bulk of the alcoholic vapours leaving the top of the analyser column are admitted to the bottom of the rectifier column. Vapours from the head of the rectifier are partially condensed and cooled as rectified spirit. The wash, freed from alcohol, is discharged from the basement of the analyser column and is termed as spentwash. After complete recovery of alcohol, the plant wastes, consisting of spentwash and fermentation washings, are thrown out of the factory as liquid waste. These wastes are described under different names such as effluents, slops, stillage, vinasse and dunder (Fig. 14.4).

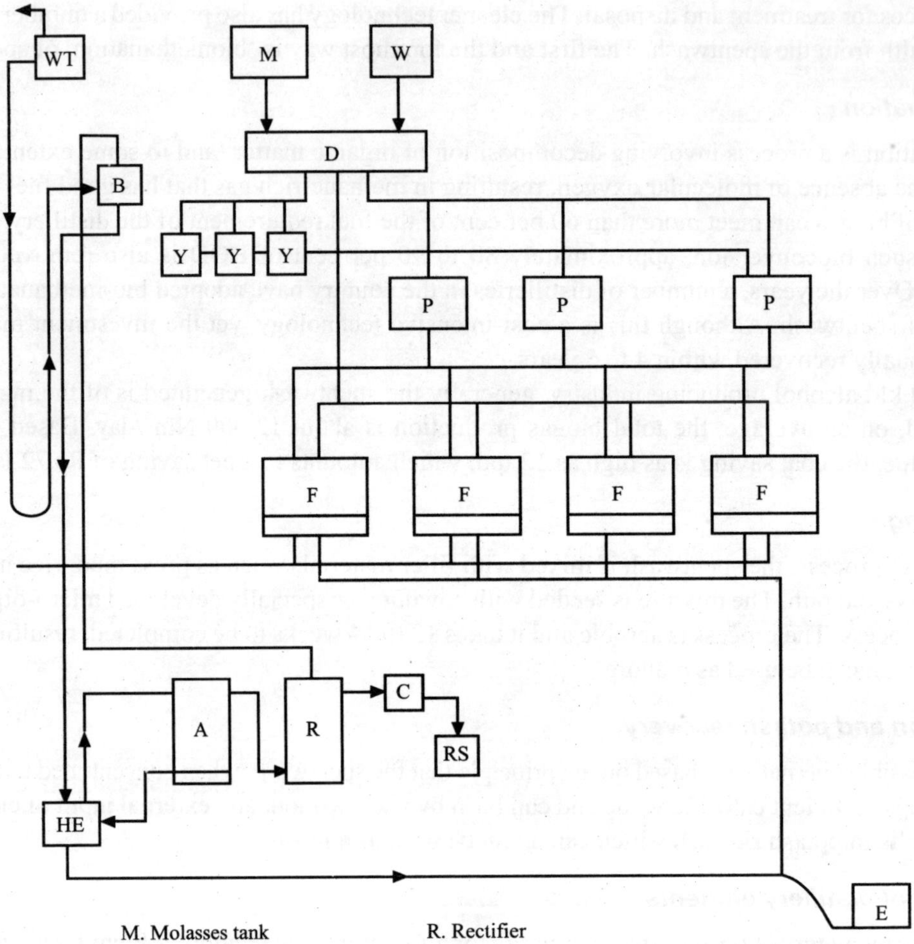

M. Molasses tank
W. Water tank
D. Molasses diluter
Y. Yeast vessels
P. Prefermenters
F. Main fermenters
A. Analyser
E. Distillery effluent or waste
R. Rectifier
RS. Rectified spirit
HE. Heat exchanger
C. Condenser
S. Yeast sludge from fermenters
WT. Wash tank
B. Beer heater

Fig. 14.4. Distillery process for ethyl alcohol from molasses.

Distillery effluents

In India, there are around 127 distilleries with a total installed capacity of 732 million litres of alcohol per annum. For every litre of rectified spirit (ethanol) manufactured, about 12–15 litres of effluents are discharged which pose a serious pollution problem, particularly due to their generation in substantial quantities. They are characterised by reddish brown to dark colour, with pH around 4.0, temperature about 90°C, having a smell of burnt or caramelised sugar.

A good deal of research work has been initiated at a number of institutes/centres; namely, National Sugar Institute, Kanpur; National Environmental Engineering Research Institute, Nagpur, on anaerobic lagooning, anaerobic digestion in closed digesters with methane recovery, ammonification, drying, incineration, potash recovery, and direct use for irrigation purposes; but the high cost involved in the treatment and recovery of by-products has been one of the main constraints.

In order to tackle this problem, there is a need to adopt an integrated system (Fig. 14.5) which may help recover energy and water for inplant utilisation, water and potash for farmlands and feed for animal nutrition, consistent with minimum pollutional hazards.

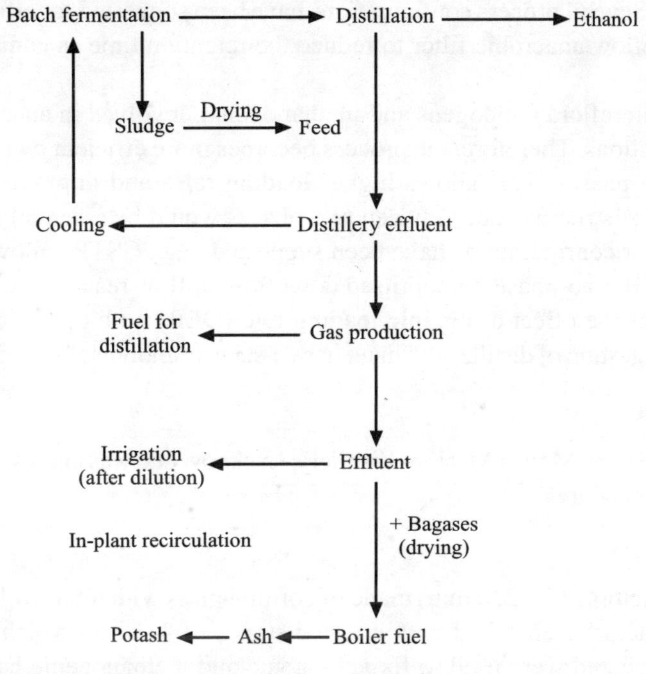

Fig. 14.5. Recycling distillery effluents for fuel, ethanol, animal feed and potash.

As such, keeping in view the prospects and problems in the utilisation of distillery wastes, it is highly important to constitute a planning group to prepare a feasibility report and a coordinated programme. Efficient utilisation of effluents may not only help provide sanitary conditions all around the factory areas but also open up a vast potential to generate new resource materials.

METHANE PRODUCTION BY TWO STAGE ANAEROBIC DIGESTION OF DISTILLERY EFFLUENT IN A FIXED FILM REACTOR

The performance of two stage fixed film reactors for methane generation from distillery effluent was investigated. Gas productivity and percentage COD removal was dependent on organic loading rates (OLR) in both the reactors.

The maximum gas productivity of 2.30 m^3/m^3d was obtained from methanogenic reactor at OLR, 9.72 kg COD/ m^3d with specific gas yield of 0.33 m^3/kg COD removed and the percentage COD removal was 73. The methane content of the biogas varied from 55–65 per cent.

The liquid effluent discharged by distilleries is one of the most polluting and environmentally unacceptable organic wastes. This high organic mineral containing waste (BOD 40,000–50,000 mg/1), if properly and adequately treated, can provide a very effective source of energy while concurrently resolving the pollution problems.

Microbial methane generation by anaerobic digestion from organic residues has gained considerable importance in recent years. Improved digester performance has been achieved by: (i) maintaining high microbial density in the reactor; and (ii) optimal operating conditions for improved microbial activity. This requires a high solid retention time (STR) and a minimum hydraulic retention time (HRT) to ensure reduced reactor volume. Several process configurations have been suggested for efficient digestion. *Gadre* and *Godbole* used an upflow anaerobic filter to reduce the retention time as compared to conventional digester.

The tow groups of microflora (acidogens and methanogens), involved in anaerobic digestion, differ in their biochemical functions. Therefore, the process becomes more efficient by providing the optimum environment in separate phases. This allows higher loading rates and improved methane yields at a shorter retention period. A strict balance between two phases would be essential for high performance. Various two-phase reactor configurations have been suggested, e.g., CSTR followed by upflow reactor CSTR followed by UASB, two-phase immobilised downflow/upflow reactor, etc.

This section describes the effect of organic loading rate (OLR) on biogas yield and per cent COD reduction by anaerobic digestion of distillery effluent in two-stage immobilised (fixed flim) upflow reactors.

Materials and Methods

The effluent was obtained from Mohan Meakins distillery, Lucknow and was analysed for its characteristics according to standard procedures.

Experimental setup

The two jacketted bioreactors (35 × 220 mm) made of corning glass with total working volume of 200 ml each were used. The reactors were filled with crushed stones of 3–5 mm which resulted in total void volume of 100 ml in each and were used to fix acidogenic and methanogenic bacteria separately. The temperature was maintained at 35°C in acedogenic reactor (RI) and methanogenic reactor (RII) separately. The upward feeding of the effluent was done by peristaltic pumps (*Watson* and *Marlow*). The gas was collected by water displacement method in calibrated columns. The schematic diagram of the set-up is shown in Fig. 14.6.

Process details

Activation of bed was carried out by circulating digested poultry manure in both reactors for about 3 weeks at a flow rate of about 30 ml/d. The two groups of microflora (acidogenic and methanogenic) were selected and fixed in reactors I and II respectively on the basis of temperature difference.

The effluent (with COD of 25920 mg/l to 97570 mg/l) was neutralised and pumped upward so as to maintain HRT of 3.3 days at different organic loading rates (OLR). The effluent from reactor I (COD, 7884 mg/l to 50,760 mg/l) was fed to the bottom of reactor II for methane generation and COD removal. At a fixed OLR the samples from both reactors were analysed for COD, volatile solids (VS), pH and generated gas was collected estimated by Orsat apparatus.

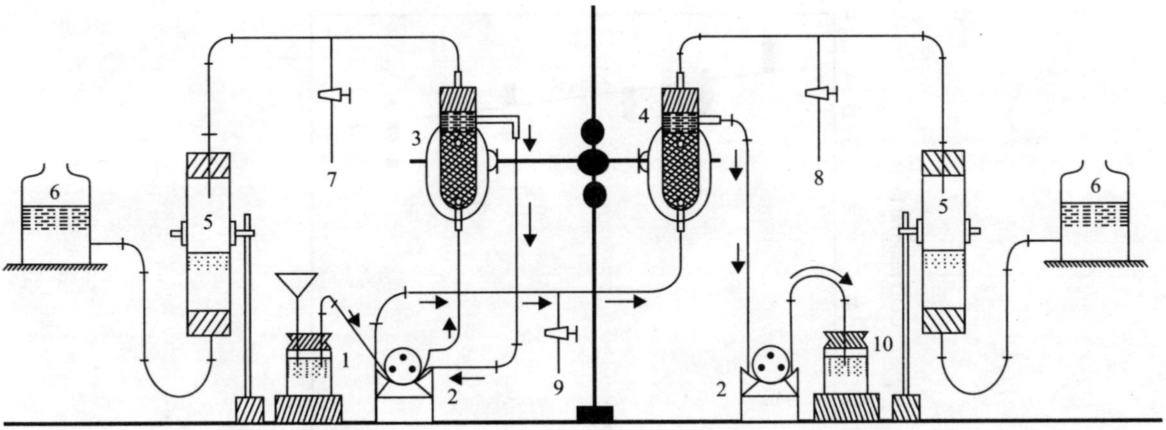

Fig. 14.6. Experimental arrangement. (1) Feed bottle; (2) peristaltic pump; (3) bioreactor (acidogenic); (4) bioreactor (methanogenic); (5) biogas (collector); (6) water collectors; (7,8,9,) sample points; and (10) treated effluent.

Results and Discussion

The complete analysis of distillery effluent is shown in Table 14.2. Since it contains high BOD (60,000 ppm) which indicates the good concentration of biodegradable constituents in the effluent, it can be a good substrate for biogas production.

Table 14.2. Analysis of distillery effluent.

Characteristics	*Value/observation*
Colour	Dark brown
Odour	Burnt sugar
Specific gravity	1.07
pH	4.5
COD, ppm	97,570
BOD, ppm	60,000
Total solids, mg/1	84,800
Volatile solids, mg/1	60,000
Ash, mg/1	24,000
Total nitrogen, mg/1	1,284
Total acidity ($CaCO_3$), mg/1	20,000

The performance of acidogenic reactor: The neutralised effluent was fed to fixed film reactors at organic loading rates ranging from 7.8 to 29.21 kg COD/m^3/d. The effluent from reactor RI was analysed for COD, volatile solids (VS), pH and gas evolved was measured. The percentage removal of COD, gas productivity (m^3/m^3d) and specific gas yield are shown in Figs. 14.7 and 14.8 respectively. The best fitting lines were drawn through the various data points. The graph (Fig. 14.7) indicates that as the OLR was increased, the percentage removal decreased from 66 per cent to 55 per cent, whereas actual values ranged from 69.5 per cent to 47.9 per cent.

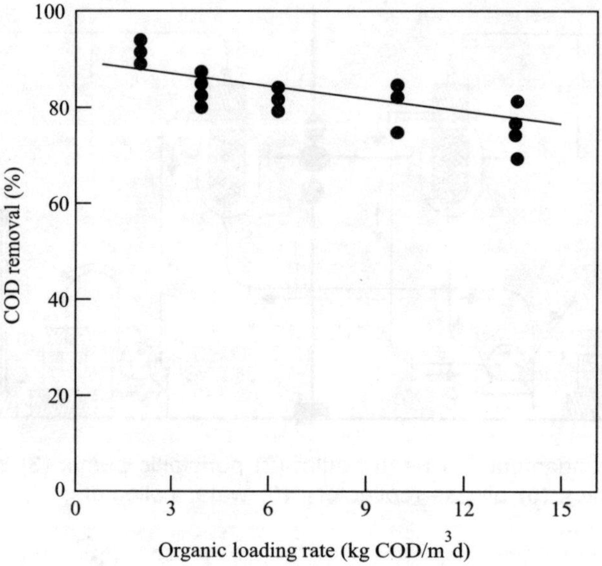

Fig 14.7. Effect of OLR on COD removal by methanogenic reactor.

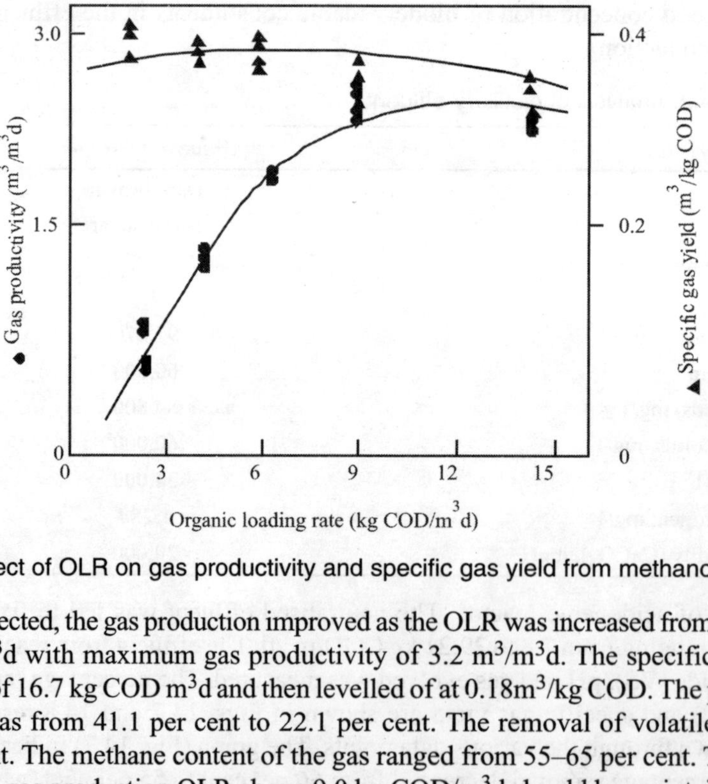

Fig 14.8. Effect of OLR on gas productivity and specific gas yield from methanogenic reactor.

As would be expected, the gas production improved as the OLR was increased from 7.8 kg COD m^3/m^3d to 29.2 kg COD/m^3d with maximum gas productivity of 3.2 m^3/m^3d. The specific gas yield increased initially upto OLR of 16.7 kg COD m^3d and then levelled of at 0.18m^3/kg COD. The percentage reduction of volatile solids was from 41.1 per cent to 22.1 per cent. The removal of volatile solids ranged from 82.5 to 85.0 per cent. The methane content of the gas ranged from 55–65 per cent. These data show that to obtain maximum gas production OLR above 9.0 kg COD/m^3d should be maintained.

Some of the data on two phase biomethanation of distiallery effluent in immobilised reactors are shown in Table 14.3 and compared with the present work. It is seen that specific gas yield and methane content in the biogas was little lower as compared to *Ghosh*. This may be improved by increasing the HRT in the second reactor and a fraction of the effluent may be recycled back for higher gas production. The effect of recirculation on the performance of two stage anaerobic process has been reported by *Joubert*. The present data compare well with those of *Kaul* and *Badrinath* who used big column reactors with a total volume of 33.41 (Table 14.3). The results obtained in this study confirm the potential of two stage fixed film reactors for biomethanaction of distillery effluent.

Table 14.3. Two phase biomethanation data for the treatment of distillery effluent in immobilised reactor.

Parameters		*Ref. 1*	*Ref. 7*		*Present work*	
			R-I	*R-II*	*R-I*	*R-II*
COD, g/1		82.0	40–80	5–45	25.9–97.5	7.8–50.7
Loading rate, kg COD/m^{3}d	R_1	39.3	2–90	2–90	7.8–29.2	2.3–15.2
	R_2	9.8				
HRT, d		7.5	0.26–11.8	0.26–11.8	3.3	3.3
Specific gas yield, m^3/kg COD removed		0.69	0.22–0.40	0.21–0.38	0.15–0.24	0.28–0.39
Gas productivity, m^3/m^{3}d		–	6.6–15	0.08–9.0	0.80–3.30	0.55–2.15
Methane content, %		75.0	51–64	50–65	45–55	55–65
COD reduction, %		70.0	36–85	28–72	47.3–69.5	73–86.9
BOD reduction, %		80.0	–	–	–	–

ENERGY FROM DISTILLERY WASTE—A NEW EXPERIMENT

The present section deals with the proposed pilot plant, for the production of biogas from distillery waste. The effect of various parameters affecting methane production such as feed water ratio, retention time, temperature, pH, digestion unit, etc., have been taken into consideration while designing the pilot plant. The basic data from various researchers have been incorporated in the planning. Suitable modifications based on experiment will be made and utilised for pilot plant study.

In recent years, energy conservation and environmental concern have increased interest in direct anaerobic treatment of organic industrial wastes. Since anaerobic digestion of any organic material results in the production of fuel in the form of biogas with simultaneous mineralisation of the organic content, special emphasis is given to this mode of waste treatment, whenever possible. There are 254 distilleries in the country, producing approximately 1400 million litres of waste alcohol/year, giving 210 billion litre of waste. The waste contains huge organic huge organic load varying from 70,000 to 130,000 mg/l of COD. Basic researches has shown that it is possible to produce good quality of biogas from the distillery waste. The present communication deals with the findings of scale reactor with an upflow filter reactor for continuous production of biogas from a distillery plant located in Pravaranagar, is proposed that a pilot plant will be set up to produce biogas from the industry waste (distillery). Although basic literature regarding the production of methane from organic load of distillery waste on bench scale is available, yet there is need of an integrated digester utilising anaerobic digestion technology to produce biogas on continuous basis. Although some work is reported about biogas generation, practically no information is available on kinetic data and optimum paramenters affecting the methane production.

Materials and Methods

Substrate characteristics

The waste from distillery is a dark brownish-black liquid. This waste comes out after the alcoholic fermentation of molasses. The fresh effluent has 90 to 100°C temperature, pH, of 4.0–5.0 and COD and BOD values of 75–100 g/1 and 35–45 g/1 respectively. The effluent is highly toxic in nature. The approximate composition is given in Table 14.4.

Table 14.4. Analysis of distillery waste.

Component	*Concentration*	*Componenet*	*Concentration*
Total solids	67	Calcium	710.0
Volatile solids	35	Silver	0.4
Total suspended solids	8	Copper	1.0
COD	80	Magnesium	180.0
BOD	40	Manganese	1.0
Alkalinity (by $CaCO_3$)	3	Iron	3.0
Potassium	4	Lead	1.0
		Zinc	5.0
		Phosphorus	45.0

Upflow anaerobic filter reactor (design and loading)

The specification of the reactor and loading scheme have been given below:

Height of reactor	117 cm
Diameter of reactor	12 cm
Effective volume	5.5 litre
Material of construction	PVC
Packing material	Red bricks
Daily organic load	40 g COD/1
Daily hydraulic load	600 ml
Hydraulic retention time	3.6 days

Proposed experimental set-up

The experimental set-up required to produce methane (Fig. 14.9) from distillery waste in continuous manner has an inlet tank where proportion of water is checked. This inlet tank in followed by a mixing tank to maintain C/N ratio. The slurry from mixing tank is pumped to anaerobic digester unit where pH and temperature is very accurately controlled for better yield of biogas.

In an experiment 75 per cent CH_4 was obtained after maintaining pH and temperature 30–35°C in fifteen days retention time accompanied by 70 per cent BOD reduction and 40 per cent COD reduction. Since impurities in the slurry are reduced in anaerobic digester hence methane production rate is accelerated in main digester which is employed just after the anaerobic digester. The experiment shows that more is the reduction in BOD and COD more will be the methane content in biogas produced. The digested slurry from main digester is partially recyled which increases the efficiency of main digester. The flow diagram of the process is shown in Fig. 14.10.

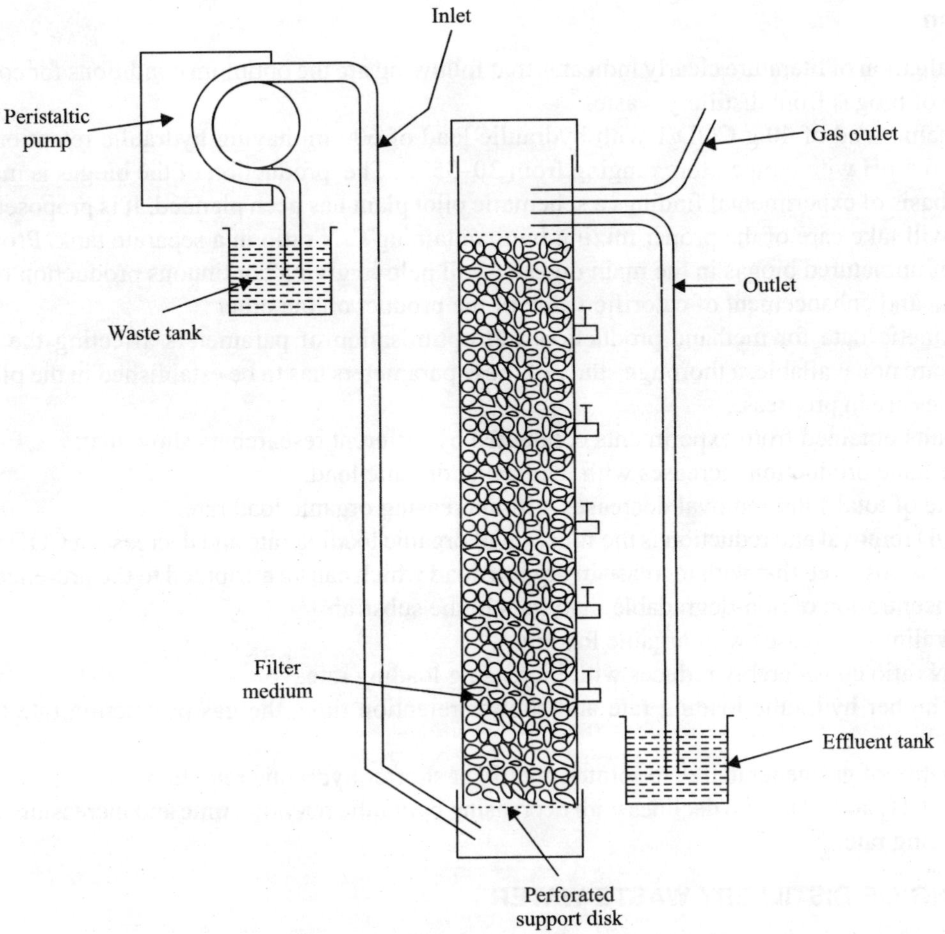

Fig 14.9. Laboratory anaerobic filter.

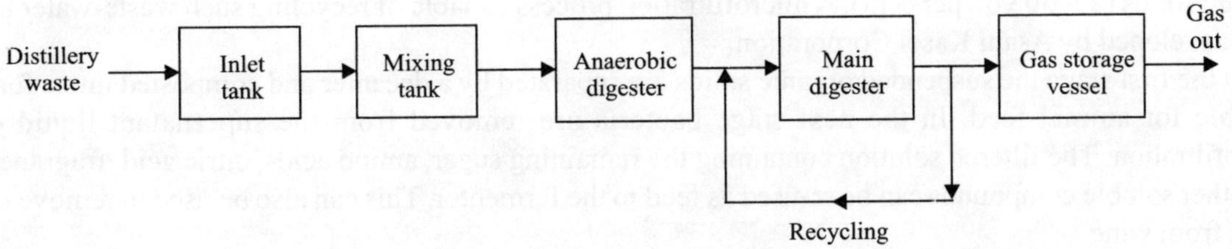

Fig 14.10. Flow process.

Sample analysis and reactor control

Sample are collected from different sampling ports which are installed along the side of reactor. The samples are analysed for methane and carbon dioxide contents using a fisher or sat apparaturs. pH and temperature are to be monitored daily. The total solids, total suspended solids, total volatile solids, alkalinity ammonia-nitrogen and COD levels of the influent and effluent have to be determined three/four times per week.

Discussion

Critical evaluation of literature clearly indicates that following are the optimum conditions for continuous production of biogas from distillery waste.

The organic load of 40 g COD/1 with hydraulic load of 600 ml having hydraulic retention time of 3.6 days at 6.5 pH with temperature ranging from 30–35°C. The production of the biogas is maximum.

On the basis of experimental findings a schematic pilot plant has been planned. It is proposed that the said plant will take care of the proper mixing by maintaining C/N ratio in a separate tank. Provision of recycling of unmatured biogas in the main digester will help in giving continuous production of the gas on one hand and enhancement of calorific value of the product on the other.

Since kinetic data for methane production and optimisation of parameters affecting the methane production are not available, a thorough study of these parameters has to be established in the pilot plant. These studies are in progress.

The results obtained from experiments conducted by different researchers show that:

1. Methane production increases with increasing organic load.
2. Rate of total solid removal decreases with increasing organic load rate.
3. COD removal and reduction is the function of organic loading rate and decrease in COD removal. It was observed that with increase in organic load which can be attributed to the presence of large concentration of non-degradable materials in the substrate.
4. Alkalinity increases with organic loading.
5. C/N ratio considerably reduces with increasing loading rate.
6. At higher hydraulic loading rate and shorter retention time, the gas production rate increases markedly.
7. Quality of gas generated deteriorates greatly at shorter hydraulic retention time.
8. The CH_4 and CO_2 ratio declines with decreasing hydraulic retention time and increasing hydraulic loading rate.

RECYCLING OF DISTILLERY WASTE-WATER

Distilleries generate about 130–150 vol. per cent waste-water. Such waste-water contains organic substances equivalent to about 50,000 mg/l of BOD (biological oxygen demand) and consists of suspended organic solids (25–30 vol. per cent). A microfiltration process capable of recycling such waste-water has been developed by Asahi Kasei Corporation.

At the first stage the suspended organic solids are separated by a decanter and composted into a form suitable for animal feed. In the next stage bacteria are removed from the supernatant liquid by microfiltration. The filtered solution containing the remaining sugar, amino acids, citric acid, fragrances and other soluble compounds can be re-used as feed in the fermentor. This can also be used to remove the dregs from wine.

15

Leather and Tannery

INTRODUCTION

The tanning industry is one of the oldest industries in India. Usually, tannery wastes are characterised by strong colour, high BOD, high pH and high dissolved salts. Disposal of these wastes into water courses or on land, with or without prior sedimentation, was not a problem perhaps decades ago.

But in recent years, the concentrated growth of this industry in certain localities has shown how such waste can cause irreversible damage to the water environment. In view of its peculiar pollution potential and the increasing industrial importance, it has become essential to treat the waste to a certain degree prior to its disposal.

Tannery waste may be classified as continuous and intermittent flow waste. Continuous flow wastes consist of wash waters after various processes and comprise a large portion of the total waste, which are relatively less polluted. Spent liquors belonging to soaking, liming, bating, pickling, tanning and finishing operations are discharged intermittently. Although these are relatively small in volume, they are highly polluted and contain varieties of soluble organic and inorganic substances.

The method of treatment of tannery waste may be classified as physical, chemical and biological:

1. Physical treatment: In this treatment, screens are required to remove fleshings, hair and other floating substances. The sludge is dried over sand drying beds and can be used as manure. No appreciable reduction of dissolved solids, BOD, COD, colour and chloride can be achieved in the physical treatment processes, but the effluent may be discharged into sewers with less chances of sewer chokage.
2. Chemical treatment: Here, chemical coagulation, with or without prior neutralisation, is followed by biological treatment and is necessary for better quality of the effluent.
3. Biological treatment: This treatment of tannery waste, in activated sludge process, after mixing with municipal waste-water in a suitable proportion, and using acclimatised microorganism is capable of reducing the BOD, COD and tannin by about 90 per cent.

CLEANER PROCESSING VIS-Á-VIS CONVENTIONAL TECHNOLOGIES

Pollution prevention by adopting cleaner technologies is the best option for tannery waste management problems. Implementation of cleaner technologies, water management, waste minimisation and recovery and reuse of chemicals in tanneries are very important for sustainable growth of the leather industry. In leather processing, large quantities of water, as much as 40 to 50 litres of water per kilogram of raw material, is used. It is also generally known that the amount of toxic material discharged in the tannery effluent arises, to a large extent, from the pre-tanning operations. Leather processing employs complex principles of biotechnology and inorganic chemistry for achieving the desired technical objectives. Over 80 per cent of organic pollution load in BOD is generated in the beamhouse, of which 10 per cent is derived from soak liquors, 70 per cent from unhairing/liming, 5 per cent from deliming and bating, and 15 per cent from the rest of the processes. Treatment of the combined waste from a vegetable tannery is shown in Fig. 15.1.

The soak water provides 60 per cent salinity and the rest comes from pickling and other operations. Even though conventional practices do allow tanners to achieve satisfactory leather quality, they suffer from disadvantages like longer processing time and large working space and environmental problems. These conventional practices are based on techniques standardised prior to the recognition of their impact on environment. Waste minimisation/low waste concept is the underlying principle for resource conservation and inplant control to achieve production consistency and reduction in pollution problems. In leather processing, chrome tanning is an important step that has raised serious ecological concerns. International regulatory laws stipulate concentrations of 0.3–2 ppm of chromium in industrial wastewater. So, several technological options have been developed to render chrome tanning eco-friendly. Among the options developed, one is the high exhaust chrome tanning system, the second option is chrome recovery and reuse, and the third is partial or complete replacement of chrome with ecologically acceptable tanning compounds.

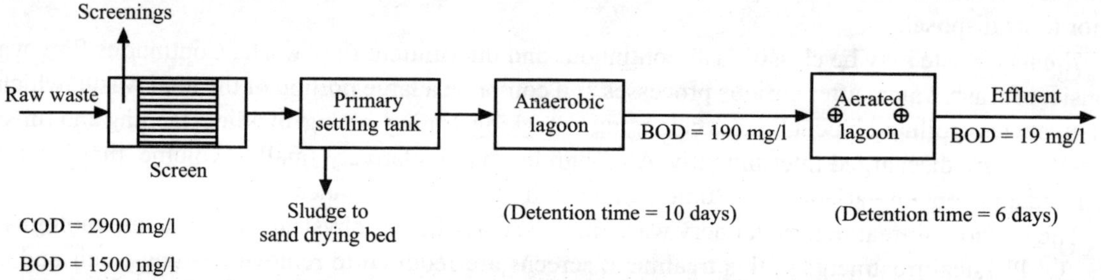

Fig. 15.1. Flow sheet for the treatment of combined waste from a vegetable tannery.

Similarly, in post-tanning operations, conventional practices require a total reappraisal in the light of current environmental regulations. The general characteristics of tannery effluent before and after treatment are presented in Table 15.1.

Table 15.1. General characteristics of effluent before and after treatment.

Characteristics	*Before treatment*	*After treatment*
pH	7.0–9.0	6.5–8.5
T.S.S.	2000–4000	10–100
T.D.S.	11000–16000	5000–10000
Chlorides as Cl	6000–9500	1600–5000
Sulphates as SO_4	2400–4000	800–2000
BOD	1000–3000	30–150

Chrome Tanning

Nearly 80–85 per cent of total finished leather produced are processed by adopting the chrome tanning system. Normally, basic chromium sulphate with following characteristics is used for tanning:

Cr_2O_3	25% (min)
Basicity	33–1/3%
pH of 10% sod	3.0–3.5

Generally, about 8 per cent of basic chromium sulphate (on pelt weight) is used for effective tanning current. But chrome tanning methods employ commercially available tanning salts which exhibit exhaustion levels of only 50–65 per cent and the rest gets discharged into the effluent. Apart from material loss, which is very significant, such poor exhaustion nature leads to alarming environmental problems. The residual discharge of chrome into the effluent on physio-chemical end-of-pipe treatment generates chrome bearing sludge and can also lead to choking of drain pipes through which it is discharged.

In the absence of viable utilisation technologies for chrome sludge, disposal becomes a serious problem. The chromium present in the tannery effluent is a hazardous pollutant. Currently used chrome tanning salts and methods employed lead to a discharge of solutions with chromium concentrations in the range of 1500–4000 ppm. But the permissible levels of chrome in the waste-water is in the range of 0.3–2.0 ppm as per the environmental regulatory agencies in many countries.

These stipulations are difficult to comply with in the present levels of exhaustion of chromium in the tanning process. Chrome tanning also contributes significantly to TDS to an extent of 25–35 per cent (both pickling and chrome tanning put together). There are a number of viable options to overcome these problems, including:

1. Chrome recovery and reuse.
2. Partial replacement of chrome tanning agent by other tanning agents.
3. High exhaust tanning systems.

Therefore, the solution to the problem rests in the reduction of discharge of chrome in the effluent by better chrome management systems.

Chrome Recovery/Reuse

In this process, the chromium present in the spent chrome liquor is first precipitated as chromium hydroxide and then separated from other unwanted chemicals.

The precipitated chromium hydroxide is then regenerated as basic chromium sulphate by redissolving with calculated amount of sulphuric acid. Chrome tanning generates a significant amount of total dissolved solids in the form of sodium chloride, sodium sulphate, etc. The major sources of total dissolved solids is given in Tables 15.2 and 15.3. The effluents coming out from a tannery processing 1 tonne of hide/skin per day, using a volume of 35,000–40,000 litres of water contain the following levels of TDS.

Table 15.2. Major sources of total dissolved solids.

Operation	*TDS (ppm)*
Soaking	25000–41000
Liming	24000–30000

(Contd...)

Operation	TDS (ppm)
Deliming	25000–60000
Pickling	34000–67000
Chrome tanning	29000–57500
Neutralisation	6000–8000

Table 15.3. Main sources of TDS discharge in tannery waste-water.

Source	Range	Average %
Soakings	30–40%	35
Beam house operations	10–20%	15
Pickling and chrome tanning	25–35%	30
Wet finishing operations	10–15%	12.5
Effluent treatment plant (alum and other chemical dosings)	5–10%	7.5

Partial replacement of chrome tanning agent by other tanning agents

The usual practice in chrome tanning is to offer 8 per cent chrome of which only 60–65 per cent exhausts and the rest gets transferred into the effluent, which increases the level of TDS and also leads to chrome bearing sludge during the treatment of effluent. In order to reduce the problems generated by using higher percentage of chrome, it is replaced to an extent of 50 per cent with other mineral salts like those of aluminium. After the pickling process, instead of using 8 per cent chrome, 4 per cent chrome can be offered and run for some time. The 6 per cent aluminium is added to the same drum and run for some time, and then basified to a pH of 3.8–4.0.

High Exhaust Tanning Systems

There are different methods of effecting higher exhaustion levels of chrome into the leather, of which chrome exhaustion by the use of additives is one. This works by modifying the affinity of chrome tanning salt and/or by introducing more reactive sites for the chrome to reach. These are different categories of chrome exhausts aids such as:

1. Compounds that are held by the substrate (hide/skin) chemically/physically, possessing suitable reactive groups, i.e. creation of additional sites for the chromium to react.
2. Compounds which can precipitate or form insoluble complexes with chromium inside the hide/skin matrix, thereby enhancing the uptake of chromium from the tanning bath.

Protein Hydrolysate (Amino Acid Mixture)

The product contains a substantial amount of dicarboxylic amino acids with reactive carboxylic groups, which offer an excellent scope for using it as a chrome exhaust aid.

DISPOSAL OF TANNERY WASTE WITH DOMESTIC SEWAGE

Tannery waste is frequently a component of the sewage of industrial cities. In very large communities the proportion present may be so small that no interference with sewage treatment processes is noted, even when the most sensitive forms of biological treatment are used. If the tannery waste content does not exceed about 10 per cent of the total flow, and the tannery discharge is regular and well distributed, no difficulty with sewage treatment processes is likely to occur. As far as digestion of sludge from sewage containing vegetable tannery wastes is concerned, the tannery solids in the sludge can amount to as much as 40 per cent of the total solids without seriously impairing digestibility. This is not at all true of sludge derived from chrome tannery waste, which accumulates nearly all of the chromium of the waste, and may carry as much as 5 to 6 per cent chromium on the basis of solids. Any substantial amount of chrome tannery sludge may be expected to interfere with normal sludge digestion.

In many small and moderate-size communities, tannery waste has been present in amounts sufficient to interfere with sewage treatment processes. The interference may be due to one or more of the following factors:

1. Excessive alkalinity or high pH produced by the lime present in the waste, and resulting in interference with biological treatment processes.
2. Hair and fleshings, which form scums on sedimentation tanks, clog sludge-removal equipment or produce mats in digesters.
3. Lime sludges and adherent deposits, which clog sewers, or interfere with proper operation of sedimentation tanks.
4. Excessive loads of organic matter, which overload the treatment units of the plant.

There are, however, a number of plants that have been designed especially for the purpose of treating sewage containing substantial proportions of tannery wastes. A treated mixture of 10 per cent domestic sewage and 90 per cent tannery wastes by sedimentation in tanks of approximately 2 hours retention. The mixed waste had a pH value of 9.0, a 5-day BOD of over 600 ppm, and a maximum hourly suspended solids content of 1,200 ppm. No values were given for the effluent from the plant. The tanks are equipped with mechanical scrapers for sludge removal and the sludge is dewatered on vacuum filters with the addition of 6.4 per cent of lime on the beds of dry solids. The filter cake is readily disposed of on farm land.

There is no reason to believe that mixtures of tannery wastes and domestic sewage cannot be adequately treated in properly designed plants, provided sufficient attention is given to the peculiar properties of the wastes. Provisions for the avoidance of sludges of waste in the discharge, for pH control and for handling a sludge heavier than that obtained from normal domestic sewage should obviously be incorporated in the design. Avoidance of sludge digestion, and possibly of secondary biological treatment, is indicated if chrome tanning wastes are present in substantial quantities. Otherwise, provision of adequate capacity of the treatment units would seem to be the only requirement

SEPARATE TREATMENT OF TANNERY WASTE

The treatment of tannery waste for disposal into water courses usually involves one or a combination of the following techniques:

1. Collection and storage of wastes, with regulated discharge adapted to the conditions of the watercourse. This technique be applied to the entire waste, or it may be restricted to one or more of the concentrated waste fractions.
2. Plain sedimentation, in basins ranging from simple earth lagoons to tanks with mechanical sludge removal. Obviously, sedimentation is also involved in (1), but it is not the primary objective, and the accumulated sludge is often discharged to the watercourse along with the wastes.
3. Coagulation with chemical, followed by sedimentation. Among the coagulants that have been used are alum, iron salts, and carbon dioxide in the form of flue gas, with or without the addition of lime.

4. Biological treatment methods, including sand filters and trickling filters. Biological methods have been confined largely to pilot plant studies, mainly because of the difficulty and expense involved in treating straight tannery wastes by such methods.
5. Present practice in waste treatment is largely a matter of adapting one or more of the first three of the techniques cited above, as is shown by the examples given below.

Storage and Sedimentation

The majority of 'tanneries which practice waste treatment of any kind have resorted to processes which are made up of various combinations of storage, sedimentation, and regulated discharge. In many instances, holding tanks are used which serve only to regulate the discharge into the stream and to extend it over a longer period of the day, making no attempt at removal of the suspended and dissolved materials. Little attention is given to sludge accumulations until it becomes necessary to remove them because the capacity of the basin is no longer adequate. In others, the weaker wastes are discharged directly and only the stronger wastes, usually the so-called intermittent wastes, are settled and subjected to regulated discharge Further refinements may be introduced, including selective mixing of components of the waste to produce mutual precipitation. Specific examples are given below.

In tanneries a simple system of 3 earth lagoons, each capable of holding an entire day's flow. These are operated in rotation on the fill-and-draw principle, filling during one day and discharging over the next 24-hour period, thus leaving the third day available for sludge removal. If desired, concrete tanks equipped with sludge collectors may be substituted, to avoid the labour involved in hand-cleaning the lagoons. In a system of this type, removal of 80 to 90 per cent of the suspended solids and of 40 to 50 per cent of the BOD may be expected.

To eliminate a larger portion of BOD, and to reduce the amount of colour in the effluent, vegetable tan liquors may be placed in separate lagoons to be discharged only during periods of high water in the stream. The design of a treatment plant of this type has been described in detail. A flow diagram is shown in Fig. 15.2. The plant, designed for 2,00,000 gallons of waste per day, has 2 earthen settling basins, each of a 5,00,000-gallons capacity, to be operated as fill-and-draw units over a 24-hour period. The basins are provided with swing draw-off pipes for collection of effluent. Sludge will be removed by gravity flow to a storage area of large capacity.

A separate lagoon for the spent tan liquors will provide storage for 200 days of normal tannery output, with controlled discharge to the stream to provide a dilution of at least 1 to 50,000. At times of very low stream flow, the 'scrub house' or bleaching wastes will also be put in the spent tan lagoon to ensure a minimum of colour in the waste discharge.

A certain amount of mutual precipitation can be secured by adding the vegetable tan wastes in controlled amounts to the alkaline beamhouse wastes and settling for 24 hours or more in fill-and-draw tanks. Eldridge has devoted considerable attention to this process, and has developed both fill-and-draw and continuous-flow designs for it.

Harnley, Wagner, and Swope the pattern have varied the pattern by putting vegetable tan liquors through an old sludge bed for colour removal by adsorption on the old sludge. The effluent is mixed with the remainder of the plant waste the bulk of which is chrome-tan waste and the mixture is settled in tanks equipped with mechanical sludge removal devices. The detention periods in these settling tanks have varied from 9 to 15 hours. About 83 to 88 per cent of the suspended solids have been removed from influents ranging from 800 to 1,000 ppm of suspended solids; 63 per cent of 423 ppm of BOD was removed in one series of tests, but only 40 per cent of 335 ppm is taken out in another series.

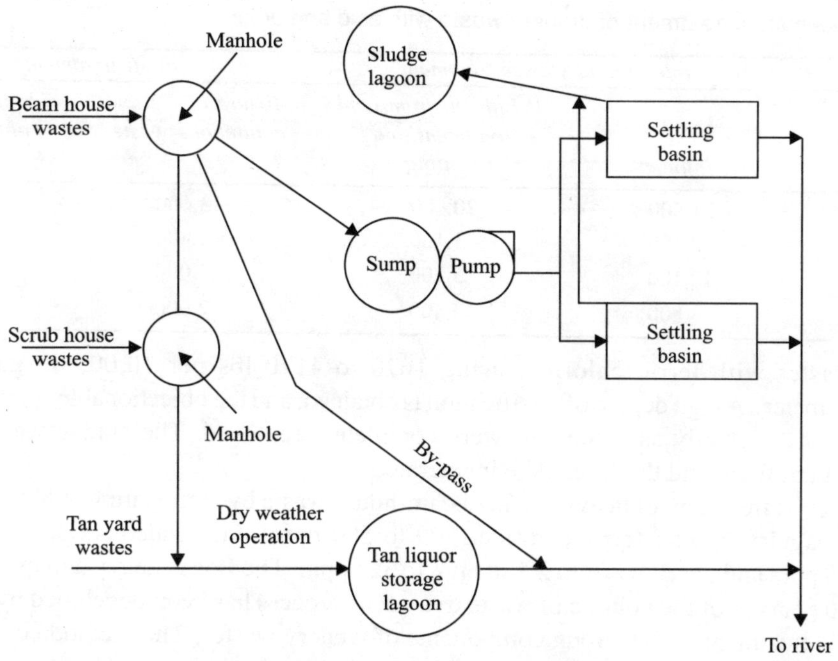

Fig. 15.2. Flow diagram of tannery waste treatment plant.

A reduction of 69 per cent of the suspended solids in a side-leather tannery waste (influent 1200 ppm, effluent 370) ppm by plain sedimentation for 3 hours. This is representative of what may be accomplished by short-period sedimentation.

Chemical Precipitation

Where the degree of treatment obtainable by plain sedimentation is insufficient some tanneries have considered coagulation by the addition of chemical. The coagulants that have been tried include sulphuric acid, carbon dioxide from flue gas, alum, and iron salts. Mechanical flocculation has sometimes been used as a part of the treatment. In general the improvements observed in the effluents as compared with those .obtained by plain sedimentation, are confined to better removals of suspended solids and possibly of colour. The additional BOD removals secured are often disappointing,

Acid and alum on a pilot-plant scale can be used. By mixing the intermittent wastes with air, adjust them with sulphuric acid to a pH of 5.5, settle the acidified wastes overnight, and added the supernatant liquor from this treatment to the continuous wastes. Then coagulate the mixture with alum and pass it to a second stage of sedimentation. The quantities of chemical used were approximately 2300 lbs of sulphuric acid and 2700 lbs of alum per l,000,00 gallons of total waste. The results obtained on a waste are briefly summarised in Table 15.4.

The volumes of sludge produced in this treatment are of interest about 11,000 gallons of sludge is secured from each 100,000 gallons of total waste treated. Of this, about 8000 gallons is from the acid pretreatment of the 43,000 gallons of intermittent waste contained in 100,000 gallons of total waste, and the remainder developed from the alum treatment of the total mixture.

Table 15.4. Experimental treatment of tannery waste with acid and alum.

	Intermittent wastes treatment		*Final treatment*	
	1. Influent	*2. Effluent from acid treatment and settling*	*Mixture of (2) and continuous waste*	*Final effluent after alum and settling*
Total solids	21,600	20,600	13,600	6,500
Suspended solids	2,500	1,100	2,800	580
Colour	12,100	5,700	3,000	850
5-Day BOD	3,800	3,400	2,000	1,030

Tannery wastes with ferric chloride, using 1670 to 4170 lbs per 10,00,000 gallons (200 to 500 g-per cubic meter). A high degree of clarification is obtained, and the objectionable sulphur compounds present in the waste, chiefly as sulphides, were completely removed. The sludge was dewatered by means of a vacuum filter, and the filter cake incinerated.

A high degree of treatment of heavy-leather beam-house waste by carbonating to 8 pH value of about 6.0 and then coagulating with ferric chloride (300 to 500 ppm). Suspended solids are reduced from 6190 ppm to 58 ppm and the BOD from 2,180 ppm to 325 ppm. The floc is heavy and quick settling, and amounted to 10 per cent of the volume of waste treated. A process has been developed by using flue gas and lime for the treatment of the strong components of tannery wastes. The method consists of mixing the intermittent wastes in a basin holding 24 to 36 hours' flow, and bubbling in due gas until the pH value of the mixture has been reduced to 6.4 to 6.7. Lime suspension is then added until phenolphthalein alkalinity appears, and the waste is settled for 3 hours The supernatant is then drawn off and subjected to a second stage of treatment identical to the first. The pilot plant, taking an influent of 13,400 ppm of BOD, 64,800 ppm of colour, and 17,000 ppm of total solids, produce a first stage effluent containing 3000 ppm BOD and 25,850 ppm of colour. The second-stage effluent had 1140 ppm of BOD, 427 ppm of colour and had 8,500 ppm total solids. The method would appear to have practical possibilities, since both flue gas and lime will ordinarily be available at most tanneries.

The waste from sheepskin tanneries, can be coagulated with lime to secure an increased removal of BOD and suspended solids. The use of fill-and-draw sedimentation tanks rather than continuous-flow tanks because of the rapid changes in the character of the waste is recommended. Some of the results are shown in Table 15.5. In this waste, coagulation with lime appears to improve the removal of BOD to a greater extent than it does suspended solids. Sludge amounts to about 5 per cent of the volume of the waste. The successful use of 300 ppm of chlorine is used to correct odour conditions due to tannery waste in a receiving stream.

Table 15.5. Treatment of sheep skin tannery waste.

Method of treatment	*Influent (ppm)*	*Suspended solids effluent, (ppm)*	*Reduction (%)*	*Influent (ppm)*	*BOD effluent (ppm)*	*Reduction (%)*
Plain sedimentation						
Continuous flow	1184	680	43	1046	537	48
Fill-and-draw	1880	461	67	1285	873	30

(Contd...)

Method of treatment	*Influent (ppm)*	*Suspended solids effluent, (ppm)*	*Reduction (%)*	*Influent (ppm)*	*BOD effluent (ppm)*	*Reduction (%)*
Lime treatment						
Continuous flow, ppm lime	918	469	46	1001	476	55
Fill-and-draw ppm lime	1980	497	73	1630	823	49

Sludge Disposal

All methods of treatment of tannery wastes involving sedimentation, with or without the addition of chemicals, result in rather formidable quantities of sludge, amounting to 5 to 10 per cent of the total volume of the waste. The disposal of this sludge in an acceptable manner frequently presents a more difficult problem than does the treatment of the waste.

Probably the most common practice is to discharge the sludge to lowlying areas of waste land or to specially prepared sludge lagoons. This is feasible only if sufficient land area can be provided. A thick sludge will require a smaller area, but thick tannery sludge is notoriously difficult to pump. Mention is seldom made of drainage from sludge disposal areas, but such drainage occurs, and may in part defeat the purpose of treatment if it reaches the stream which the treatment is designed to protect, The partially dried material from these disposal areas has some fertilising value and, in addition, contains about 10 per cent of lime. In many places it is taken by farmers as fertiliser and soil-conditioner.

A somewhat more satisfactory method of sludge disposal is the provision of drying beds of sand or cinders, with or without underdrains. The Winslow Brothers and Smith plant, described previously, Sand beds are provided with skimming pipes arranged so that the top water from the beds could be drawn back into the primary sedimentation tanks. This practice is recommended by other engineers who have concerned themselves with tannery wastes. Siebert has recommended that beds be provided on the basis of 1 sq ft of area per lb of hide processed per day. The wet sludge depth on the beds should not exceed 2 ft and that the sludge can be removed from the beds at a moisture content of 80 per cent.

Vacuum filters have been used in a few places for dewatering tannery sludges, but have not been particularly successful. The sludges appear to clog the filter media quite readily, and the usual sludge-conditioning chemicals do not effect much improvement. There is a possibility that inexpensive filter aids or waste products from the tannery, such as buffer dust and shavings might improve the filtrability of the sludge and make the use of vacuum filters practicable.

Secondary Treatment

Treatment of tannery waste beyond the level of plain sedimentation or chemical precipitation has heretofore been seldom employed, and authorities in charge of pollution control have been reluctant to force it on the tanneries because of the great expense involved. However, some form of secondary treatment of tannery waste is necessary in situations where the degree of dilution available in the stream is small. The following steps are involved:

1. Mixture of spent tan liquors and lime sludge followed by 24-hour sedimentation.
2. Treatment of the remaining mixed tannery waste, plus the effluent from step (1) with 10 grains of copper as per gallon followed by sedimentation for 4 hours.
3. Treatment of the effluent from (2) on trickling filters using screened cinders or coke as medium, at rates of 0.1 to 0.3 million gallons per acre per day.

4. Treatment of the affluent from (3) on sand beds at rates of 0.29 to 0.43 mgad.
5. Drying of sludge from steps (1) and (2) on sand beds.

In any case it is evident that proper secondary treatment can produce from tannery waste an effluent approximating that obtained from a plant providing complete treatment for domestic sewage except for the residual colour, which appears difficult to remove. The emission factors for various operations are shown in Table 15.6.

Table 15.6. Emission factors for various operations.

Parameter	*Soaking*	*Liming*	*Deliming*	*Pickling*	*Vegetable tanning*	*Chrome tanning*	*Dyeing and fatliquoring*	*Composite (incl. washing)*
BOD	83.0–18.8	17.5–35.0	1.5–4.5	0.3–0.5	9.0–27.0	0.5–1.2	1.5–3.0	35.0–105.0
COD	22.5–45.0	35.0–87.5	3.8–10.5	0.8–2.3	22.5–60.0	1.5–3.8	3.8–10.5	87.5–280
Total solids	262.5–415.5	105.0–175.0	6.0–15.0	26.3–52.6	37.5–90.0	45.0–90.0	6.0–15.0	525.0–875.0
Dissolved solids	240.0–360.0	84.0–105.0	3.8–9.0	25.5–50.3	31.5–75.0	43.5–86.3	5.1–13.5	455.0–735.0
Suspended solids	22.5–52.5	21.0–70.0	2.3–6.0	0.8–2.3	6.0–15.0	1.5–3.8	0.9–1.5	70.0–140.0
Chlorides as Cl	112.5–225.0	14.0–28.0	1.5–3.0	15.0–3.8	1.5–3.8	23.0–38.0	0.8–1.5	210.0–332.5
Total chromium as Cr.	–	–	–	–	–	3.0–7.5	–	3.5–8.8

Pollution abatement

Considerable reduction in pollution can be achieved if cleaner tanning processes are adopted. In Table 15.7, a series of options have been indicated for different stages of tanning process which result in significant reduction of pollution. Capacity to absorb and adopt these will depend on a variety of factors — such as, machinery and equipment, monitoring systems, process control equipment, technical personnel available and motivation—to cite a few. Suffice to say that with appropriate policy initiatives, positive and negative motivation could be provided to encourage tanners to adopt these progressively.

Table 15.7. Reduction of pollution obtained in cleaner production method.

Subject	*Results*
Curing	
Fresh hides and skins	No salt pollution in soaking
Cooling	No salt pollution in soaking
Partial salt elimination	Less salt pollution (10–15%)
Antiseptics	Lower toxicity

(Contd...)

Subject	*Results*
Soaking	
Antiseptics	Lower toxicity
Fleshing	Upgrading waste and lower chemicals consumption
Unharing-liming	
Enzymatic treatment	Lower toxicity and organic pollution
Hair saving methods	Lower nitrogen and organic pollution
Direct recycling	Lower chemicals and organic pollution
Splitting of limed pelts	Upgrading waste and lower chemicals
Deliming	
Carbon dioxide	Less nitrogen (–40%)
Weak acids	Less nitrogen
Tanning	
Pickling float recycling	Less salt pollution (–20%)
Solvent free degreasing	Lower toxicity (75%) and COD
Wet white production	Upgrading waste
Tanning float recycling	Chromium recovery (90%)
Partial recycling	Partial chromium recovery (60%)
Precipitation method	Total chromium recovery (99%)
Improvement in exhaustion	Good chromium uptake (95%)
Other mineral tanning	No chromium tanning
Three compartment drums	Improvement in chromium recovery
Vegetable tanning	
Dry drum	Good vegetable tanning uptake (95%)
Close circuit in pits	No effluent process and lower COD
Dyeing	
Environmentally acceptable chemcials	No heavy metals and lower AOX
Three compartment drums	Better uptake of chemicals
Through feed machine	No dyes in effluents
Finishing	
Water based finishes	No VOC emissions
Roller coating and curtain coating	Better uptake of chemicals

Set of discharge norms prescribed by MINAS for tannery effluent in India may be seen at Table 15.8.

Table 15.8. Minimal national standards (MINAS) for tanneries.

Parameter	*Limits not to exceed*
pH	6.5–9.0
BOD_5, 20°C, mg/l	100
Total suspended solids Mg/l	100

(Contd...)

Parameter	*Limits not to exceed*
Sulphides, mg/l	2
Total chromium, mg/l	2
Oil and grease, mg/l	100

Note : (For effluent discharge into land waters, BOD limit may be made stricter to 30 mg/litre by the concerned State Pollution Control Boards).

One saving grace of this industry is that there is no serious gaseous emission. It may however be mentioned that in some states where the treated effluent is discharged in surface water, more stringent norms are prescribed, particularly with reference to TDS and chloride.

Rationale for discharge norms

The rationale for discharge norms for effluent will depend on the nature of recipient. Depending on whether the recipient is sewer, sea, surface water or marine coastal water, the discharge norms vary. In some countries the discharge norms vary from location to location, depending on the carrying capacity of the recipient. Though in many developing countries very stringent norms are prescribed, in developed countries a much more rational view is taken. Environment protection is a paramount duty of any government. At the same time, employment generation and economic development are also equally important. One at the cost of other is not thinkable in modern society.

Leather industrial sector in India is employing over 1.7 billion persons and its annual turnover is estimated at over hundred billion rupees. Export from this sector is estimated to be over Rs. 60 billion, with potential to grow fruther. In other developing countries too, these vital economic benefits of the industry are realised. Accordingly, a fine balance must be struck. In developing countries, a gradual approach towards stringent norms will yield encouraging results.

Situation in India

Though tanneries in India were late starters in the area of environment protection, of late considerable development has taken place. For example, today, Tamilnadu boasts of having no tannery without access to an effluent treatment facility. In Uttar Pradesh, Kanpur and Unnao tanneries have also overcome this problem. So also the tanneries in Jullundur and Bangalore. Calcutta is the only location where the tanneries still operate without proper effluent treatment facility. It is hoped that sooner than later, here too, tanneries will be able, to overcome this problem when the Calcutta Leather Complex is completed.

Common effluent treatment plants, to treat effluent generated by a cluster of small scale tanneries, have been set up in large numbers in different parts of India. These pose interesting technical and managerial challenges but the Indian experience in this regard is noteworthy.

Residual problem

Solution of one problem gives birth to another. Disposal of treated effluent and sludge (classified as hazardous due to presence of chromium) is increasingly posing a problem. Experiments are underway as to how best to use the treated effluent for irrigation purposes and sludge for any useful purposes such as making brick for compost. Tamil Nadu Agricultural University, Coimbatore is working closely with All India Skin and Hide, Tanners and Merchants Association, Madras and UNIDO's Regional Programme Office at Madras, on these projects. What appear as major problems today will soon become golden opportunities for societal development. From being a perceived foe, leather industry soon will become a sought after friend of society.

A word of caution

It has been established that solid and liquid wastes generated by tanneries can be safely and cost effectively handled and disposed. However, this is not possible without full cooperation of the tanners, small and big. Introduction of cleaner tanning processes must not be looked upon as an unnecessary burden but as a necessary process modification to attain sustained growth of the industry. Any lethargy today could spell doom for tomorrow. Likewise, effluent treatment plants are like any other manufacturing unit. An effluent treatment plant receives waste water as raw material and produces relatively cleaner water through its processes. In operation and maintenance of this plant, as much care is needed as in any other manufacturing plant. Tanners must treat the effluent treatment plant very much as part of the tannery and not as a separate and outside entity.

Future

With the currently available technology and resources, it is eminently feasible to treat the tannery effluent to required standards. And cost effectively too. Continuous research work is going on how best to reduce cost, introduce natural treatment systems and find beneficial uses for the resultant wastes. UNIDO through its regional programme for pollution control in the tanning industry in South East Asia has been grappling with many of these new challenges. It has been well understood that any country or region which takes a lead in this regard will be able to take a leap forward compared to others who fail to do so.

Sustained growth of this industrial sector anywhere in the world presupposes effective addressing of this problem. Indian leather industry, as of today, seems to have more or less got an upper hand in dealing with this problem. With sustained efforts, it will be able to totally overcome this problem and thus lay the base for continued growth of this industrial sector.

DEVELOPMENT OF NEUTRALISING AGENT FOR CHROME TANNED LEATHER FROM THE WASTE LIQUOR OF PENTAERYTHRITOL MANUFACTURE

Waste liquor from pentaerythritol manufacture containing huge amount of sodium formate as by-product was modified to develop neutralising agents for chrome tanned leathers. Chrome tanned leathers were treated with the neutralising agent, dyed, fat liquored and finally dried. Dried leathers were found to be fuller and brighter in colour than the leather treated with ordinary alkali-sodium bicarbonate as neutralising agent and found similar result in comparison to the leather neutralised with a standard neutralising agent available in the market.

The waste liquor of pentaerythritol manufacture contains a considerable amount of pentaerythritol and sodium formate which is not possible for further recovery from the economical point of view. Also due to pollution effect the accumulation of the waste becomes a greatest headache for the industries having pentaerythritol plant. Therefore it becomes necessary to find out a viable solution to utilise or treat this waste to control the pollution effect for the survival of these industries. However, there is possibility to utilise this for the development of various leather chemicals. On the other hand, neutralisation of leather—particularly of chrome tanned leather prior to dyeing and fat-liquoring is of great importance if satisfactory results are aimed at. It has been studied that chrome salts bound by the hide-protein continue to undergo changes due to hydrolysis accompanied by the formation of free acid in the leather while the chromium compound itself becomes increasingly basic. In wet chrome leather this process of hydrolysis is a continuous one.

The main object of neutralising chrome leather is to reduce this excessively positive charge to such an extent that the negative dyestuff ions, and also the similarly charged particles of the fatty emulsion do not combine on the surface of the leather too rapidly and unevenly. However, neutralisation of chrome tanned leather is done by using various neutralising agents such as sodium formate, borax, sodium acetate etc.

Since waste liquor of pentaerythritol production contains huge amount of sodium formate as by-product which is difficult to recover from economical point of view as well as due to the pollution problem caused by this waste liquor, the same can be utilised for developing neutralising agent for chrome tanned leather.

Therefore, this section deals with the utilisation of waste liquor from pentaerythritol manufacture to develop neutralising agents for chrome tanned leather.

Principle

The reactivity of a number of organic materials with chromium in solution has been studied which are used to control the chrome tanning process. Such materials are called masking agents because they modify the chromium tanning reaction to give improved tannage characteristics. The sensitivity of a tannage to pH variations can be greatly decreased by the use of these masking agents. These are additives to the chrome and at a slightly high pH than in an unmasked tannage. The conditions of the chromium salts in the solution can be represented by the following three stages:

Solution

The basic chromium sulphate dissolves and as the sulphate ionises, a cationic chromium complex is formed.

Masking with sodium formate

The formate ion in the solution forms a basic chromium formate complex with the displacement of some of the sulphate from the chromium complex.

Fixation

The cationic chromium complexes react with the anionic carboxylic groups of the acid amino acids of the hide protein in the initial tanning reaction.

The above three stages can be represented as:

Solution

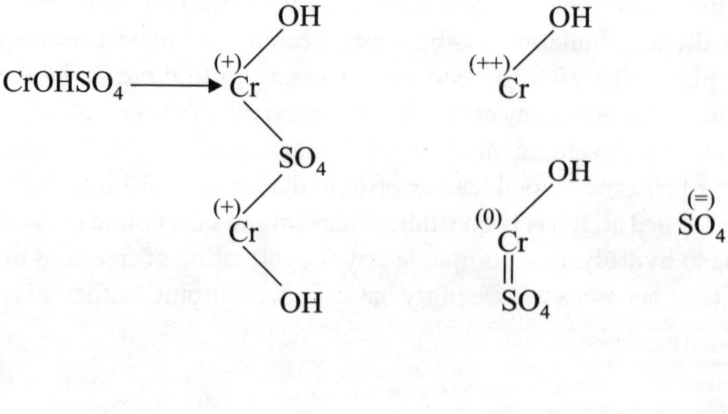

Masking with sodium-formate

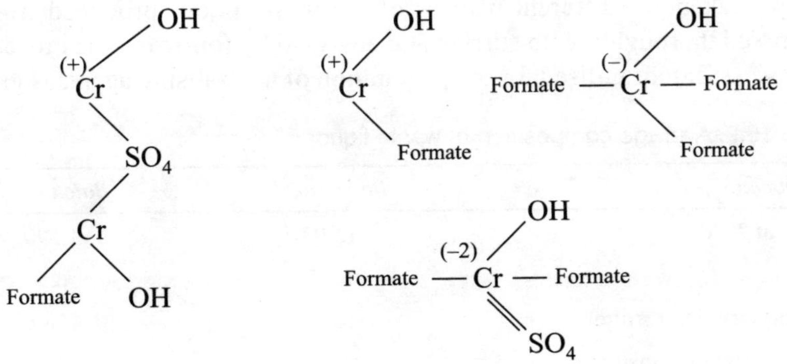

Fixation

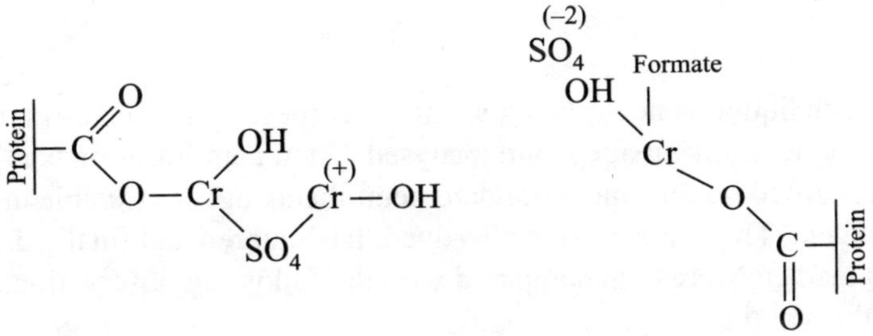

The fixation of chrome tanning material as related to masking and pH is given in Fig. 15.3. The fixation of unmasked chrome tanning material by the hide is very sensitive to pH. Above a critical point precipitation takes place and fixation is sharply decreased. Out of two masking agent, the reaction is less pH depended with formate masking.

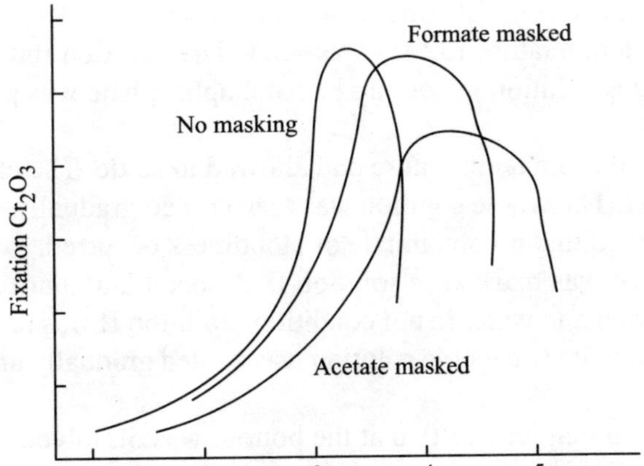

Fig. 15.3. Effect of masking on the fixation pH characteristic of chrome tanning.

Analysis of Waste Liquor

A number of sample taken for different batches of the waste liquor produced from pentaerythritol manufacture was mixed thoroughly with stirring and analysed by following the procedure. The average composition of the waste liquor utilised for the preparation of neutralising agents is given in Table 15.9.

Table 15.9. Average composition of waste liquor.

Component	*Batch no. 1*	*Batch no. 2*
Sp. gr. at 20°C	1.292	1.290
pH	6.55	6.5
Pentaerythritol, gms/litre	164.3	128.6
Sodium formate, gms/litre	417.24	404.8
Solid, wt. %	60.1	56.5

Experimental

To see the effect of waste liquor as neutralising agent for chrome tanned leathers, the waste liquor was modified to be used as a neutralising agent and analysed. Then neutralisation was done on the chrome tanned leathers with modified liquors and a standard neutralising agent available in the market and with ordinary neutralising agent. The leathers were also dyed, fat-liquored and finally dried to make finished leathers. The finished leathers were then compared with the following observations:

1. pH of the leathers at the surface and centre.
2. Fullness of leathers.
3. Brightness and uniformity of colour of finished leathers.

Preparation of modified waste liquor as neutralising agent

A fixed amount of waste liquor taken in a 4 necked round bottom flash was dissolved by adding water and formic acid gradually with constant stirring. It was then heated to 50–55°C to obtain clear solution. pH of the solution at room temperature (32°C) was 4.25. The solution thus prepared was marked as solution I. 23 per cent aqueous solution of sodium hexametaphosphate was prepared by dissolving it in hot water with stirring.

The solution was cooled to room temperature and allowed to settle. The clear solution was decanted and poured into the solution I. The whole solution was then cooled gradually to about 15°C and kept for abut 2–3 hours. It was observed that no precipitate or cloudiness occurred.

The solution thus prepared was marked as solution II. A specific amount of phthalic anhydride was dissolved in caustic lye solution and water in hot condition. Solution II was then heated to 45°C and then the above solution was added to it. The whole solution was cooled gradually and become cloudy at about 40°C.

On further cooling all the precipitate settled at the bottom was dissolved by adding a mixture of 1:4 formic acid and water. The pH of the solution was then raised to about 6 by adding 46 per cent caustic lye. The materials required for the preparation of two batches of modified waste liquor required are given in Table 15.10.

The experimental set-up for the preparation of modified waste liquor is given in Fig. 15.4.

Table 15.10. Materials required for preparation of neutralising agent.

Materials required	*Batch no. 1*	*Batch no. 2*
Waste liquor	200 gms	200 gms
Formic acid	42 gms	35 gms
Sodium hexametaphosphate	6 gms	6 gms
Caustic lye	6 gms (46.5% conc.)	7 gms (46% conc.)
Phthalic anhydride	6 gms	6 gms

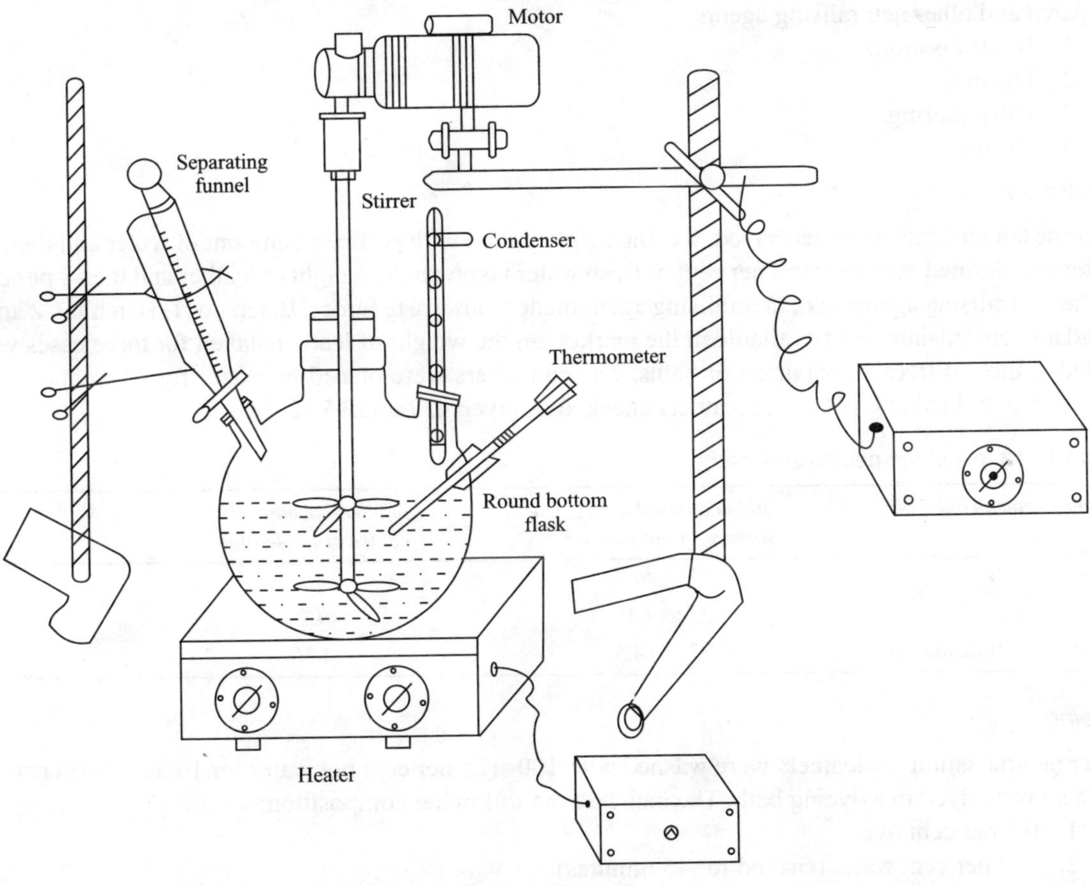

Fig. 15.4. Experimental set-up for preparation of neutralising agent from waste liquor.

Analysis of neutralising agent prepared

The analysis of modified waste liquor was done by following the procedure as reference and the analytical results in given in Table 15.11.

Table 15.11. Analytical result of modified waste liquor.

Component	*Batch no. 1*	*Batch no. 2*
pH	6.0	6.1
Sodium formate (wt. %)	32.7	29.4
Pentaerythritol (wt. %)	12.3	12.2
Solid content (wt. %)	44.45	44.4

Applications of the neutralising agents on the chrome tanned leathers

The following operations were performed to make finished leather by applying the modified waste liquor prepared and other neutralising agents:

1. Neutralisation.
2. Dyeing.
3. Fat-liquoring.
4. Drying.

Neutralisation

Chrome tanned leathers were shaved, weighed and washed with sufficient amount of water and then the water was drained out. 100–150 per cent of fresh water taken on the weight of leather and then 3 per cent of the neutralising agents viz., neutralising agent made from waste liquor, Batch no. 1, Batch no. 2 and a standard neutralising agent available in the market, on the weight of leathers taken for three cases were added in three different containers or baths. The containers were placed in shaker for 45 minutes. The baths were drained out, pH of the leathers checked as given in Table 15.12.

Table 15.12. pH of the neutralised leathers.

Leather neutralised with	*pH decreased at the surface of the leather*	*pH decreased at the centre of the leather*
Batch no. 1	7.275 to 4.3	4.1
Batch no. 2	7.25 to 4.45	4.2
Standard neutralising agent	7.375 to 4.5	4.35

Dyeing

After neutralisation the leathers were washed with 100–125 per cent hot water for 10 minutes. Then the leathers were dyed in a dyeing bath. The bath had the following composition:

1. 0.5 per cent dye.
2. 100 per cent water (shaked for 45 minutes).
3. 0.5 per cent formic acid (for fix-up).

The leathers were taken up from the dye bath and the cut portion of the leathers were checked for dye penetration.

Fat-liquoring

After dyeing the leathers were washed with 100–150 per cent hot water and then fat-liquoring was done.

The baths used for fat-liquoring had the following composition:

1. 7 per cent fat-liquor.
2. 100 per cent water (shaked for 45 minutes).
3. 0.5 per cent formic acid (added gradually).
4. Drying.

The leathers were then dried. The same operation was done with ordinary alkali, sodium bi-carbonate.

Result

Dyed leathers were found fuller and brighter in colour in comparison to the leather neutralised with ordinary alkali viz., sodium bicarbonate, and found similar result in comparison to the leather neutralised with other standard organic salts (masking salts) available in the market.

Discussion

It has been found that in normal tanning procedure sodium salts are occasionally added during neutralisation and as well as in the tanning process of the chrome tanned leather. Hence, as waste liquor from the pentaerythritol plant contains a considerable amount of sodium formate, the same can be utilised for leather industry.

Different types of ordinary neutralising agent viz., sodium bicarbonate, borax, etc., does not make the leather so full, whereas 'neutralising agent' prepared from the waste liquor neutralise uniformly and make leather full. Further after neutralising with the modified waste liquor, the leathers were dyed and the dyed leathers were found brighter and uniform in colour after drying.

Thus the neutralising agent prepared from the waste liquor has a beneficial point of view that at the time of neutralisation of the chrome tanned leather, no further addition of any other organic salts are necessary, which are usually added by the tanners in their normal tanning procedure to make leather full and more even substance.

Further from the present experimental work it is found that the pollution effects of the waste liquor from the pentaerythritol manufacture can be controlled successfully by utilising the waste liquor for the development of neutralising agent which would be an useful industrial chemical for leather industry also.

UTILISATION OF WASTE LIQUOR FROM PENTAERYTHRITOL MANUFACTURE AS REDUCING AGENT FOR CHROME TANNING MATERIAL

This section deals with the development of chrome tanning material by utilising the waste liquor from pentaerythritol manufacture as a reducing agent and application of the same on pickled pelt leather. For these, standard B and C liquor and B and D liquor with waste liquor were prepared by the reduction of sodium dichromate and analysed. The chrome liquor made with waste liquor was found to have a better penetration point of 6.5 than the standard basic chrome liquor for which it is 4.9. The experimental results show that the chrome liquor prepared with waste liquor will be quite beneficial to the leather industry because of having several advantages like — quicker tannage as evidenced by boiling test of the tanned leather, uniform penetration of the liquor, no chance of case hardening and considerable saving in the amount of masking agent. The chrome tanned leather was also found to be fuller than the leather treated with standard chrome liquor.

Pentaerythritol [$C(CH_2OH)_4$], is manufactured from acetaldehyde and formaldehyde using soda lye as catalyst. The by-product of this manufacture is sodium formate. Though most of the pentaerythritol and sodium formate are recovered from its mother liquor by successive evaporation, crystallisation and centrifuge, the waste liquor still contains a considerable amount of these two materials which is not possible for further recovery so far as economy is concerned. Moreover, due to acidic in nature, it causes a pollution or disposal problem to the industries having pentaerythritol plant. Therefore, it is necessary to utilise or to treat this waste to control the pollution effect for the survival of these industries.

The syrapy waste liquor from the manufacture of pentaerythritol can be used for the preparation of various leather chemicals viz., chrome tanning material, fat-liquoring component etc. Therefore, in the present study efforts has been made for the development of chrome tanning material to be used in leather industry by the reduction of Cr^{6+} with waste liquor as reducing agent.

Analysis of Waste Liquor

A number of samples taken from different batches of the waste liquor produced from pentaerythritol manufacture was analysed by following the procedure, the results of which are given in Table 15.13. The samples were then well mixed in equal proportion with stirring thoroughly. Besides above, the waste liquor contains polypentaerythritol, methylether, formals of pentaerythritol, gums and formose sugar. *Lichvar* also reported that the waste liquor contains C_{3-8} hydroxy-aldehyde and hydroxy ketones.

Table 15.13. Analytical results of waste liquor.

Sample no.	*Sp. gr. at 20°C*	*pH*	*Pentaerythritol gms/litre*	*Sodium formate gms/litre*	*Solid wt. %*
1	1.290	6.5	126.80	398.80	58.0
2	1.286	6.45	102.39	413.82	52.5
3	1.287	6.5	109.60	409.86	54.9
4	1.292	6.55	164.33	417.24	59.4

Principle and Procedure for Preparation of Chrome Liquor

Chrome liquor which contains basic chrome sulphate $Cr(OH)SO_4$ is prepared by acidifying dichromate of sodium or potassium and subsequently reducing the chemicals by an organic and inorganic reducing agent viz., glucose, cane-sugar, molasses, sodium bisulphite etc. The term reduction means the change of acidic hexavalent chromium (Cr^{6+}) to basic trivalent chromium (Cr^{3+}) which forms the basic salt reacting with acid used. The basic salt in the chrome liquor is used for tanning hides and skins. The chrome has a definite basicity and pH under definite condition. The following reaction which takes place when sulphuric acid and sodium dichromates are used:

$$Na_2Cr_2O_7 + 2H_2SO_4 = 2H_2CrO_4 + Na_2SO_4 + SO_2 + O$$

$$2H_2CrO_4 + SO_2 + O + H_2SO_4 + \text{Reducing agent} = 2Cr(OH)SO_4 + Na_2SO_4 + 3H_2O + 3O$$

Since syrapy waste liquor of pentaerythritol plant contains a considerable amount of sugar materials which can be utilised as reducing agent for the above reaction. Further, a number of different organic materials viz., sodium formate, sodium acetate etc. are generally used to control the chrome tanning process. Such materials are called masking agent because they modify or mask the chromium tanning reaction to give improve tannage characteristics. The sensitivity of the tannage to pH variations can also be greatly decreased by the use of these masking agents. Therefore, the waste liquor containing a huge amount of sodium formate can be utilised as additive to the chrome tanning bath.

Experimental

To see the effect of waste liquor on the reduction of dichromate for preparing basic chrome (B and C) tanning material, and standard B and C liquor and a B and C liquor with the waste liquor were prepared, analysed and applied on pickled pelt leathers. The performance of these basic chrome tanning materials was examined by the 'precipitation figure'. Finally, the chrome tanned leathers were compared by the 'boiling test'.

Preparation of B and C liquor

Standard B and C liquor

75 grams of sodium dichromate was dissolved in water. The 75 grams of conc. H_2SO_4 (98 per cent) was added as a thin stream with stirring thoroughly. Glucose as reducing agent was added occasionally with constant stirring and heated to about 85°C. An additional amount of glucose was added until the liquor become greenish blue colour.

B and C liquor with waste

The liquor was prepared by the same method as above where waste liquor was used in place of glucose as reducing agent.

Analysis of basic chrome liquor

The analysis was done by following the procedure as mentioned above in analysis of waste liquor and the results given in Table 15.14.

Table 15.14. Analytical results of B and C liquor.

Component	*B and C liquor (standard)*	*B and C liquor (waste)*
Acidity (gms. SO_4 per gm liquor)	0.216	0.129
Basic chromium (gms. Cr^{3+} per gm. liquor)	0.114	0.0829
Basicity (%)	31.6	44.1
pH (10% solution)	2.35	3.5
Sodium formate (wt. %)	0.0	0.9
Solid content (wt. %)	68.3	61.2
Precipitation figure at pH	4.9	6.5

Precipitation figure

The fixation of chrome tanning material by the hide is very sensitive to pH. Above a critical point precipitation takes place and fixation is sharply decreased. The fixation of masked chrome tanning material is less sensitive to pH than unmasked one. The precipitation figure which can be regarded as a measure of astringency of a chrome liquor, can be defined as the amount of alkali which can be added to a chrome liquor to just bring about a cloudiness.

For this, 10 per cent (wt/wt), chrome liquor solutions were prepared and heated to 40–45°C. It was then titrated slowly with standard caustic solution until slight turbidity occurred. The result is given in Table 15.14.

Application of Chrome Liquor on Pickled Pelt Leather

Procedure

A certain amount of leathers were placed in two separate containers containing 5–6 per cent common salt and 80 per cent water based on the weight of leathers in each case. The containers were placed in shaker and shaked for about 1 hour. Then a total of 16 per cent of standard B and C and B and C with waste liquor based on the weight of leathers were added in two times separately in two containers and again shaked for about 5 hours until the cut edge of the leather were observed to be fully soaked with basic chrome liquors. For both cases, the fixation of Cr_2O_3 inside the different layers of leathers was also marked at this point of observation. About 1.5–2 per cent of sodium bicarbonate solution on the weight of leathers tanned with standard B and C liquor was added to adjust the final pH of the tanning bath to 3.3 so that the leather stand the boiling test. The final pH of the tanning bath containing leathers tanned by B and C liquor made with waste liquor was found to be 3.5 without addition of any salt solution.

Comparison of Chrome Tanned Leathers by Boiling Test

Two samples were cut out from the tanned leathers for the above two cases and their boundaries were outlined on a paper with a pencil. The samples were placed in a bowl of water kept at boiling point first for 2 minutes and then for 5 minutes after raising the pH to 3.3 of tanning bath containing leathers tanned with standard basic chrome liquor. Then the samples were taken out, cooled and placed on the tracing to check if any shrinkage of the leathers were observed as evidence by the contraction of the boundary lines for both cases. The results of this test is given in Table 15.15.

Table 15.15. Results of leather tanned with standard and waste liquor.

Leather tanned with	*pH of tanning bath*	*Boiling time (min.)*	*Shrinkage (%)*
B and C liquor (standard)	3.1	2	17.5%
B and C liquor (with waste)	3.5	2	No shrinkage
B and C liquor (standard)	3.3	5	No shrinkage
B and C liquor (with waste)	3.5	5	No shrinkage

Thus from the above experimental work, it has been found that the chrome tanning material prepared from the waste liquor of pentaerythritol plant showed a very encouraging results as compared to the performance of standard B and C liquor and it will be beneficial to the leather industry because of having the following advantages:

1. It is not necessary to basify the chrome liquor.
2. Tannage is quicker.
3. Since waste liquor contains a great amount of sodium formate, so considerable savings in the amount of masking agent required for satisfactory tannage.
4. The liquor uniformly penetrate into the leather and given fuller leather than the leather treated with usual chrome liquor.
5. No chance of 'case hardening' resulting from non-fixation of chrome salts on the grain surface which is in the case of non-masked chrome liquor.

16

Pesticides

INTRODUCTION

Pesticide is a substance, organic or inorganic, used to destroy or inhibit the action of plant or animal pests, the term thus includes insecticides, herbicides, rodenticides, miticides, etc. Virtually all pesticides are toxic to man to a greater or lesser degree. They vary in biodegradability. The noteworthy feature of our pesticide consumption is the predominance by insecticides (75 per cent), followed by fungicides (15 per cent), herbicides (7 per cent) and others (3 per cent) while it is 23 per cent, 51 per cent, 24 per cent and 2 per cent respectively in developed countries where emphasis is on fungicides.

The development countires consume about 81 per cent of the total pesticide production in the world, whereas consumption by develping countries is only 19 per cent.

India's share is meagre 2 per cent. One improtant factor for the projected decrease in the quantity of pesticide consumption in developed countries is the switchover from first generation pesticides to second and third generation types.

Developed countries have newer products and formulations which are safer, more effective and biodegradable. Among the main reasons for the poor growth of the pesticide market in developing countries is the high cost of pesticides as compared with other agricultural inputs. The low return on investment in small farm holdings directly affects the use of the pesticides.

Farm holdings in the developed countries are very large and more pesticides are used as farmers get higher returns.

The pesticide formulation industry in the country falls under two categories:

1. Large scale units in the organised sector, including multinationals.
2. Small scale industiral (SSI) units formulating different products for large scale/multinational units or marketing under their own brand name.

There are about 600 SSI units in the pesticide sector, of which about 400 are presently operational. The SSIs can be further categorised into three groups:

1. Units working mostly on a job-order basis for MNCs and large industries.
2. Units marketing products on their own, through a network of dealers and distributors.
3. Mixed units, catering partly to the job-order market but also marketing their own products.

The pesticide formulation industry is seasonal in nature and its operation depends on the requirements of the farm sector.

PROCESS DESCRIPTION

There are four common generic products in the pesticide formulation sector.

1. Granule-based formulations, such as phorate and butachlor.
2. Dust powder, for example, BHC, methyl parathion, fenvalerate, etc.
3. Liquid formulations, namely, malathion, fenvalerate, cypermethrin, endosulphan, monocrotophos, etc.
4. Wettable powder, for example, mancozeb, malathion, etc.

The process basically consists of mixing (blending) the basic pesticide (technical), filters (raw granules, pulverised soapstone, china clay, solvents, etc.) and adjuvants (stabiliser, emulsifier, etc.) as per the product recipe, to a targetted technical pesticide content, commonly known as active ingredient (AI), in the product. The formulation is essentially a batch process.

The formulation recipe is prepared considering the following factors: (i) targeted active ingredient; (ii) guaranteed shelf life; (iii) purity of technical; (iv) batch size; and (v) marketing size.

Normally, the targeted AI content is slightly higher than the prescribed values.

GRANULE FORMULATION

The main product in the granule formulation category is the 10 per cent phorate granule formulation. The formulation is carried out in a rotating drum called a blender. The most prevalent capacity of a blender is 1,000 kg/batch; however, blenders of 800 kg and 1,200 kg are also used. The time required for one batch varies from two hours to four-and-a-half hours. The variation depends on the technique of technical dosing.

Raw Materials

The raw materials used are:

1. Bentonite clay granules—carrier media.
2. Technical—basic pesticide.
3. Diethylglycol (DEG)—stabiliser/deactivator.
4. Rosin—encapsulating agent.
5. Solvent (mineral turpentine oil—MTO or *o*-xylene) — to dissolve rosin.

Process Steps

The main process steps are:

1. Raw granules weighing and charging.
2. DEG weighing and charging.
3. Technical weighing and charging.
4. Encapsulation (rosin—solvent weighing, mixing and charging).
5. Drying.
6. Reverse blending.
7. Unloading.
8. Quality check.
9. Packing.

The raw granules, after weighing, are transferred either manually or mechanically (conveyor/bucket elevator) to the blender via a charge hopper. The weighed quantity of DEG is transferred manually (using buckets) to a spray container and is charged into the blender using an air displacement pump. The requisite quantity of technical is then transferred from the drum to the spray container either manually or mechanically by a pump. The technical is sprayed in the blender through nozzles by forced or gravity spray technique. The rosin and solvent are weighed separately and mixed in a mixer. The required quantity of mix is charged into the blender through the same arrangement as that for DEG.

After charging all the ingredients, the blender is run for about 60–75 minutes. The phorate granules are then unloaded into a storage bin/big bags for a quality check. Two or three grab samples and one composite sample from each batch are analysed for the AI percentage in product and other relevant parameters.

In case the bulk product fails to meet the quality criteria, it has to be reprocessed. For reprocessing, two batches of low AI and high AI are mixed and charged to the blender for a homogeneous product. The quality of the reprocessed batch is checked again.

The product is packed manually or by automatic machine in the requisite packings.

LIQUID FORMULATIONS PLANT

Emulsion concentrates (EC) and spraying liquids are made under this category.

Raw Mateirals

The raw materials used are:

1. Technical—basic pesticide.
2. Solvents—(aromax, C-9, acetone, cyclo-hexane, etc.).
3. Emulsifier and adjuvant.

The formulation is carried out in a churner called a blender. The capacity of the blender varies from plant to plant, ranging from 1 kl to 5 kl.

Process Steps

The main process steps are:

1. Solvent weighing and charging (major quantity).
2. Technical weighing and charging.
3. Initial blending.
4. Emulsifier, additives weighing and charging.
5. Balance solvent charging.
6. Final blending.
7. Quality check.
8. Unloading.
9. Packing.

The solvent is transferred from the storage tank/drum to a weighing container either manually or mechanically. After weighing, a major portion is charged into the blender. Similarly, the liquid technical is charged into the blender. In case the technical is solid, it is charged directly into the blender manually. Subsequently, the emulsifier/adjuvants are charged. All solid materials are charged direclty to the blender and liquid materials through technical charging assembly. Finally, the balance quantity of solvent is charged to flush the line. After all the ingredients have been charged, blending is carried out. The blending duration depends upon the product and ranges from 30 minutes to two hours. The formulated bulk product is transferred to a holding container/tank for a quality check.

A composite sample of the formulated product is drawn and analysed in the plant laboratory for quality check of AI percentage and specific gravity.

If the product fails the quality check, a fresh recipe, based on the laboratory analysis, is prepared for reprocessing the batch.

The product is packed in requisite sizes either manually (volumetric filling) or on semi-automatic volumetric type filling machines. Subsequently, a leak test is carried out and finally shrink packing is applied wherever necessary.

In case of a switchover from one product to another, the entire equipment and accessories are cleaned with the appropriate solvent to avoid contamination of the next batch.

DUST/WETTABLE FORMULATION PLANT

Dust powder and wettable dust powder are made under this category. The main operation is carried out in a blender, generally of 1 or 2 tonne capcity. The batch duration depends upon the product and on the capcity of the blender and is normally about two to four hours.

Raw Materials

The raw materials used in the dust and wettable formulation plants are:

1. Pulverised soapstone/China clay/precipitated silica—filler.
2. Technical—basic pesticide.
3. Adjuvant.

Process Steps

The main process steps are:

1. Soapstone weighing and charging.
2. Additives weighing and charging.
3. Technical (soild/liquid) weighing and charging.
4. Manual pre-mixing.
5. Pre-mix pulverising.
6. Manual charging of pre-mix to blender.
7. Blending.
8. Unloading.
9. Quality check.
10. Packing.

The most common filler is soapstone. Most units have pulverising facilities while some purchase pulverised soapstone itself. For pulverising, the big lumps of soapstone are crushed in a primary crusher to bring down the size of lumps to about 1″–2″. The crushed lumps are stored in a silo. The weighed quantity of soapstone with (after manual pre-mixing) or without solid technical is charged to the pulveriser. Implex pulveriser or Raymond mill are the commonly used pulverisers. The pulverised material is pneumatically conveyed to a cyclone separator. This is collected in a cyclonic hopper, whereas the fines remain entrained in the carrier air.

The major part of this air is recirculated and the balance is vented through a bag filter. The objective of partial recirculation is to avoid the building up of fines which is not desirable from the product quality viewpoint. The pulverised material from the cyclone hopper is fed to the blender (directly or through a screw converyor). In some units, specially those with a Raymond mill, a rotary air lock valve is installed to isolate the blender from the pressurised air circuit. In units using pulverised soapstone/china clay, these are directly charged to blender, usually manually.

The liquid technical, after weighing, is charged into the blender through a pump nozzle assembly. The nozzles are fitted in a detachable pipe to facilitate their cleaning/replacement. After all the material has been charged, the blender is run for one to two hours, depending upon the product.

The blended material, which is the final product, is drawn at the discharge end of the blender directly into the bags, either manually or through a screw conveyor, and checked for quality.

Before sealing the bags, a composite sample of each batch is checked for quality, specially the AI percentages. In case the formulated product fails to meet the quality criteria, it is reprocessed.

For reprocessing, a fresh recipe is prepared as per analysis and accordingly the entire operation is repeated. When switching over from one product to another, the entire assembly line is cleaned to avoid contamination of the new product. The blender is cleaned with soapstone, the technical charging assembly with the appropriate solvent and the pumps with some cheaper solvent.

WASTE/EMISSION LEVEL

In pesticide formulation, mainly fugitive air emissions occur which are difficult to quantify. However, since the process is batch type, the losses can be worked out by input and output material balance comparisons. The losses take place because of fugitive emissions and spillages. The type of pollution and pollutants generated in different formulation plants are listed in Table 16.1.

Table 16.1. Types of pollution in the pesticide formulation sector.

Type of pollution	*Pollutant*	*Plant*
Air pollution (dominantly fugitive in nature)	Dust—soapstone, china clay, granule	Dust plant Granule plant
	Fumes—basic pesticide, solvents,	Dust plant Liquid plant Granule plant
Water pollution	None except container washing, wherever practised	–
Hazardous waste	Blender cleansing waste	Granule plant Dust plant

The levels of pollution from different types of formulation plants are given below:

Granule Formulation Plant

The fugitive emission occur during raw material weighing, charging to blender, blending, unloading of blender and packing operations. Empty drums of technical containing residual material also causes fume generation and emissions.

Both technial (phorate) and solvent (MTO and *o*-xylene) contain volatile organic compounds (VOC). During blending, vapours are generated due to natural evaporation. The normal evaporation rate measured for basic pesticide and MTO is 1.3 gm/hr.m^2 and 12 gm/hr. m^2, respectively, from stationary surface at ambient temperature. The actual evaporation rate might be different due to slightly higher temperature in blender and higher turbulence in liquid spray. The phorate concentrations in fugitive emissions from blender are given in Table 16.2.

Table 16.2. Typical phorate concentration levels in fugitive emissions from blender in granule plant.

Stage of operation	*Phorate concentration (mg/m³)*	*Time (min)*
Before material loading	182	12
Phorate spray and mixing	300	40
DEG spray and mixing	264	20
Rosin and MTO spray and mixing	210	40
Weighted time average	249 mg/m³	–
Total flow	250 m³/hr	–
Batch time	2.5 hrs	–
Total phorate emissions	156 gm/batch	–

As mentioned earlier, direct quantification of overall pollution level from different sources is not feasible but losses can be worked out by material balance. Typical overall losses from 1tonne blender is given in Table 16.3.

Table 16.3. Typical losses in granule formulation plant (product phorate 10% G).

Capacity of blender (kg)	*Inputs—raw material (kg)*	*Ouput—product (kg)*	*Losses**	
			kg	*%*
1000	1010.0	998.0	12.0	1.2

* Indicates total losses inclusive of fugitive emissions.

Dust/Wettable Formulation Plant

The fugitive emissions occur during raw material weighing, charging to pulveriser, pulverisng, blending, unloading of blender and packing operations. Empty drums of basic pesticide containing residual material also cause fume generation and emissions. Typical losses during formulation of 2 per cent methyl parathion dust powder are given in Table 16.4.

Table 16.4. Typical losses in manufacture of dust powder formulation (product: methyl parathion 2% DP).

Capacity of blender (kg)	*Inputs—raw material (kg)*	*Ouput—product (kg)*	*Losses**	
			kg	*%*
1000	816.2	753–803	13–63	1–2.6
2000	2023–2040	2000–2010	23–40	1–2

* Indicates total losses inclusive of fugitive emissions.

Liquid Formulation Plant

Fumes are generated during raw material weighing and charging into blender due to natural evaporation and spillages. Although the blender remains closed while carrying out the operation, the churning of liquid in the blender increases the temperature and turbulence which accelerates the natural evaporation rate. The losses for a typical batch of 35 per cent endosulphan EC and 76 per cent DDVP EC formulation are given in Table 16.5.

Table 16.5. Typical losses in liquid formulation.

Product	*Batch size (kg)*	*Inputs—Raw materials (kg)*	*Ouput—product (kg)*	*Losses*	
				kg	*%*
Endosulphan	2000	2020.4	2005.85	14.65	0.72
DDVP	1300	1325.4	1312.58	12.82	0.97

Apparently, the losses and hence the pollution from the pesticide formulation industry is low, ranging from 1 per cent to 3 per cent. However, the basic pesticide being highly toxic, even at this level the health and safety impact of pollution is severe, specially in and around the factory area. The different types of waste/emissions along with the source/activity for different plants is given in Table 16.6.

Table 16.6. Types of waste/emissions and source/activity.

Plant	*Types of waste/emissions*		*Source/activity of waste/emissions*
Granules	Spillages		
		Granules	Raw material—storage, weighing and charing Product—unloading, transfer to silo, packing
		Technical	Transfer from drum to spray container
		Rosin mix	Transfer from mixer to spray container during mixing of rosin-solvent
	Leakages		
		Technical	Valve fitted in pipe to transfer from drum to spary container techinical transfer pump
	Fumes		
		Techinical	Weighing container, spray container, blender
		Solvent	Weighing, mixing with rosin and blender
	Fine dust		Charging of granules, unloading of blender, floor sweeping
	Residual technical		Empty drums
	Filter and residual impurities		Filtering of technical
	Packing losses		Packing
Dust/Wettable	Spillages		
		Soapstone (lumps)	Transfer from primary crusher to silo, weighing, charging to pulversier
		(powder)	Weighing, charging to blender
		Technical	Weighing, pre-mixing with soapstone, charging to pulveriser
		(solid)	Transfer from drum to spray container, weighing
		(liquid)	
		Product	Unloading, filling, weighing
	Leakages	Technical (liquid)	Valve fitted in pipe to transfer from drum to spray container technical transfer pump
	Fumes	Technical (liquid)	Weighing container, spray container, blender,

(Contd...)

Plant	Types of waste/emissions		Source/activity of waste/emissions
	Fine dust		Primary crushing, material transfer, charging to pulveriser, blending, unloading, product weighing
	Residual technical		Empty liquid technical drums
	Filter and residual impurities		Filtering of technical (liquid)
	Packing losses		Packing
Liquid	Spillages	Solvent	Manual transfer from storage container to weighing container
		Technical (solid) (liquid)	Weighing, charging to blender Transfer from drum to weighing container
		Product	Unloading, filling, packed container tippling
	Leakages	Technical (liquid)	Valve fitted in pipe to transfer from drum to spray container Technical transfer pump
		Solvent	Transfer pump
		Product	Transfer valve and pump, filled bottles/container
	Fumes	Technical (liquid)	Weighing container, spray container, blender, packing
		Solvent	Natural evaporation—blender
	Residual technical		Empty liquid technical drums
	Filter and residual impurities		Filtering of technical (liquid) and product
	Packing losses		Packing
Factor common to all plants	Packing material		Packing material of incoming raw materials, outer packings, rejected product packing materials
	Indirect factors	Off-specification product	Non-adherence to operational practices, absence of preventive maintenance, lack of supervision
		Expired product	Unpredictable rains, improper production schedule
		Inventory management	Improper data recording and data interpretation

In terms of quantity and economics, waste generation in the pesticide formulation industry is not significant. However, when considered from the point of view of impact on the environment, health of workers and safety considerations, the wastes, which are generally highly toxic, assume significance and need to be curtailed.

FUTURE TRENDS

Conventional formulation, as discussed earlier, despite having the advatnge of low cost and simple processing, is associated with severe impact on the environment and major health hazards. Mainly due to the risk to the environment associated with these conventional formulation methods, industires in developed countreis have already shifted to newer formulation types. With stringent emission norms coming up in the near future in India, the future trend here also is to move away from conventional formulations which are mostly water-based, therefore resulting in improved handling and bio-efficacy.

NEW FORMULATIONS

The new formulations can be broadly categorised as follows:

1. Suspension concentrates.
2. Water dispersible granules.
3. Flowable concentrate for seed treatment.
4. Micro emulsions.
5. Suspo emulsions.
6. Extruded type granules.
7. Micro capsule suspension.
8. Controlled release formulations.

A brief description of some of the new formulation types listing their processing techniques, flow diagrams, advantages/disadvatages are given below:

Suspension Concentrates (SC)

Pesticides which are crystalline and insoluble in water can be formulated as suspension concentrates. The physical requirements of basic pesticides are:

1. High melting point (> 90°C).
2. Chemically stable in water.
3. Low solubility in water.

The formulation normally consists of the basic pesticide (10 to 80 per cent), a dispersing agent (0 to 10 per cent), anti-caking agents (0 to 5 per cent) and water for the balance percentage.

Process steps

The basic process steps for SC formulation are:

1. The basic pesticide is made into a slurry with water in the presence of the dispersing agent.
2. The slurry is ground through bead mills to give very fine particles (about 1/1000th of a millimetre).
3. De-gassing of the ground slurry.
4. Final blending along with the addition of thickening and anti-settling agents to prevent caking.

Careful selection of the dispersing agent is required to ensure that formulation does not thicken on storage. Good suspension concentrates readily dilute in the spray container to give a suspension which will stay for long periods. A lot of fungicides, triazine and urea herbicides are now formulated as suspension concentrates in advanced countries. The process flow diagram is shown in Fig. 16.1.

Micro Capsule Suspension (CS)

Micro encapsulation is the packing of liquids/solids/gases in small containers (capsules) with particle sizes ranging from 5–50 microns. There is increased use of CS formulation in advanced countries to reduce pesticide levels caused by extended residual effects. A typical composition of micro capsule suspension is given below:

Oil phase		*Water phase*	
Active ingredient	10–25%	Emulsifier	1–3%
Solvent	0–15%	Water-soluble monomer	2–10%
Oil-soluble monomer	2–10%	Anti-foaming agent	1%
		Water	35%

Process steps

The processing of CS is done *in situ*. The basic process steps are:

1. The basic pesticides containing the oil-soluble monomer are emulsified in water—Monomer (A).
2. The water-soluble monomer is dissolved in water in a separate mix tank, along with the emulsifier and anti-foaming agents—Monomer (B).
3. Monomer (A) and (B) are mixed in a tank (fitted with a high shear stirrer), where the monomer meet and react at the pesticide/water interface, forming a polymer shell around the pesticide droplets.

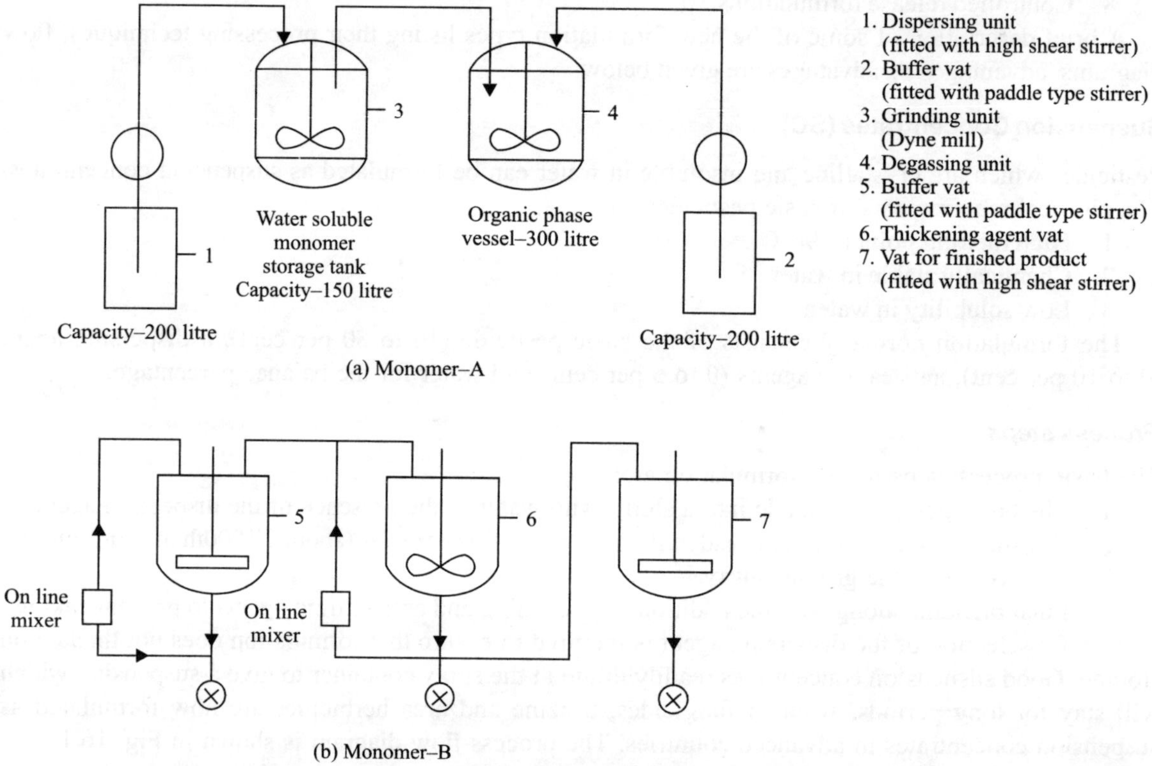

Fig. 16.1. Flow sheet for suspension concentrate.

During the mixing of the oil phase and the water phase, the pH is adjusted to 7.5 by adding neutralising agent. Anti-settling and wetting agents are also added for quick wetting and making a good suspension. The process flow sheet is shown in Fig. 16.2.

Water Dispersible Granules

This newer type of formulation is very convenient to use as it is in free-flowing form but readily disperses when added to the spray container. The composition is similar to that of the wettable powder but there is an extra step to form the granule either by extrusion or agglomeration.

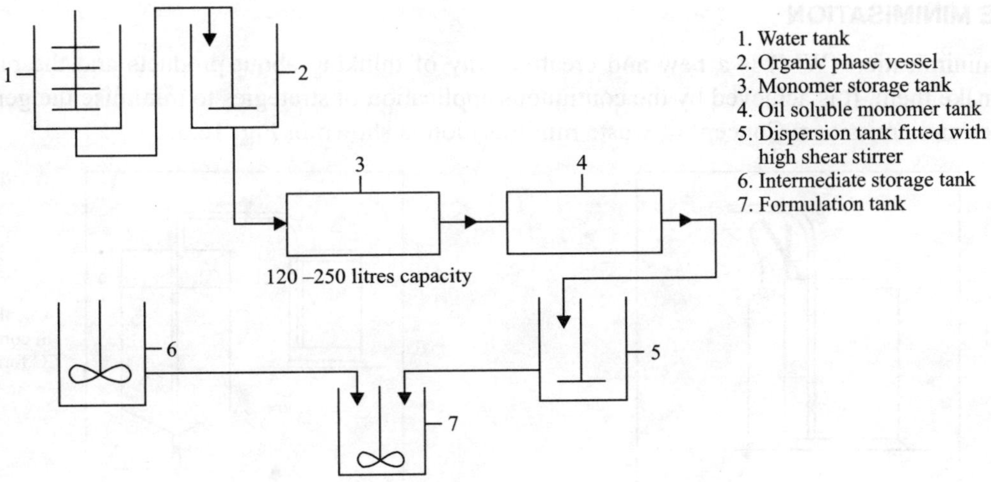

Fig. 16.2. Process flow diagram for microcapsule suspension.

Process steps

The process steps are as follows:

1. Raw material mixing.
2. Dry grinding.
3. Granulation.
4. Drying.

Granulation methods

There are different granulation methods, namely agglomeration, pressure compactation, dehydration of pre-mix or dispersion. Each granulation technique has its own advantages and disadvantages and the correct choice depends upon the physio-chemical properties of the basic pesticide and other raw materials, safety aspects and market requriement.

Controlled Release Formulation (CR)

Controlled release formulations are being developed to provide slow release of the basic pesticides in the field and to minimise the toxic exposure risk during formulation and application. In this type of formulation, the basic pesticide is coated with or physically trapped by polymers which slow down its rate of release.

The CR formulation has several distinct advatages over the conventional formulations.

Some of these are:

1. Low flammability.
2. Satisfactory activity effect for a longer period of time.
3. Same or higher efficacy at lower dose.
4. Longer application intervals.
5. Reduced ground water contamination risk.

In the end, it can be concluded that continuous efforts should be made to catch up with current trends in the developed countries and to venture out for a timely entry into the era of new formulations which are relevant to our needs and environment.

WASTE MINIMISATION

Waste minimisation (WM) is a new and creative way of thinking about products and the processes which make them. It is achieved by the continuous application of strategies to minimise the generation of wastes and emissions. Concept of waste minimisation is shown in Fig. 16.3.

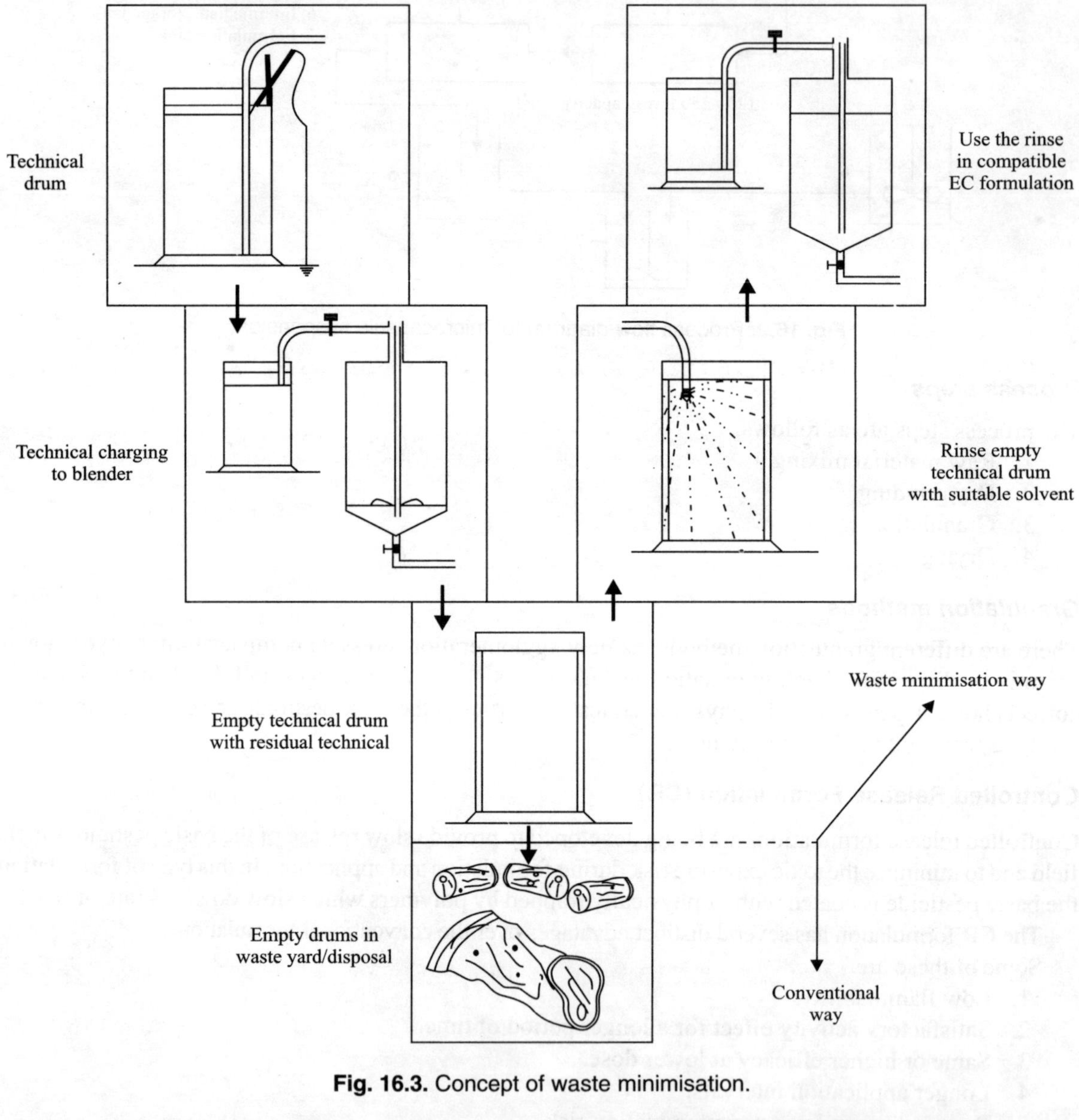

Fig. 16.3. Concept of waste minimisation.

The waste generated from the pesticide formulation industries is extremely toxic and hazardous. Hence WM should start from the waste generation source itself. After exhausting the source reduction opportunities, attempts should be made to recycle the waste within/outside the unit. The techniques which are available in these two areas, along with a few illustrative examples, are given below.

WASTE MINIMISATION TECHNIQUES

Recycling

On-site recovery—re-use

1. Rinsing of empty technical drum with suitable solvent to recover residual pesticide.
2. Controlled recycle of blender cleansings.
3. Recycle of fines collected from bag filter.

Product Change

1. Water-based emulsion concentrate.
2. Use of water soluble packing material.

Technology change

1. Use volumetric filling M/C instead of manual filling.
2. Use forced spray instead of gravity spray for technical dosing in granule plant.

Source Reduction

Good housekeeping

1. Prevention of leakages by regular preventive maintenance of pumps and timely replacement of worn out gaskets/seals.
2. Avoidance of fume generation by keeping the liquid containers closed when not in use.

Input material change

1. Use higher purity technicals.
2. Use granules with low fines content.
3. Use of less toxic solvents.

Process Change

Better process control

1. Optimise number of spray nozzles in dust and phorate blender.
2. Use accurate balance for weighing.
3. Optimise blending time.

Equipment modification

1. Introduce thin steel plate lining to reduce sticking of material in phorate blender.
2. Insulate inlet/outlet openings of shrink packing machine.
3. Consider open air circuit system instead of closed loop air circuit for impex pulversier.
4. Install an air lock rotary valve at the outlet of the cyclone of impex pulveriser/Raymond mill.

Waste Minimisation Need

Highly toxic and hazardous chemicals are dealt with in the pesticide formulation industry. The wastes generated are also mostly hazardous. The most important need for WM is, therefore, one of reducing toxic and hazardous waste generation. With the enactment of the hazarouds waste mangement and disposal rules, the treatment and disposal of hazardous wastes is going to be mandatory, and expensive, and it would be prudent to prevent or at least minimise its generation. The need for WM from other aspects is discussed below.

Pollution control

Although presently no emission norms have been prescribed by pollution control authorities, the prescription of such norms is imminent. Most of the pollution in this sector is in the form of fugitive emission, and controlling it by conventional air pollution control systems is extremely expensive and difficult. The adoption of WM renders residual emissions relatively lower, simpler and cheaper to treat, and would enable the industry to approach the forthcoming standards.

Health and safety

Safety norms have to be complied with as per the Companies Act, as do the occuptional health norms. These regulations can be effectively and economically met through the application of WM. Deterioration in safety and occuptional health standards is mainly due to fugitive emission and minimising the same would, by itself, enable approaching the prescribed norms.

Raw material conservation

With the increasing cost of input materials, namely, technicals, solvents and fillers, no industry can afford to lose them in the form of waste. The specific loss of input materials can be considerably reduced by adopting WM measures.

Environmental pressures

Public awareness in the field of environmental protection is growing day by day. Over the last few years, several non-governmental enviromental protection organisations have come up, not only for creating awareness in this field, but also to act as watchdogs against polluters. The industry can no longer keep itself isolated from such pressure groups and has to take positive steps to mitigate environmental pollution caused by it. WM could be one such positive step. In addition, today's industrial workers are much more aware of their right to a clean work environment. A sizable number of skilled workmen, who are not easily available, now consider environment as an important factor in an organisation. Ensuring a clean environment through WM could be a definite way of attracting, retaining and even motivating good workmen. The product manufactured in the clean and safe environment can lead to a broader market within the country and can also open up export markets.

It could thus be concluded that the implementation of WM measures serves the multiple objectives of reducing raw material losses, improving safety levels, improving shopfloor air quality and reducing the cost of compliance. Once WM becomes a part of the company's activites, it can be used as a marketing tool for developing market opportunities.

POTENTIAL FOR WASTE MINIMISATION

Small pesticide formulation units in India operate under constraints such as obsolete and inefficient technology, small scale of operation, dependence on large mills for most important raw materials such as technicals and solvents as well as untrained manpower. The market is unpredictable and small adverse conditions are enough to take the company into the red.

Accordingly, minimising losses during formulation, how-so-ever small, is important for the company's profitability. These losses include fugitive emissions, natural evaporation losses of liquid materials, spillages, leakages, filling and packing losses. It is surprising to find that such losses have come to be accepted as part and parcel of the process, and values in the range of 1 per cent for liquid formulation and 3 per cent for dust/wettable formulations are considered normal. Since the profit margin in this sector is estimated to be around 10 per cent, bringing down the losses to half the above value would mean considerably enhanced profits. Most of the losses occur due to poor housekeeping, sub-optimal operating practices, lack of exploitation of recycling opportunities, and similar controllable reasons. Obviously, the judicious applications of simple techniques and tighter process control would easily help in bringing down the losses. Some of the potential areas of WM are given in Tables 16.7 and 16.8.

Table 16.7. Potential of waste minimisation in small pesticide formulation units.

Granule formulation plant

Area	*Present waste*		*Expected waste minimisation potential*	
	Quantum (T/year)	*Value (Rs.'000/year)*	*(%)*	*(Rs. '000/year)*
Technical left in empty drums	0.17–0.21[a]	25–31[b]	> 80	20–25
Blender cleansings	0.50–0.60[c]	0.5–0.6[d]	> 90	0.45–0.54
Technical spills and leaks	0.015–0.022[e]	2.2–3.3[b]	80	1.8–2.6
Fugitive emission	Not quantifiable	Not quantifiable	Not quantifiable	
Granules spillage	4–5	3–4[d]	65	2.0–2.5

Expected total savings = Rs. 25,000 to 30,000/year.

[a] @ 250–300 ml technical left/drum, for about 700 drums use/year.
[b] Calculated at phorate cost.
[c] @ 50 kg/blender (1T)/month, for about 150 working days/year, for two blenders.
[d] Calculated at the granules cost.
[e] @ 100–150 ml/d/plant, for 150 working days/year.

Table 16.8. Potential of waste minimisation in small pesticide formulation units.

Dust formulation plant

Area	*Present waste*		*Expected waste minimisation potential*	
	Quantum (T/year)	*Value (Rs.'000/year)*	*(%)*	*(Rs. '000/year)*
Technical left in empty drums	0.12–0.15[a]	36–45[b]	> 80	30–36
Technical spills and leaks	0.015–0.022[c]	4.5–6.6[b]	80	3.6–6.0
Fugitive emission	Not quantifiable	Not quantifiable	Not quantifiable	
Fluid coupling to avoid waste due to power failure*	Not quantifiable	35.0	100	35.0
Soapstone spillage	4–5	3–4[d]	65	2.0–2.5

* Physical quantum of the waste generated due to power failure is very low but the monetary loss due to loss of production is substantial and is taken as the value of the present waste.
Expected total savings = Rs. 70,000 to 80,000/year.
[a] @ 250–300 ml technical left/drum, for about 450 drums use/year.
[b] Calculated at methyl parathion cost.
[c] @ 100–150 ml/d/plant, for 150 working days/year.
[d] Calculated at the soapstone powder cost.

It may be noted that though losses/emissions are low in quantitative terms, once dispersed/spilled in the working area, they pose a severe health hazard within and around the premises of the unit, as all pesticides are highly toxic. In view of the loss of expensive materials and the potential health hazard posed by such losses, it is important to eliminate/reduce the losses to the extent possible, through innovative, sustainable WM techniques.

The expected annual savings for a pesticides formulation company with one granule and one dust formulation plant amounts to between Rs. 94,000 and Rs. 110,000. In India, most SSIs operate about two to five plants, and thus the total savings potential might easily be a couple of lakhs of rupees annually. Apart from direct savings, the indirect savings through reduction in the discharge of even smaller quantities of toxic emissions on the shopfloor can make an immense impact on the health and safety of the persons working with you (including your ownself).

WASTE MINIMISATION OPPORTUNITIES

The experience of several pesticide formulation units shows that there are several proven waste minimisation (WM) techniques which could be implemented to tap the potential discussed earlier. Most importantly, these measures enable reduction in the adverse environmental hazard/impact of formulation activities within and around the unit.

In this chapter, various WM measures have been described. For the sake of clarity, the WM opportunities have been discussed separately for the three different types of formulation plants, namely, granule formulation plant, dust/wettable formulation plant and liquid formulation plant. Some general WM measures have also been identified and are as follows:

1. Stricter supervision and control at each formulation step.
2. The display of the operational sequence for each formulation step in the local language at appropriate places.
3. The provision of timer with timing of blending operation.
4. The recording of actual input to the process as well as actual output from the process and the time taken for each batch.
5. Recording the maintenace carried out with cost and the next schedule maintenance.
6. Proper inventory management, particularly for costly items such as technical.
7. Training the operators and supervisors.

For each measure (except the general ones listed above), the technical requirements, economic viability, environmental aspects and anticipated benefits have been discussed.

The WM opportunities have been grouped as per the following WM techniques.

Source Reduction

1. Good housekeeping.
2. Process change:-
   (a) Input material change.
   (b) Better process control.
   (c) Equipment modification.
   (d) Technology change.

Waste minimisation measures in the granule formulation plant are shown in Tables 16.9 to 16.11.

Table 16.9. Waste minimisation measures in the granule formulation plant.

Waste minimisation measures	*Anticipated benefits*	*Technical requirements*	*Economic viability*	*Environmental and safety impact*	*Remarks*
Source reduction *Good housekeeping*					
Regular checking of technical transfer pumps, valves, rosin mixer churner, and replacement of worn out gaskets and washers, etc.	Reduced spillages Reduced exposure to toxic fumes Clean shopfloor area	Process: Routine maintenance	I = Negligible S = Rs. 2,000 P = Nil	Reduced risk of contact with hazardous materials Reduced exposure to toxic fumes	The prevailing breakdown maintenace practice to be changed to preventive maintenance Savings have been estimated based on the assumption that about 100 gm technical material spills/day for about 150 working days/year
Process change *Input material change*					
Use of better quality gunny bags	Low spillage in handling Low fugitive emissions Enhanced recycle life	Tie up with suitable supplier	I = Rs. 1 lakh S = Rs. 6,000 P = > 16 years	Low exposure of workers to dust emissions	At present used fertiliser bags are utilised Savings have been estimated based on the assumption of 0.5% spillages batch at an average of 1500 batches per year
Better process control					
Screening the raw granules	Low fines will result in reduced fugitive emission Less material sticking to blender walls Reduced chances of off-specification product Better quality product Reduced hazardous waste generation Better housekeeping	*Equipment* Screen Extraction and exhaust system *Process* Additional enclosed space —5 × 5 metre required *Manpower* At least two unskilled workers are required	I = > Rs. 1 lakh S = NQ P = NQ	Low exposure of workers to toxic fumes Reduced hazardous waste generation	Maximum 1–2% fines are permissible and guaranteed by suppliers Investment beyond the capability of small units Resultant fine material is not hazardous and can be disposed off for land filling

(Contd...)

Waste minimisation measure	*Anticipated benefits*	*Technical requirements*	*Economic viability*	*Environmental and safety impact*	*Remarks*
Quality check of all incoming raw materials	Accurate recipe of formulation Reduced chances of off-specification product	Routine analysis	I = Nil S = NQ P = Nil	Reduced reprocessing waste	In most of the units, the purity of technical is tested on an adhoc basis
Blender cleaning	Reduced sticking of material to blender wall Reduced risk of off-specification product	Process: Blender to be cleaned at least once a fortnight	I = Nil S = NQ P = Nil	Low generation of hazardous waste Reduced reprocessing waste	
Technical nozzle spray maintenance	Uniform technical spray and blending Reduced risk of off-specification product	Process: Replace at least once in a day, the technical spray nozzles set with cleaned nozzles	I = Nil S = NQ P =Nil	Reduced reprocessing waste	The nozzles are replaced in most of the units but only when wet product or failure of batches in quality check is observed
Complete blender unloading	Reduced risk of off-specification product	Process: Unload blender completely after final blending	I = Nil S = NQ P = Nil	Reduced reprocessing waste	Due to negligence of operators, the blender is not unloaded completely and offsets the composition of next batch
Equipment modification					
Use of blender lined with non-sticky or low sticky material such as stainless steel	Reduced sticking of material to blender wall Reduced risk of off-specification product Reduced cleaning frequency	Equipment: Lining material	I = Rs. 3000 S = Rs. 150 P = > 15 years	Low generation of hazardous waste	The option is under implementation in one unit. The impact can be assessed only after implementation Savings have been estimated based on the assumption that 50% waste, equal to 150 kg/season/blender, can be avoided
Use of forced spray technique for technical charging	Better and uniform spray Reduced batch cycle time Better quality product	Equipment: Positive displacement pump Nozzles Accessories	I = Rs. 25,000 S = Rs. 66,000 P = < 1 year	No adverse impact on shopfloor environment, provided fume extraction from blender is	For forced spray the blender operation to be accordingly optimised Savings have been estimated assuming 20% increase in

		Technology: Indigenously available		effective	production (200T/year) and formulator's profit as 10% of the product market price
Technology change					
Raw granules movement from storage area in enclosed trolleys and mechanised charging system	Low spillages Better product quality Reduced risk of off-specification product Optimum production	*Equipment* Trolleys (5–6 nos) Bucket elevator Enclosed charing hopper *Process* Space: About 2 × 2 metre for charging operation	I = Rs. 1.25 lakh S = Rs. 6,000/ season P = 20 years	Reduced fugitive emissions Improved work environment	The option is under implementation in one unit. Full impact can be analysed only after implementation
Extraction of fumes from spray container and blender	Avoidance of dispersion of toxic fumes in shop-foor	*Equipment* Extraction hood and ducting Fan and motor *Process* Additional space to install fan and motor—1 × 1 meter	I = Rs. 15,000 S = Nil P = NA	Reduced exposure of workers to toxic fumes and fine dust emissions	In some units the extraction and exhaust system exists but needs improvement
Scrubbing and detoxification of fumes and fine dust emissions before discharging them into the atmosphere	Elimination of toxic fumes discharge into the atmosphere	*Equipment/Chemicals* Scrubber 0.1 N NaOH—scrubbing liquid *Space* Scrubber—2 × 2 metre	I = Rs. 10,000 S = Nil P = NA	Reduced health hazards	Phorate technical gets detoxified in the presence of alkali in a week's time
Extraction of fumes from rosin and solvent mixer	Reduced exposure of workers	*Equipment* Extraction hood and ducting Fan with motor *Space* Space requirement—1 × 1 metre	I = Rs. 5,000 S = Nil P = NA	Improvement in working environment	
Recycling: on-site recovery/ re-use	Recovery of spilled material	*Equipment*	I = Negligible S = Rs. 6,000	Reduced risk of contact and expo-	In most of the units, the trays are used but collected

(Contd...)

Waste minimisation measure	*Anticipated benefits*	*Technical requirements*	*Economic viability*	*Environmental and safety impact*	*Remarks*
Collection and recycling of drippage/leakage from equipment such as technical transfer valve and pump, rosin mixer and pump	Reduced toxic fumes exposure	Collection tray beneath the concerned equipment	P = Nil	sure to toxic material	material is rarely recycled
Collection and recycling of leftover liquid material from hand pump	Recovery of technical material Reduced exposure to toxic fumes	*Equipment* Pump holder	I = Rs. 1,000 S = Rs. 2,500 P = < 1 year	Avoidance of contact risk with toxic material	
Recovery of the residual technical material from empty drums	Recovery of technical material	*Process* Empty drum to be kept in inverted position for some time (about five minutes) to drain and collect the residual material	I = Nil S = Rs. 11,500 P = Nil	Reduced toxic fumes emission during empty drum storage	Saving have been claculated based on the recovery of about 100 gm technical (phorate)/drum and for the use of 750 drums/year
Rinsing of empty technical drum with MTO and recycling wash MTO	Recovery of technical material Detoxification of empty drum for safe recycle/ disposal	*Process* The empty drum to be washed with about 5 ltr of MTO. The wash MTO to be recycled for rosin preparation	I = Nil S = Rs. 13,500 P = Nil	Safe diposal of contaminated technical drum	Savings has been calculated for the rinsing of empty phorate drums with MTO
Recycle the waste generated in blender cleaning	Recycling of toxic waste Disposal cost eliminated	*Process* Controlled quantity of waste recycle in new batch formulation	I = Nil S = Rs. 150 P = Nil	Significant reduction in hazardous waste	The jobwork units send back waste to parent company Savings have been estimated assuming 150 kg waste generated after blender lining is recycled as granules Hazardous waste disposal cost not included since presently the waste is being dumped without any treatment

Note: NQ = Not quantifiable; and NA = Not available

Table 16.10. Waste minimisation measures in the dust/wettable formulation plant.

Waste minimisation measure	*Anticipated benefits*	*Technical requirements*	*Economic viability*	*Environmental and safety impact*	*Remarks*
Source reduction *Good housekeeping* Periodic checking and replacement of wornout gaskets, bearings of pumps, transfer valves, sealings, pipe fitting, etc.	Reduced leakage losses	Nil	I = Nil S = Rs. 2,000 P = Nil	Improved work environment Reduced toxic emissions from the spills Reduced health hazards	Shift from breakdown maintenance practices to preventive maintenance practices is required Savings have been estimated based on the assumption that about 100 gm technical (malathion) leaks/spills/day for about 150 working days/year
Usage of industrial vacuum cleaner for floor cleaning	Reduced dust settlement on the equipment and thus less maintenance problems Possibility of recycling of collected material	*Equipment* Vacuum cleaner	I = Rs. 25,000 S = NQ P = NQ	Cleaner work environment	The effectiveness of vacuum cleaner in avoiding dust re-entrainment is much more than the normally used brooms
Equipment modification Installation of a rotary air lock valve	Reduced fugitive emissions Reduced health hazards	*Equipment* Rotary air lock valve	I = Rs. 6,000 S = NQ P = NQ	Reduced toxic fumes emission Reduced health hazards Improved work environment	Savings could not be quantified as the losses are of fugitive type and consisting of various components which are difficult to quantify
Installation of a vent pipe in blender to avoid pressure build-up	Reduced fugitive emissions Reduced health hazards	*Euipment* A small pipe (100 mm × 300 mm length) Small bag *Process* Provide a vent pipe of above dimensions and attach a filter bag to it	I = Rs. 500 S = NQ P = NQ	Reduced toxic fumes emission Reduced health hazards	

(Contd...)

Waste minimisation measure	*Anticipated benefits*	*Technical requirements*	*Economic viability*	*Environmental and safety impact*	*Remarks*
Installation of a simple bag clamping arrangement along with laminar plates at blender discharge	Reduced fugitive emissions Reduced product spillages	*Equipment* Bag clamping arrangement		Reduced skin contact with toxic materials Improved work environment Reduced health hazards	Needs further study as the bag clamping arrangements are not available indigeno-usly
Process change *Technology modification*					
Volumetric weighing and mechanical handling of liquid technical	Reduced fugitive emission Reduced spillages	*Equipment* Volumetric weighting pump *Process* Variations in the specific gravity of technical are not tolerable	I = Rs. 25,000 S = Rs. 2,500 P = > 10 years	Improved work environment Reduced toxic fumes emission Reduced health hazards	Fine control required since the specific gravity of technical changes from type to type and from batch to batch Operators need to be trained to operate the system
Installation of a fluid coupling assembly	Avoidance of waste generation in case of power failure Reduced worker exposure to toxic material Reduced loss of production time	*Equipment* Fluid coupling assembly with motor	I = Rs. 40,000 S = Rs. 33,000 P = < 2 years	Improved work environment Reduced health hazards	The option is environmen-tally very attractive. Although the investment is very high, the savings are also large, as worked out for an average loss of one batch (1 T)/week of methyl parathion over 150 days/ season
Installation of a premixer	Reduced spillages	*Equipment* Premixer Motor	I = Rs. 75,000 S =NQ P = NQ	Reduced contact between skin and the toxic materials Improved work environment Reduced health hazards	Premixing needed only while handling viscous technical Feasible only if quantity of products requiring pre-mixing is sufficiently high

Extraction of fumes/fine dust missions from technical handling area, spray container, blender, charge hopper, discharge end, packing area, etc.	Avoidance of dispersion of toxic fumes in the shopfloor area	*Equipment* Extraction hoods and ducting Fan with motor *Space* Additional space to install fan and motor 1 × 1 metre	I = Rs. 25,000 S = Nil P = Not applicable	Improved work environment Reduced health hazards	In some units extraction systems for some sources exist but need improved
Recycle/recovery					
(On-site) recycle Recycling the fine dust collected in bag house	Use of the technical left in the fine dust Reduced waste quantity	Nil	I = Nil S = Rs. 16,000 P = Nil	Reduced waste quantity	Controlled quantity collected material is to be recycled Savings have been calculated for the waste of 15 kg/batch for 2,250 batches/year at the cost of soapstone
Re-using the solvent used for pumps cleaning	Solvent saving	*Process* Use of same solvent for at least three to four cleanings	I = Nil S = Rs. 100 P = Nil	Reduced waste disposal problem	Savings have been estimated assuming 50% reduction in the solvent requirement
While product swithch over cleaning of spray nozzles and charging the cleaning solvent (if used) to blender in new batch formulation of same product	Reduced risk of off specification product Reduced waste disposal cost	Nil	I = Negligible S = NQ	Reduced waste disposal problem	Cleaning of the spray nozzles is already in practice
Off-site recycle Use contaminated solvent (used in cleaning of technical spray container and transfer line) in compatible EC formulation	Technical and solvent saving	The compatibility needs to be checked	I = Nil S = NQ P = Nil	Reduced waste disposal cost	Care shall be taken that the contaminated solvent does not have an impact on product quality
Recovery Recovery of the residual technical from the empty technical drums	Saving of technical	*Process* Keeping the empty technical drum in inverted position for some time (about five	I = Nil S = Rs. 2,050	Reduced toxic fumes emission during storage	Savings have been calculated based on the recovery of about 100 gm technical (methyl parathion) and for the use of 170 drums/year

(Contd...)

Waste minimisation measure	Anticipated benefits	Technical requirements	Economic viability	Environmental and safety impact	Remarks
		minutes) to drain and collect residual technical			
Recovery of the technical adhered to empty technical drum walls	Recovery of the adhered technical Reduces the disposal cost of the empty containers	*Process* Wash the empty drum with 5 litres of suitable solvent, collect the wash and recycle it in the formulation	I = Nil S = Rs. 2,050	Reduced toxic fumes emission from the drum during storage	Savings have been calculated based on average cost of technical (Rs. 150/kg) and about 170 empty drums/year

Note: NQ = Not quantifiable; and NA = Not available

Table 16.11. Waste minimisation measures in the liquid formulation plant.

Waste minimisation measure	Anticipated benefits	Technical requirements	Economic viability	Environmental and safety impact	Remarks
Source Reduction *Good housekeeping*					
Regular checking and replacement of wornout gaskets, sealings, valves of transfer valves, flanges, pipe fittings, etc.	Reduced leakage losses Reduced loss of production time	Nil P = Nil	I = Nil S = Rs. 4,500 Reduced toxic emissions from	Improved work environment preventive maintenance is required spills Reduced health hazards	Shift from breakdown maintenance practice to pumps, transfer Savings have been calculated based on the assumption that about 100 gm technical (endosulphan) leaks/day for about 150 working days
Process change *Equipment modification*					
The technical transfer drum opening can be closed by a loosely-fitting cork through which the transfer pipe should be inserted	Reduced evaporation losses	Minor modifications in the existing equipment	I = Negligible S = Rs. 150 P = Nil	Reduced toxic fumes emissions Improved work environment Reduced health hazards	Savings are calculated based on the evaporation rate (of endosulphan) of average exposure time of three hours/day for 150 days/year

Closing the open top filling container with an appropriately designed cover	Reduced evaporation losses	*Equipment* Stainless steel cover Gauge glass	I = Rs. 1,500 S = Rs. 4,500 P = Nil	Reduced toxic fumes emission Improved work environment Reduced health hazards	Savings have been estimated based on the natural evaporation rate of product endosulphan 35% EC for an average drum opening of 10 hours/day for about 200 working days Gauge glass may develop cracks resulting in more waste
Provide a mesh between the conveyor and the heating region of the shrink packing m/c	Reduced bottle damage Avoiding explosion possibilities	*Equipment* Stainless steel mesh	I = Rs. 500 S = NQ P = NQ	Reduced fumes Improved safety aspects	
Reducing the gap between links of the shrink packing m/c by winding the links by an asbestos rope of 5 mm diameter	Avoids bottle breakage Reduced heat loss	*Equipment* Asbestos rope	I = Rs. 500 S = NQ P = NQ	Reduced fumes Improved safety aspects	From the space between the links of the link conveyor, smaller size bottles fall down and break
Closing the inlet and outlet door of shrink m/c with movable Hysil board with a provision for increasing/decreasing the opening as per bottle size	Reduced energy loss	*Equipment* Hysil board of 5 mm thickness	I = Rs. 250 S = NQ P = NQ	NA	
Re-using the solvent used for the cleaning of transfer pumps	Reduced solvent requirement	*Process* Recycle the solvent for at least three to four cleanings	I = Nil S = Rs. 100 P = Nil	Reduced waste disposal problem	Savings have been estimated assuming 50% reduction in the solvent requirement
Extraction of fumes from the blender, packing section etc.	Improvement in shop quality Better health of the employees	*Equipment* Extraction duct Blower with motor	I = Rs. 30,000 S = Nil P = NA	Significant improvement in the shop air quality Reduced health hazards	No economic returns but substantial improvement in the shop air quality Subsequent recovery chemicals is not feasible due to multiplicity of products and high cost of recovery system

(Contd...)

Waste minimisation measure	*Anticipated benefits*	*Technical requirements*	*Economic viability*	*Environmental and safety impact*	*Remarks*
Technology modification					
100% leak testing of containers before filling with the product the disposal of leaking	Reduced leakages Operation of recovering the product from the leaking containers and Drier containers avoided	*Equipment* Pressure testing machine Water storage tank Tables/fittings P = NA *Manpower* Five workers required *Space* Leak test area: 3×3 metre	I = Rs. 90,000 op cost: Rs. 31,000 S = Rs. 18,000 Reduced health	Reduced toxic fumes emission Improved work environment 35% EC hazards	Savings are claculated based on 1–2% product loss from the 2–5% leaking bottles of endosulphan
Recycle/recovery					
Collection and recycling of leftover material from hand pump	Recovery of technical material Reduced exposure to toxic fumes	*Equipment* Pump holder	I = < Rs. 1,000 S = Rs. 2,500 P = < 1 year	Improved work environment Reduced health hazards	
Recycling the solvent used for technical charging assembly in formulation and switching over the other products	Technical saving	*Process* The contaminated solvent can be collected and stored and should be used in the same formulation	I = Nil S = NQ P = Nil	Reduced waste disposal problem	
Recovery of the residual technical from the empty technical drums	Technical saving	*Process* Empty drums to be kept in inverted position for some time (about five mts.) to drain and collect the residual material	I = Nil S = Rs. 14,500 P = Nil	Reduced toxic emissions during empty drum storage	Savings have been calculated based on the recovery of about 100 gm technical (endosulphan)/drum and for the use of 440 drums/year
Recovery of the technical adhered to the empty technical drum's walls	Recovery of the technical material	*Equipment* Container for collecting the rinse *Process* Wash the empty drum with 5 litres of suitable solvent Collect the rinse and recycle it in the next batch	I = Nil S = Rs. 20,000 P = Nil	Reduced toxic emissions during empty drum storage	Savings have been calculated for the rinsing of empty endosulphan technical drums with acetone

Note: NQ = Not quantifiable; and NA = Not available

Recycling

1. On-site recovery/re-use.
2. Creation of useful by-products.

Product Change

The economic viability is experessed in terms of the investment (I); annual savings (S) worked out for the duration of season (taken as five months); and payback period (P) in years. In some cases it has not been possible to quantify the savings (shown as NQ) because of the following reasons:

1. The loss is in the form of fugitive emissions which are difficult to quantify.
2. In some cases, the impact can be assessed only after the implementation of the WM measure.
3. In some cases it is very difficult to pinpoint a single reason for a particular loss (for example, off-specification product) and so the savings.
4. Some WM options yield intangible benefits, such as reduced maintenance requirement, for which the savings quantification is very difficult.

In addtion to the above the listed WM opportunites, another area could be WM through 'product change'. The two WM opportunities identified in this category are:

1. Use of newer type of packing material.
2. Change over to newer type of formulations, for example:-
   (a) Substitution of solvent-based formulations with water emulsion suspensions.
   (b) Replacement of dust-based products with water-soluble grains and slow active release granules.

Both of the above mentioned WM opportunities have not been included in the earlier WM opportunities list as these are yet in the development stage in the country.

Salient features

Salient features of both of these opportunities are discussed below:

1. Use of newer type of packaging material. Using newer type of packing material, which could be more easily disposed, for example, water-soluble packs for wettable powder and water-dispersible granules. Use of these types of packing materials will results in:
   (a) Avoiding hazardous waste disposal problems.
   (b) Reduced toxic material exposure to end user.
   (c) Change over to newer type of formulations.

Advantages

The major advantages of these product types over conventional formulations are listed below:

1. Safer for field application and use.
2. Produce less waste during formulation.
3. Less skin and eye toxicity.
4. Higher efficacy at lower dose.
5. Low cost due to low solvent cost.
6. More amenable to reprocessing on expiry of shelf life.

BARRIERS TO WASTE MINIMISATION AND ENABLING MEASURES

It may be better to clarify that the barriers as well as enabling measures are specific to individual enterprises and no general prescription can be evolved.

For the sake of clarity and easy understanding, the barriers have been classified into the following categories:

1. Lack of commitment and involvement.
2. Lack of data.
3. Technical constraints.
4. Economic constraints.

Unit to evolve own guidelines to overcome the barriers.

Lack of Commitment and Involvement

Non-involvement of employees

In many units, a majority of workers are temporary or contract workers and, therefore, do not participate in WM activities until and unless ordered by the chief executive/owner.

Lack of recognition

Employees in small-scale industries are suppose to do whatever work is assigned to them. The immense impact of other forms of motivation, that is, public recognition, awards, etc. is often not realised. The employees, therefore, lack initiative in taking up new and challenging assigments.

Fear of failure

There is a complete dearth of success stories in the sector which could instil confidence in the management to try out WM concepts which include new formulations, material handling and packing techniques in their respective units. The fear of failure dissuades the management from implementing WM programmes. This also gives rise to the NMF (Not Me First) syndrome wherein people would be willing to try out an idea provided it has been successfully implemented elsewhere.

Lack of Data

Poor record keeping and reporting

In the absence of records and systematic interpretation of records such as product-wise raw material consumption, daily log sheets of input and output, it is very difficult to assess how much waste is being generated, when and in which product category. In the absence of environmental norms, environmental records, for example, quantity and quality of emissions, are just not available. Data generation, therefore, is time consuming and effort-intensive.

Ad hoc production planning

Pesticide formulation is demand dependent. The units undertaking job work normally receive the tentative production schedule in advance. However, changes in plans are frequent. Further, due to climatic changes, the production schedule changes on a day-to-day basis. This hampers any systematic work—be it baseline data collection or impact assessment of implemented measures.

Technical Constraints

Obsolete technology

In small-scale pesticide formulation industries, the risk potential is very high. From the point of view of operational safety, the prevalent technology is obsolete. Most workers are unskilled and without a technical background; therefore it is difficult to retrofit any additional equipment/practice.

Lack of systems for professional upgradation of employees

There is a near total lack of systematic training of employees to upgrade their job skills. The employees are, therefore, unable to comprehend new subjects such as WM.

Large number of raw materials

A large number of raw materials with different chemical and physical characteristics (specially technical-basic pesticides) and in different proportions are to be handled for different formulations. Even the purity of the same technical (basic pesticide) varies from one manufacturer to the other and between batches. Such variations stop small-scale formulators from switching from manual material handling systems to mechanical.

Limited access to technical information

Generally these units have limited access to information and success stories on WM and new formulations and packaging techniques within the country as well as overseas. Further, there is a near total absence of relevant technical literature. The information available from abroad is neither directly relevant nor tailor-made for the small-scale pesticide formulation sector.

Economic Constraints

Cost and availability of technical (basic pesticides)

As mentioned earlier, small-scale formulators are totally dependent on large scale units for the supply of technical material. Technical materials are very expensive and the sale proceeds are realised towards the end of the season. This causes a cash flow problem. These factors dampen or delay the impetus for implementing WM measures that need investment for want of funds.

In the absence of environmental norms, waste treatment and environmental costs in the sector are non-existent and therefore, the economics of WM measures is computed mainly on the basis of direct economic returns and short-term financial gains. The benefits accruing from reduced pollution are not accounted for. As the saving in environment cost is often a major saving, its exclusion disfavours acceptance of the WM measures.

Units to Formulate Own Guidelines to Overcome Barriers

In fact, each unit has to evolve its own prescription for overcoming the barriers.

Involve employees in the WM programmes

The success of the WM programmes, specially its long-term sustainability, depends upon the involvement of employees. There is no better asset than a motivated employee willing to actively participate in programmes aimed at improving the working environment in the company.

Reward the 'achievers'

One of the easiest ways of extracting the best out of employees is to motivate them by rewarding their efforts. The rewards need not necessarily be monetary. Even small gestures such as appreciation in public or a pat on the back in front of the others puts an employee on a high pedestal and he would put in more effort to be recognised more often. Make him the champion of the cause and not the cause of a failure.

Improve record-keeping and reporting

The management should introduce simple record-keeping and reporting procedures to enable data compilation. The emphasis should be on collecting basic but reliable data which could be used for conducting WM programmes.

Develop manpower

To upgrade job skills of employees and to enable them to comprehend WM concepts, systematic training should be provided. The involvement of external experts could be considered to provide this training.

Include environmental costs in economic analysis

The management should realise that sooner or later it will have to abide by environmental regulations and incur costs on meeting stipulated requirements. Environmental costs, even though imaginary presently, should therefore be included in the economic analysis so as to give a total picture of the proposed WM measure.

Plan the investment

The management should take up only those measures for which it has enough resources for complete implementation. If phase-wise implementation is requied, it should be properly planned so that the remaining phases do not languish.

17

Paper Industry

INTRODUCTION

The paper industry is one of the oldest industries in India, dating back more than a century. Agro residue-based mills are more prevalent in developing countries, which have predominantly agricultural economies. The major agro-pulp producing countries are China, India, Mexico, Brazil, South Africa and Indonesia.

The capacity of the agro residue-based pulp and paper industry in Asia generally ranges from 15 to 60 tonnes/day. Constraints imposed due to limitations of availability of raw material within economically transportable areas, frequent operational interruptions causing heavy losses in large capacity machines, and the high cost of transporting low-quality paper producible from agro residues have been responsible for the relatively small size of plants.

There are over 400 small pulp and paper mills in India, of which about 100 are purely agro residue-based, whereas others utilise a mixed feed of waste paper and/or agro residues. The characteristic features of these mills are summarised below:

1. Most of the mills are in the size range of 10 to 40 tonnes/day capacity.
2. Most of the mills follow the soda chemical pulping process and do not have any facility for chemical recovery. Thus, compared with large mills, their chemical consumption—and consequently the pollution load—is very high.
3. The mills continue to use old technology, with little emphasis on modern instrumentation and process control.
4. The availability of raw material varies from season to season and accordingly the mills have to make frequent adjustments to be able to use the diverse raw material. The product quality also gets affected and the mills are only able to produce low-quality paper. Some of the mills have started using a small proportion of long fibre, thus improving the quality of paper.

Thus, while the Indian agro residue-based pulp and paper industry faces the challenge of improving its efficiency and cost of production, it also has the opportunity to grow, due to the abundant availability of agro residues.

PROCESS DESCRIPTION

The various production sections and their operations/processes are listed in Table 17.1.

Table 17.1. Major sections and associated unit operations/processes.

Section	*Unit operations/processes*
Raw material preparation	Dedusting, depithing, leaf removal
Pulping section	Cooking, beating, pulp washing, refining, bleaching, cleaning and thickening
Stock preparation	Blending, pulp conditioning
Paper machine	Refining, centricleaning, dewatering, drying of paper
Utilities	Steam generation, electricity, raw water supply

A detailed flowsheet of the manufacturing process is presented in Fig. 17.1.

Paper manufacturing is an energy and water intensive process. The main sources of energy are fuels (coal, rice husk, bagasse) for boilers, electricity, and diesel for captive power generation. The specific energy consumptions for electrical and steam energy are 800 to 1200 kW/tonne of paper and 4 to 6 tonnes/tonne of paper, respectively. The specific water consumption ranges from 180–280 m^3/tonne of paper produced. The agro residue-based raw material (RM) is procured by the mills from nearby farms and in some mills raw material is screened at the site. The dust from the screening section is disposed of as solid waste along with municipal waste. In small mills bagasse is generally not depithed. The prepared agro raw material is then conveyed by a belt conveyer to spherical digestors.

Pulping Section

Cooking

The agro residue is chemically digested in a spherical digestor at 150–160°C and 6–7 atm pressure for about six hours. Charging and discharging takes another 1.5 hours and 0.5 hours, respectively. The cooking in small agro-based units is done with caustic soda (NaOH) and steam. The quantity of NaOH charged, varies from 6 to 14 per cent of RM, depending on the type of agro residue used. For every tonne of agro residue, about 1.5–2.0 tonnes of steam is used, depending on the type of pulp required—hard cooked or soft cooked. During digestion, a solid to liquid (bath ratio) in the range of 1:3 to 1:4 is maintained.

Blow tank

After cooking, the content of the digestor is discharged, under pressure, either into a blow tank where the pressure is released or directly into potchers. Water is added to reduce the pulp consistency from an inlet value of 12–14 per cent to about 3–4 per cent, so that it can be pumped to the washing and cleaning section.

Washing

The pulp is then pumped to the washers (potchers) for washing with back water (recycled water from succeeding stages). The washing operation takes about four to six hours. The wash water called black liquor, which has a total solids (TS) content of around 1 per cent due to residual alkali and lignin, is discharged into the drains.

Screening

The washed pulp contains sand and uncooked agro residue as impurities. The impurities are removed through screening and centricleaning.

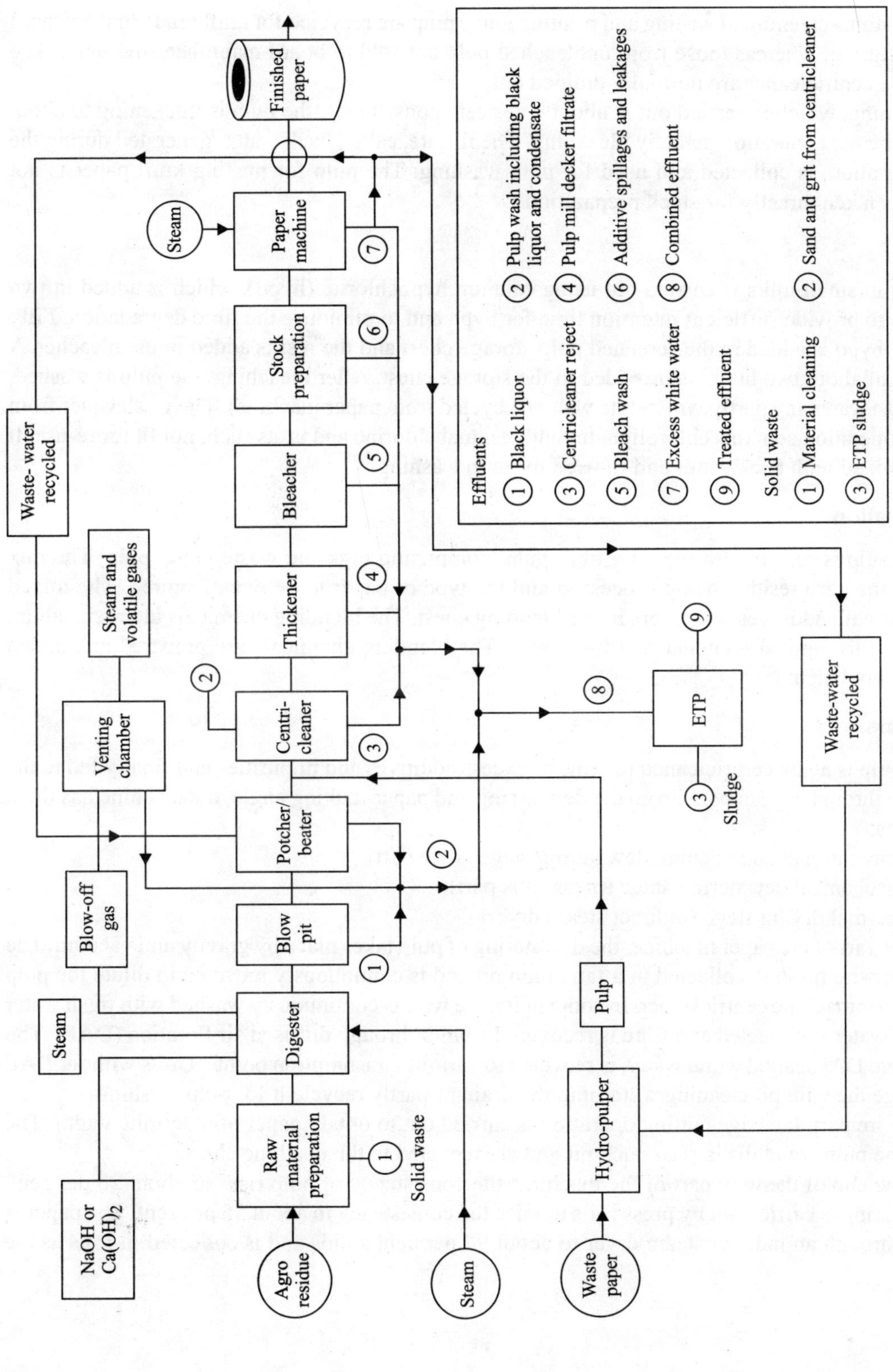

Fig. 17.1. Generalised process flowsheet of agro-based paper mills and sources of wastes.

The rejects from screening of writing and printing grade pulp are recycled for kraft grade (unbleached) paper manufacturing, whereas those from unbleached pulp are sold to board manufacturing units. The rejects from the centricleaner are normally drained out.

After screening, which is carried out at about 1 per cent consistency, the pulp is thickening to about 4 per cent for the next operation, namely bleaching. The filtrate, called back water, generated during the thickening operation, is collected and used for pulp washing. The pulp for making kraft paper is not bleached and is taken directly for stock preparation.

Bleaching

The bleaching in small mills is carried out using calcium hypochlorite (hypo), which is added in two stages in order to provide sufficient retention time for hypo and to minimise the fibre degradation. Fifty per cent of the hypo is added in the screened pulp storage chest and the rest is added in the bleacher. A retention time of about two hours is provided in the storage chest. After bleaching, the pulp is washed, partly with fresh water and partly with white water (recycled from paper machine). The wash water from the bleaching operation contains chlorolignates and residual chlorine and is, as such, not fit for re-use. It is, therefore, mixed with back water and re-used for pulp washing.

Stock Preparation

The bleached pulp is mixed with the long fibre pulp, comprising rags and waste paper pulp. The mix depends upon the agro residue being processed and the type of paper to be manufactured. The mixed pulp is blended with additives and fillers in the blending chest. The blending chemicals are rosin, alum, talc, dye (optional), optical whitener and high gum. The blending chemicals are prepared and added manually in every batch.

Paper Machine

The blended pulp is again centricleaned to remove excess additives and impurities and finally fed to the paper machine through a head box. From the dewatering and paper-making angle, the machine has three principle stages:

1. The gravitational and vacuum dewatering stage (wire part).
2. The mechanical dewatering stage (press rolls part).
3. The thermal drying stage (indirect steam dryers).

On the wire part of the paper machine, the dewatering of pulp takes place by gravity and vacuum. The water from the wire mesh is collected in a fan pump pit and is continuously recycled to dilute the pulp fed into the paper machine centricleaner. In some units, the wire is continuously washed with fresh water showers. This water is collected and fibre is recovered from it through diffused air flotation (DAF). The clear water from DAF, called white water, is recycled to various consumption points. Units without DAF either discharge the wire pit cleaning water into the drain or partly recycle it for pulp washing.

After the wire part, the edge cutting operation is carried out to obtain paper of a definite width. The edge cuts of the pulp web falls in the couch pit and are recycled to the machine chest.

Towards the end of the wire part of the machine, the consistency of pulp rises to about 20 per cent. Further dewatering is carried out by press rolls to raise the consistency to about 55 per cent. The paper is finally dried through an indirect steam dryer to about 94 per cent solids and is collected in rolls as the final product.

Utility Section

The utility section comprises water supply, boiler house, and electrical power supply.

The total requirement of raw water is generally met by the mills, own bore wells; about 200–300 m^3 of water per tonne of paper is used.

The steam requirement is met by boilers of capacities ranging from 5 to 10 tonnes of steam per hour. The maximum steam pressure rating is usually 10 kg/cm^2. The process steam consumption is about 2 tonnes per tonne of paper at a pressure of 4 kg/cm^2 in the paper machine and 4 tonnes per tonne of the pulp at a pressure of 6 kg/cm^2 in the pulp mill. The paper machine condensate is partially returned to the boiler. The make-up water requirement for the boiler is met by a softener plant.

The flue gas from the boiler is discharged through an induced draft (ID) fan into a stack. Control systems—multiclones or bag filter are provide to control the particulate emissions.

Some units have diesel generating (DG) sets to meet the electrical power requirement in case of non-availability of power from the grid.

STATUS OF WASTE GENERATION

The agro residue-based pulp and paper mill generates wastes in the form of waste-water discharges, air emissions and solid wastes. The most prominent form of waste discharge is the waste-water, followed by air emissions and solid wastes.

Waste-Water

Table 17.2 depicts sources of waste-water generation from various sections/equipment.

Table 17.2. Sources of waste-water generation.

Section/Equipment	*Sources*
Raw material preparation section	Wet depithing of bagasse
	Wet cleaning of straw, elephant grass, etc.
Pulping section	
Digestor house	Leaks and spills of black liquor gland cooling water from refiners, etc.
Pulp washing	Unbleached pulp wash
Centricleaners	Rejects containing high concentration of fibres, grit or sand
Decker thickener	Filtrate of pulp thickening
Hydropulper potcher	Wash water
Bleaching	Chlorination stage waste-water containing chlorolignin
	Hypochlorite waste-water
Stock preparation section	Leaks and spills of chemicals/additives and floor wash
Paper machine	
Centricleaner rejects	Rejects containing fibres, grit or sand
Wire pit	Waste-water containing fibre
Fan pump pit	Fan pump pit overflow
Couch decker	Filtrate of couch decker containing fibres, talc and sizing chemicals
Utilities	
Boiler	Boiler blow-down
Softener plant	Softener regeneration discharges

The section-wise and combined waste-water specific pollution load factors of agro residue-based mills is presented in Table 17.3.

Table 17.3. Waste-water pollution load.

Parameter	*Pulp mill section*	*Paper machine*	*Factory combined*
Waste-water flow rate (m^3/T)	110–168	60–80	162–230
BOD (kg/T)	128–210	27–41	161–245
COD (kg/T)	705–1140	80–140	865–1215
TSS (kg/T)	118–262	61–109	192–362
TS (kg/T)	872–1292	205–394	993–1408

Air Pollution

In small paper mills, air pollution occurs mainly from two sources—digesters and steam boilers. A third possible source is the captive power generation facility.

In agro residue-based mills, after the raw materials digestion with caustic soda is completed, the pressure in the digester is released. During this process, about 1.4 tonnes of steam per tonne of pulp escapes into the atmosphere. The escaping gases contain volatile organic compounds. Unlike the conventional sulphate mills (kraft mills), these mills do not generate hydrogen sulphide and mercaptans (organic sulphides), as sodium sulphide is not used in pulping. The pollution is mostly confined to the surroundings of the mill.

Agro residue such as paddy husk and bagasse are the most commonly used fuel for generating steam. Besides agro residue, coal and diesel oil, etc. are also used as fuel in some mills. The air pollutants of concern are:

1. Suspended particulate matter (SPM).
2. Sulphur dioxide.
3. Oxides of nitrogen.

Mills with new boilers generally employ particulate removal devices, such as a multi-clone or bag filter. The gaseous emissions from boilers as well as from DG sets are usually taken care of by dispersion through stacks of the appropriate height.

Solid Waste

Solid waste is basically generated from the process, and from pollution control facilities. The sources and quantities of solid waste generated are given in Table 17.4. The amount of waste (rejects) from centricleaner and screenings is very small, and is just a small fraction of the total waste generated during raw material handling.

It can be seen that small paper mills have to handle 0.3 to 0.4 tonnes of organic solid waste and 0.7 to 1.1 tonnes of boiler ash per tonne of paper produced. At present, the solid wastes, particularly boiler ash, are disposed of in low-lying areas. After natural solid drying, the organic solid waste is either disposed of by burning, or sold to secondary users, such as brick kilns.

The waste generation in pulp and paper is quite significant. It not only represents a loss of resources, but also demands high treatment and disposal costs. Waste minimisation efforts are, therefore, likely to yield the dual benefits of better resource utilisation and reduced pollution control costs.

Table 17.4. Sources and quantities of solid waste.

Source	*Specific waste generation (kg/T)*
Production process	
Raw material cleaning	70–100
Johnson screen	20–60
Centricleaner rejects (from trap and riffler)	
Pulp mill	10–20
Paper machine	8–15
Treatment facility	
Effluent treatment sludge	
Primary clarifier	60–90
Secondary clarifier	100–150
Dust control system in boiler	500–750
Utilities	
Boiler bottom ash	175–350

WASTE MINIMISATION (WM)

In order to be able to tap the available potential of waste minimisation (WM) in the pulp and paper industry, let us first get familiar with the concept of waste minimisation.

Waste minimisation is a new and creative way of thinking about products and the processes which make them. It is achieved by continuous applications of strategies to minimise the generation of wastes and emissions. Waste minimisation is best practised by reducing the generation of waste at the source itself. After exhausting the source reduction opportunities, attempts should be made to recycle the waste within the unit. The type of techniques which are available in the two areas along with a few illustrative examples are given below:

WASTE MINIMISATION TECHNIEUQES

Recycling

1. Recycle back water and white water in pulp washing, bleaching and pulp dilution actitives.
2. Recycle couch pit pulp in machine chest.
3. Recover and recycle steam condensate.
4. Recover and recycle fibre from white water by installing efficient save-all systems.

Creation of useful by-products

1. Manufacture lignosulphate from black liquor.
2. Use short/waste fibre for board-making.
3. Generate biogas from organic waste.
4. Use dust from raw material cleaning as fuel in boiler.

Product Change

1. Produce high yield variety of paper.
2. Produce unbleached paper instead of bleached paper.

Technology change

1. Modify pulping processes, for example, by using anthraquinone in cooking.
2. Consider alternative pulping processes such as urea pulping.
3. Modify washing and dewatering process, for example, by using twin wire belt press.
4. Use environment-friendly bleaching processes, such as oxygen bleaching, ozone bleaching.

Source Reduction

Process change

Better process control

1. Optimise cooking process.
2. Practise hot stock refining.
3. Refine at highest possible pulp consistency.
4. Use retention aid chemicals for maximising the utilisation.

Equipment modification

1. Install efficient showers.
2. Have blow tank of adequate size to avoid spillage of pulp.
3. Provide broke pulper.
4. Use high pressure fibre saver in centricleaner.
5. Use capacitors for power factor improvement.
6. Provide couch pit and couch pit pump of adequate size.
7. Install consistency regulator.

Input material change

1. Use non-toxic dyes.
2. Hydrogen-peroxide bleaching.

Good Housekeeping

1. Repair all leakages.
2. Keep taps closed when not in use.
3. Cover vibratory screens to avoid spills.
4. Prevent shower from being blocked.

Need for Waste Minimisation

The conventional end-of-pipe pollution control in small scale agro residue-based pulp and paper mills is very expensive. It has been estimated that the investment cost of effluent treatment alone could be as high as 20 per cent of the plant and machinery cost. The annualised operating cost could be upto 10 per cent of the turnover. A better approach would be to first exploit the opportunities of waste minimisation and control only the residual pollution. This approach not only leads to resource savings and hence lower production cost, but also, reduces the effluent treatment cost. Waste minimisation is necessary from other aspects also, as described below.

Need from regulatory point of view

Meeting progressively more stringent discharge standards has been an major problem faced by the pulp and paper industry. As discussed earlier, conventional effluent treatment is highly expensive. Additionally, it is very difficult to achieve the prescribed standards with such treatment facilities, unless one goes in for advanced treatment methods which could be beyond the technical and financial capabilities of this sector. The adoption of waste minimisation techniques renders the residual effluents relatively easier, simpler and cheaper to treat and would enable mills to approach the prescribed standards.

Need from the point of view of raw material conservation

Given the increasing cost of input materials—raw material, chemical and energy—no industry can afford to lose these resources in the form of waste. The specific consumption of these raw materials can be considerably reduced by adopting waste minimisation measures, for example, by recycling some waste streams during the manufacturing process.

Need from the point of view of water conservation

Water has increasingly become a scarce commodity, and several industries are experiencing acute water shortages, specially during lean periods. Most small scale agro residue-based pulp and paper mills are not located close to water bodies with abundant water supplies, and therefore have to depend on ground water. The continuous lowering of the ground water table leads to water shortage, increasing pumping costs. Instances have been reported where the industries had to curtail their production levels for want of adequate amounts of water. Efforts made towards conserving water not only get paid back through reduced costs, they also ensure continuous availability of water, thus sustaining desired production levels.

Need from the point of view of energy conservation

The pulp and paper manufacturing is an energy intensive industry with energy accounting for more than 25 per cent of production costs. It has been found that, there is a potential for saving energy to the tune of 20 per cent by simple, low-cost measures. For the industry to retain its competitiveness and also to reduce its energy intensiveness, it is essential that it take steps to reduce energy consumption, which can be achieved through a concentrated waste minimisaton effort.

Need from the point of view of operating cost of ETP

As mentioned earlier, unlike large scale plants, small scale plants can hardly afford to install capital intensive waste-water treatment plants which require heavy investments and entail high recurring operational costs. Waste minimisation techniques help in reducing the overall pollution load and hence the cost of effluent treatment.

Need from the point of view of environment pressure groups and worker's awareness

The public awareness in the field of environment protection is growing day by day. Over the last few years, several non-government environmental protection organisations have come up, not only to create awareness in this field but to also act as watchdogs against polluters. The industry can no longer keep itself isolated from such pressure groups and has to take positive steps to mitigate environmental pollution. Waste minimisation could be one such positive step. In addition, today's industrial workers are far more aware of their right to a clean work environment. Number of skilled workmen, now consider the environment as an important factor in an organisation. Ensuring a clean environment through waste minimisation could be a definite way of attracting, retaining and even motivating good workmen. It also inculcates, among the workers, a mentality of controlling waste, which is of prime importance in getting the competitive edge. It could thus be concluded that the implementation of waste minimisation measures serves the dual objectives of reducing raw material, energy and chemical losses as well as brings down the investment and operating costs of pollution control systems.

Potential for Waste Minimisation

Small pulp and paper mills in India operate under constraints, such as obsolete and inefficient technology, a small scale of operation, poor grades of raw material which changes from seasons to season, low profit margins and untrained manpower. Further, the technology for chemical recovery from black liquor is still not developed for this sector, resulting in higher chemical consumption as well as higher pollution load. This leads to high levels of waste generation and decreased profitability. The potential of waste minimisation in this sector is, therefore, very high. Table 17.5 illustrates the potential of waste minimisation in quantitative and monetary terms.

Table 17.5. Potential of waste minimisation in small agro residue-based pulp and paper mills*.

Area	*Unit*	*General performance/ consumption range*	*Potential of waste minimisation*	
			Quantity	*Value*
Fibre	Yield %	40–44	40–100 kg/T	250–600 Rs./T
Caustic	kg/T	60–120	10–15 kg/T	100–150 Rs./T
Additives	kg/T	11–15	1–2 kg/T	5–10 Rs./T
Steam	Tonne/T	4.5–6.5	0.5–1.5 T/T	125–400 Rs./T
Electrical energy	kW/T	850–980	60–120 kW/T	150–300 Rs./T
Overall capacity	Per cent	5–20	75–300 kg/T	75–300 Rs./T
Water consumption	m^3/T	180–280	60–120 m^3/T	25/50 Rs./T
Total potential				730–1800 Rs./T

* These figures are an indication of the magnitude of waste minimisation (WM) in average agro-based pulp and paper mills of 30 TPD.

The overall potential of waste minimisation and subsequently enhancing profit margin in the pulp and paper sector is about Rs. 730 to 1800 per tonne of paper produced. In addition, waste-water treatment costs get reduced from a level of Rs. 1000–1250 per tonne to Rs. 500–700 per tonne of paper made.

It can be seen from Table 17.6 that about 40 per cent of waste minimisaton potential can be tapped by implementing just the minor ones.

Table 17.6. WM potential of minor, moderate and major measures.

Area	*Units*	*Tapping of WM potential*		
		Minor measures	*Moderate measures*	*Major measures*
Fibre recovery	kg/T	16–40	10–25	14–35
Caustic saving	kg/T	5–7	5–7	0.1
Additives saving	kg/T	0.5–1.0	0.5–0.8	0–0.2
Steam conservation	T/T	0.2–0.6	0.1–0.3	0.2–0.6
Electricity saving	kW/T	18–35	18–35	24–50
Capacity increase	kg/T	15–60	22–90	38–150
Water conservation	m^3/T	30–60	15–30	15–30

Let us now have a look at the various waste minimisation opportunities in greater detail.

WASTE MINIMISATION OPPORTUNITIES

The experience of number of agro residue-based pulp and paper mills show that there are several proven WM techniques which could be implemented to tap the WM potential already discussed. The impact of WM measures often extends beyond the point of application and has a cascading effect on other sections/ operations as well.

Better raw material preparation has its beneficial effect on pulping, washing, paper-making and even on paper quality. Most importantly, these measures enable reduction in environmental costs due to reduced end-of-pipe pollution control requirements.

In this section, various WM measures have been described. For the sake of clarity, the WM opportunities have been grouped under the four major production sections—the raw material preparation section, pulping section, stock preparation section and paper-making section. The technical requirements, economic viability, environmental aspects and anticipated benefits for each measure are also discussed.

Raw Material Preparation

The fibrous raw material used by the industry contains 8 to 10 per cent dust and fines and upto 25 per cent pith in the case of bagasse. These fines (impurities) do not contribute to the pulp value. Instead, they consume costly cooking chemicals and steam. In addition, they end up as pollution load in terms of COD and TS.

The impurities adversely affect the washing efficiency and overload the centricleaner. Those impurities which escape the centricleaner cause specks (black spots) on the finished paper. They also increase the wear and tear (maintenance) of mechanical equipment (refiner, pumps, etc.) Most importantly, the drainability of pulp in the machine wire part is adversely affected, resulting in reduced paper machine speed and consequent loss of production.

The appropriate preparation of raw material is therefore not only important from the economic standpoint, but also from the environmental, equipment efficiency and product quality points of view. Proper raw material preparation can result in savings (Table 17.7).

Table 17.7. Benefits of proper raw material preparation.

Potential area	*Savings/improvement quantity*	*Potential value/tonne of paper*
Chemical saving—caustic	12–18 kg/T paper	Rs. 120–180
Steam requirement for cooking	0.1–0.2 T/T paper	Rs. 25–50
Increase pulping capacity	6–8%	Rs. 25–100
Total		Rs. 170–330
Reduction in pollution load	8–10%	Rs. 95–120
Total		Rs. 265–450

Unfortunately, in most agro residue-based pulp and paper mills, the raw material preparation section is conspicuous by its absence. A myopic view might indicate that there is, therefore, no waste generated. However, as discussed earlier, this has a serious deleterious effect in all downstream sections.

Pulp Mill Section

This section is the single largest polluting section in a pulp and paper mill, accounting for about 80 per cent of the total pollution load. At the same time, it also has scope for the largest number of WM interventions, ranging from raw material substitution of technology modifications, and recycling.

The overall potential of benefits accruing from waste minimisation in the pulp mill section is given in Table 17.8.

Table 17.8. Benefits of proper pulping operation.

Potential	*Savings/improvement quantity*	*Potential value/tonne paper*
Chemical saving—caustic	3–12 kg/T paper	Rs. 30–120
Fibre recovery	20–45 kg/T paper	Rs. 120–240
Water conservation	30–60 m^3/T paper	Rs. 12–25
Reduced steam requirement	0.2–0.6 T/T paper	Rs. 50–150
Reduced electrical energy	20–50 kWh/T paper	Rs. 50–125
Increased pulping capacity	5–7%	Rs. 20–75
Total		Rs. 282–735
Reduction in pollution load	25–30%	Rs. 300–360
Extra fuel value	25–40 kg/T paper	Rs. 15–25
Total		Rs. 597–1120

The raw material preparation techniques discussed in Table 17.9 emerge as important WM measures due to their impact on waste reduction in downstream sections. The various waste minimisation measures which would enable tapping of this potential are discussed in Table 17.10.

Stock Preparation Section

The major operation in the stock preparation section is the blending of pulp to make it suitable for making paper. Additives such as sizing chemicals, fillers, dyes, etc., are blended with the pulp. As such, there is no significant waste generation from this section. Waste minimisation opportunities, as discussed in Table 17.11, predominantly relate to chemical suostitution and process optimisation which would have an impact on waste in the paper-making section.

Paper-Making Section

The paper-making section is the largest consumer of fresh water in the entire mill, accounting for more than 60 per cent of the total water consumption. The total amount of water in circulation is still higher, as large quantities are recycled as tray water. Consequently, the waste water generation is also the largest in this section. As a general rule, the pulp enters the section at 4 per cent consistency, which improves upto 20 per cent after mechanical dewatering, resulting in the generation of 20 m^3 of waste-water per tonne of paper. Loss of fibre along with waste-water is a serious loss, but most fibre loss minimisation measures are economically viable. Enhancing the productivity of the paper machine also simultaneously affects the generation of waste.

The paper machine is highly energy-intensive, in terms of both electrical energy and steam energy, and substantial savings can be obtained by cutting down energy waste. The waste minimisation measures are based upon water recycling, fibre recovery, energy conservation and improvements in paper machine productivity. The various WM measures in this section are discussed in Table 17.12.

Utilities

In the utilities section, the steam boiler is the major source of waste generation. It generates solid, liquid and gaseous wastes. Quite often, the role of the boiler departments is limited to supplying steam at the desired pressure and in desired quantities. The efficiency of steam generation is rarely taken seriously, resulting in increased waste generation and higher energy losses. The waste minimisation measures discussed in Table 17.13 pertain mainly to boiler efficiency improvement.

Housekeeping Measures

The general housekeeping in most of the small agro residue-based pulp and paper mills is far from satisfactory. Good housekeeping is, in fact, the first step in any waste minimisation programme. The expenditure is small and the savings are immediate, with very attractive pay-back period—usually less than a year. A well kept mill has tremendous psychological impact on the employees and helps in building the right attitude for waste minimisation.

Other side benefits such as better working conditions, reduced risks and hazards, etc., further motivate the employees. Housekeeping-related waste minimisation measures are mostly unit specific and can be easily identified through a careful inspection of the production sections. A range of such measures found in some mills, albeit to varying degrees, is given below. The technical requirements are usually minimal and have therefore not been discussed in detail. The economics (expenditure and savings) is mill-depended and it is difficult to give a general range. The measures given below should be taken as guidelines, and specific measures should then be evolved from case to case.

Suggested housekeeping measures

1. Installation of appropriate chutes, to collect screening rejects. The collected rejects should be handled and disposed with minimum spillage.
2. Repair of raw material converyor, to prevent spillage of raw material and contamination of cooked pulp.
3. Modification of the digester loading chute, to prevent the spillage of raw material.
4. Covering of all vibratory screen and chemical dosing tanks by proper lids to prevent spills.
5. Provision of dykes in rag pulp dumping area, to contain and channelise the flow of black liquor into the drain.
6. Installation of spring-actuated self-closing valves in all water hose pipes, to minimise water wastage.
7. Avoidance of spillage of lime sludge in the hypo section, by proper containment, handling and disposal.
8. Proper collection and storage system for dedusting rejects, to contain their spillage.
9. Provision of an appropriate discharge system for unloading cooked pulp in potchers (in cases where the pulp is directly discharged in potcher), to avoid spillage of pulp and black liquor.
10. Control of leakages and spillages in the handling and preparation of chemicals and additives.
11. Avoidance of pump gland leakages through proper pump maintenance.
12. Timely repair/sealing of water and steam leakages from pipes, valves, flanges.

Table 17.9. Waste minimisation measures in the raw material preparation section.

Waste minimisation measure	*Anticipated benefits*	*Technical requirements*	*Economic* viability*	*Environmental impact*	*Remarks*
Dry dedusting of straw for removal of fines and dust	Reduction in consmn of: Steam 7–10% Caustic 7–10% Electrical (marginal) Reduction in screening rejects by 10–15% Increased pulping capacity Fines and dust can be used as fuel Enables re-use of black liquor due to reduced silica content Better runability of paper machine Reduced wear and tear of pumps and refiners	Equipment: Vibratory/rotary screen Conveyor Dust handling system Manpower: No additional skilled manpower requirment Process: Application for RM having moisture <15%	I = Rs. 2–3.5 lakh P = < 1 year	Reduction in TS by 6–8% and in COD by 5–10%	All equipment/technology available indigenously. The measure is easily implementable.
Wet cleaning for removal of fines and dust	Reduction in consmn of Steam 7–10% Caustic 8–12% Reduction in screening rejects by 10–15% Enables re-use of black liquor due to reduced silica content Increased pulping capacity Better pulp quality Better runability of paper machine Reduced wear and tear of pumps and refiners	Equipment: Cleaning pond Screw conveyor Showers Dewatering equipment Process: Additional space reqmt of about 5,000 sq. mt Manpower: No additional skilled manpower reqmt. Technology: Indigenous dewatering equipment not found to be successful	I = Rs. 15–20 lakh S = Rs. 3–4 lakh P = > 5 years	Reduction in TS by 8–10% and COD by 8–12%	The measure, though environmentally attractive,requires high capital investment and needs additional space. Additional quantities of wastewater are generated.

Installation of disk cutter for removal of fines, dust and non-cellulosic material	Reduction in consmn of: Steam 15–20% Caustic 15–20% Reduction in screening rejects by 15–20% Enables re-use of black liquor due to reduced silica content Increased pulping capacity Fines and dust can be used as fuel Better pulp quality Better runability of paper machine Reduced wear and tear of pumps and refiners	Equipment: Air blower Screw conveyor Disk mill Cyclone Collection chutes Process The measure is flexible to suit different raw materials, provided moisture is < 15% Technology: Disk mill not available indigenously Manpower: Skilled manpower required	I = Rs. 10–15 lakh S = Rs. 2–3.5 lakh P = > 4 years	15–20% reduction in TS and 8–15% in COD	Recommended mainly for rice straw. For other agro residues, the fibrous RM loss could be high (upto 35%) Additional fuel
Using 25% top cut elephant grass (the top 25% is mainly leaves and doesn't contribute to fibre value)	Improved packing density Increased pulping capacity Better pulp quality Reduced caustic reqmt. by 3–7% Reduced steam reqmt. Easy RM transportation Better runability of paper machine Dried out leaves can be used as fuel	Equipment: Cutter Conveyor Technology: Indigenous cutter are not effective and require additonal unskilled manpower	I = Rs. 1–2 lakh S = Rs. 2–3 lakh P = < 0.5 years	Redution in TS load by 8–10% and in COD load by 10–12%	A good technology, yet to be demonstrated
Baling of wheat and rice straw	Reduced storage area Reduced fibre degradation Better housekeeping Reduced loss due to entrainment in air Easy handling	Equipment: Baling machine Bale breaker Process: Yet to be demonstrated Manpower:	I = Not available S = Not yet quantifiable	Reduced air-borne particulate. Reduced pollution load due to degraded raw material	The measure is feasible only if mechanical harvesting is practised

(Contd...)

Waste minimisation measure	*Anticipated benefits*	*Technical requirements*	*Economic* viability*	*Environmental impact*	*Remarks*
	Only feasible with mechanical harvesting	Additional unskilled manpower required Technology: Not indigenously available			
Depithing of bagasse	Reduced consmn of: Steam 10–15% Caustic 15–20% Electricity (marginal) Enables the use of BL due to reduced viscosity Improved pulp yield Increased pulping capacity Removed pith can be used as fuel Better runability of paper machine	Equipment: Depither Allied equipment Technology: Indigenously available Manpower: No additional manpower is required	I = Rs. 4–6 lakh S = Rs. 5–6 lakh P = < 1 year	20–25% reduction in TS and 15–30% in COD	The measure is easily implementable

* The assumed capacity of paper mill for all listed options is 30 TPD.

I = Investment; S = savings; and P = payback period.

Note: The following abbreviations means: RM = Raw material; reqmt. = requirement, consmn. = consumption

Table 17.10. Waste minimisation measures in the pulping section.

Waste minimisation measure	*Anticipated benefits*	*Technical requirements*	*Economic viability*	*Environment impact*	*Remarks*
RM impregnation with recycled liquor	Improved packing density (7–10%) Increased pulping capacity by 7–10% Reduced cooking time by 5–7% Reduced caustic Reduced steam reqmt. Enables low bath ratio Uniform pulp quality	Equipment: Belt conveyor Liquid feed tank Showers Black liquor feed tank Paddle mixer Process: Suitable for all agro residues Manpower: No additional manpower is required	I = Rs. 1–15 lakh S = Rs. 0.5–1 lakh P = < 2 years unquantifiable	Reduction in pollution load	The measure is easily implementable black
RM impregnation with caustic	Improved packing density Increased pulping capacity Reduced cooking time Reduced caustic consmn. by 3–7% Reduced steam consmn. Enables low bath ratio Uniform pulp quality	Equipment: Belt conveyor Showers Liquor feed tank Paddle mixer Caustic collection tank Process: Suitable for all agro residues Manpower: No additional manpower is required	I = Rs. 5–8 lakh S = Rs. 3–4 lakh P = < 2 years	Marginal reduction in pollution load	The measure is easily implementable
Multiple loading of digester	Increased digester capacity by 10–15% Reduced steam consmn. Increased pulp yield	Nil	Savings due to reduced steam consmn. to be quantified on case-to-case basis, other savings difficult to quantify	10–15% reduction in digester vent gases	The extra time required for double loading will be more than offset by the reduction in number of batches for same pulp production
Reduction in bath ratio	Reduced consmn. of: steam by 5–10% cooking time	Nil	I = Nil S = Rs. 3–6 lakh	No direct impact	No financial involvement. Measure is easily implementable

(Contd...)

Waste minimisation measure	Anticipated benefits	Technical requirements	Economic viability	Environmental impact	Remarks
Using anthraquinone or anthraquinone-plus-molasses as cooking aid	Higher pulp yield Improved delignifiction Reduced washing reqmt. Reduced water consmn. Improved bleachability Better pulp quality Better runability of paper machine	Technology: Availability of good quality anthraquinone in domestic market is uncertain	S = Unquantifiable	Reduction in TS and COD load in effluent and better quality of pulp	Increase in chemical cost will be compensated by high yield
Neutral ammonium sulphite cooking	Enables use of black liquor as fertiliser Requced bleaching chemical reqmt.	Process: Not yet successfully demonstrated	S = Not yet quantified	80% reduction in pollution load, as BL can be used for agricultural applications	Due to the disadvantages, such as corrosion of digestor, higher cooking temperature and time, and emission of ammonia gas, this measure needs further investigation
Alkaline sodium sulphate cooking	Reduced washing reqmt. Reduced water consmn. Improved bleachability Reduced bleaching chemical reqmt. Higher paper strength	Nil	S = Unquantifiable	Reducion in colour intensity of effluent	If proper pH conditions are not maintained, it could lead to corrosion of digestor
High yield pulping, that is, mechano-chemical pulping for unbleached paper	High pulp yield 8–15% Reduced cooking chemicals	Equipment: Mechano-chemical digestors Technology: Available indigenously	I = Rs. 10–15 lakh S = Unquantifiable	Reduction in TS and COD by 10–15%	Increase in steam and electrical energy consmn.
Optimisation of cooking process (that is, operating at required steam pressure, temperature, chemical dosing)	Better pulp quality Reduced uncooked and screening rejects Reduced refining reqmt. Reduced chemical consmn. Reduced steam reqmt. Better pulp yield	Equipment: Temperature indicator Pressure indicator Control panel Manpower: Skilled operator required	I = Rs. 1–1.5 lakh S = Unquantifiable	Reduced overall pollution load	Strongly recommended for all mills

Screw press for removal of concentrated black liquor from pulp	Easy handling and treatment of black liquor Enables recycling of BL for using residual caustic Hot BL recycle reduces steam reqmt. Enables better impregnation of RM with BL Reduced washing time by 50–60% Water conservation in washing by 50% Reduced energy reqmt. in effluent treatment Enables solar drying of black liquor Reduced fibre loss at washing stage	Equipment: Recirculation tank Blow tank High consistency pump Dillution pump Screw press Piping Process: Unwashed pulp conveying system to be modified Technology: Indigenously available but needs modification Manpower: Additional skilled manpower required	I = Rs. 1.2–3 lakh S = Rs. 1.5–3 lakh P = < 1 year	Reduction in effluent volume by 40–50%	The measure is at pilot scale and is being implemented in some mills
Twin-wire belt press for pulp dewatering	Easy handling and treatment of black liquor Enables recycling of BL for using residual caustic Hot BL recycle reduces steam reqmt. Enables better impregnation of RM with BL Elimination of separate batch washing Water conservation in washing by 80% Reduced energy reqmt. in effluent treatment Enables solar drying of black liquor Reduced fibre loss at	Equipment: Twin-wire belt press and circulation pumps Blow tank High consistency pump High density tower Process: Washing system almost eliminated Technology: Available but expensive Manpower: Skilled manpower is required	I = Rs. 10–15 lakh S = Rs. 6–10 lakh P = < 2 years	Reduction in effluent volume by 60–70%	Versatile piece of equipment, usable for any agro residue raw material

(Contd...)

Waste minimisation measure	*Anticipated benefits*	*Technical requirements*	*Economic viability*	*Environmental impact*	*Remarks*
	washing stage Low power consmn.				
Vacuum washers for washing of pulp: With barometric leg With vacuum pump	Easy handling and treatment of black liquor Enables recycling of BL for using residual caustic Hot BL recycle reduces steam reqmt. Enables better impregnation of RM with BL Reduced washing time due to continuous wash Water conservation in washing by 50% Reduced energy reqmt. in effluent treatment Enables solar drying of black liquor Reduced fibre loss at washing stage	Equipment: Tall civil structure High density tower High consistency pump Barometric leg or vacuum pump Vacuum washers Process: Counter-current multi-stage washing to be incorporated Technology: Indigenously available but expensive Manpower: Skilled manpower required	I = Rs. 60–80 lakh S = Rs. 6–10 lakh P = 8–10 years	Reduced effluent volume	This option recommended for medium-sized mills (35 TPD and above). The measure would be attractive when taken up in conjuction with solar evaporation measure
Segregation of initial thick black liquor from potchers and its recirculation to the extent possible	Easy handling and treatment of remaining BL Re-use of residual caustic due to BL recycle Hot black liquor will reduce steam reqmt. Enables better impregnation of RM and BL	Equipment: Collection sump Pumps and pipes Process: Separation drains required for thick and weak black liquor	I = Rs. 0.5–0.8 lakh S = Rs. 1.5–2.5 lakh .P = < 1 year	Reduction in sodium load in effluent	The measure is simple and easy to implement
Solar evaporation of concentrated black liquor generated by implementation of any of the four preceding measures	Use of BL cake as fuel Reduced treatment cost Reduced colour of effluent Improves achievability of discharge standards	Equipment: Briquetting machine Civil work for solar evaporation ponds with appropriate lining, etc.	I = Rs. 15–40 lakh S = Rs. 20–40 lakh P = < 1 year (investment mainly for land cost)	30–40% reduction in overall pollution load	Availability of land in nearby areas is a must

	BL cake can be briquetted with RM dust and both to be used as fuel	Process: Space reqmt. of about 50,000 sq mt for solar ponds			
Anaerobic treatment of black liquor	Biogas generated can be used as fuel Reduced treatment cost Reduced colour of effluent	Equipment: Digester (anaerobic) Gas holder Burners Pressure piping Technology: Proven and available but expensive Manpower: Skilled and trained manpower requried	I = Rs. 125–175 lakh S = Rs. 20–45 lakh P = > 5 years	30–40% reduction in overall pollution load	The measure is economically not attractive for a typical small mill
Chemical recovery from black liquor	Caustic recovery 60–80% Reduced pollution load	Equipment: Recovery boiler and associated system Pumps Multi-stage evaporator Causticising plant Lime sludge recovery or disposal system Technology: Not yet proven Manpower: Skilled manpower required	I = Rs. 300–600 lakh S = Rs. 15–30 lakh P = > 20 years	Upto 50% reduction in COD load	Technological problems of high silica content, high viscosity (due to high % of fines) and low solids and lignin content in raw BL (thus lesser heat generation in boiler) yet to be overcome. The measure is not economically viable
Hot stock refining	Reduced electrical energy consmn. Reduced fibre degradation Better pulp quality Reduced screening rejects Improve runability on wire	Equipment: Refiner Process: Sequence of washing and refining to be modified	I = Rs. 1–2 lakh S = Difficult to quantity	Reduced solid waste	Well-proven technology. Can be easily implemented

(Contd...)

Waste minimisation measure	Anticipated benefits	Technical requirements	Economic viability	Environmental impact	Remarks
Use of screening rejects for unbleached paper-making	RM saving Cooking chemical saving	None	I = Nil S = Rs. 1–2 lakh	Reduced solid waste	The measure is valid only for mill making both bleached and unbleached paper
Conversion of lignin in black liquor to lignosulphonate	Reduced pollution load and effluent colour Additional by-product which is an import substitute	Equipment: Plant and equipment for manufacture of lignosulphonate Technology: Not provent for agro residue-based black liquor	I = Rs. 300–400 lakh S = Rs. 15–20 lakh P = 20 years	Upto 50% reduction in COD load	It is a very expensive measure and virtually amounts to setting up another industry. Trial samples of product not accepted by market
Recovery of ligning for land application	Reduced pollution load and effluent colour Additional by-product Increased crop yield	Equipment: Plant and equipment for lignin recovery Sulphuric acid and other chemicals Process: Sophisticated layer separation required	I = Rs. 8–10 lakh S = Yet to be established due to undeveloped market	Upto 50% reduction in COD load	Market of lignin as a soil conditioner and fertiliser yet to be developed. Impact of sodium on soil not established
Buffering with caustic soda before bleaching	Reduced fibre degradation and thus higher pulp yield Reduced bleaching time Better bleaching and hence brighter pulp Better paper machine runabiliy	Process: Precise pH control (at least 8.5) required	Difficult to quantify	Not yet established	Suggested only for mills making high quality paper
Vacuum bleacher for pulp bleaching With barometric leg: With vacuum pump qu-	Reduced water reqmt. Enables bleaching at higher consistency Enables multi-stage counter-current washing, thus improving washing efficiency and reducing fibre loss Increased bleaching capacity	Equipment: Barometric leg or vacuum pump Tall civil structure High density tower High consistency pump Vacuum bleachers Process:	I = Rs. 60–80 lakh S = Rs. 10–15 lakh (estimated) P = 5–6 years	Reduced AOX fromation Reduced COD load. Enables total recycle to bleach wash water	The measure is highly cost intensive. Applicable only for mills making high ality bleached paper

	Reduced chemical consmn.	Clear high vertical space required for installing equipment			
Bleaching with sodium hypochloride	Better pulp quality Less sludge generation from bleaching chemical preparation section Smaller chemical preparation are reqmt. due to higher content of chlorine in bleach liquor	Equipment: NaOCl stroge tank	The measure is not economically viable due to the high cost of NaOCl	Handling and disposal of solid waste is reduced	The mesure can be implemnted if market price parity betwen NaOCl and $CaOCl_2$ is brought about
Bleaching with hydorgen peroxide (H_2O_2)	Better pulp quality Significant environmental benefits due to: No chlorine in shop air No chlorine in effluent No AOX generation No sludge generation in bleaching chemical preration section Chemical prepration area is smaller Elimination of hazard in chlorine handling	Equipment: H_2O_2 handling and dosing equipment Process: Care and safety reqmt. for handling	I = Rs. 0.5 lakh	Chlorine-free shop air. No chlorine in effluent. No AOX generation. Elimination of sludge in bleach liquor preparation	The measure yet to be demonstrated in agro residue-based mills
Oxygen-assisted bleaching (OCEoP)	Better pulp quality Significant environmental benefits due to: Less chlorine in shop air Less chlorine in effluent Less AOX generation No sludge generation in bleaching chemical preparation section Chemical preparation area is smaller	Equipment: Oxygen preparation equipment Process: Safety reqmt. in oxygen handling	Not yet quantified	Less chlorine in shop air and in effluent. Less AOX generation	The measure is yet to be demonstrated in agro residue-based mills. Apparently, it is not economically viable

(Contd...)

Waste minimisation measure	*Anticipated benefits*	*Technical requirements*	*Economic viability*	*Environmental impact*	*Remarks*
	Reduction of hazard in chlorine handling				
Bleaching with ozone	Better pulp quality Significant environmental benefits due to: No chlorine in shop air No chlorine in effluent No AOX generation No sludge generation in bleaching chemical preparation section Elimination of hazard in chlorine handling	Equipment: Ozonation plant Process: Safety reqmt. for ozone preparation and handling Technology: Not available indigenously Yet to be proven for agro residue mills	Not yet quantified Ozonation plants are very expensive	Chlorine free shop air No chlorine in effluent No AOX generation. Elimination of sludge in bleach liquor	The measure is yet be demonstrated in agro residue-based mills. Import of ozonation plant required preparation
Provision of hot water for maintaining bath ratio in digester recovered from DG set waste heat	Reduction in cooking time 30–45 minutes Reduction in steam reqmt. 0.2–0.5 tonne per tonne paper	Equipment: Piping and pumps Waste heat recovery unit Technology: Indigenously available	I = Rs. 8–10 lakh S = Rs. 10–12 lakh P = < 1 year	Not direct impact	Option is valid for units having captive DG set and operating if for at least 40% of time
Insulation of digester	Reduced steam reqmt.	Equipment: Insulation material	I = Rs. 1–1.5 lakh S = Rs. 2–3 lakh P = < 1 year	Nil	Care needs to be taken to avoid sagging and deterioration due to the spillage of BL

I = Investment; S = savings; and P = payback period.

Note: The following abbreviations means: RM = Raw material; reqmt. = requirement, consmn. = consumption

Table 17.11. Waste minimisation measures in the stock preparation section.

Waste minimisation measure	*Anticipated benefits*	*Technical requirements*	*Economic viability*	*Environmental impact*	*Remarks*
Using poly-aluminmum silicate-sulphate (PASS) as sizing chemical, instead of alum	Improved drainability Increased filter retention Enables manufacture of high grammage paper	Nil	I = Not quantified S = Not quantified	Reduced TS in effluent	The option not yet tried in agro-based mills. The effectiveness of size is increased when PASS is used with cationic starch
Adoption of alkaline neutral sizing to enables making high ash content paper	Lower pulp consmn. per tonne of paper Lower production cost and higher profitability Reduced pollution load due to lower pulp reqmt.	Technology: Yet to be demonstrated in agro-based mills Process: The process is applicable only when pulp temp. in blending chest is < 38°C	I = Negligible S = Yet to be quantified	No direct impact	Applicable only for mills making writing and printing paper
Consistency indicator	Consistency regulation becomes easy Variation in grammage avoided Reduction in paper breakage Enables uniform paper drying	Equipment: Consistency indicator	I = Rs. 15–20 lakh S = Rs. 1.2–2.5 lakh P = < 1 year	Marginal reduction in TS load	Reduces dependence on human judgement
Consistency regular	Automatic consistency regulation Variation in grammage avoided Reduction in paper breakage Enables uniform paper drying	Equipment: Consistency regulator Technology: Indigenously not available Experience of some large mills not satisfactory	I = Rs. 15–20 lakh S = Rs. 2–4 lakh P = > 7 years	Marginal reduction in TS load	More suitable for continuous pulp-making. The measure is very expensive and not economically viable
Substituting existing dyes with less toxic or non-toxic dyes	Reduced toxicity of effluent Reduced toxic content in product Environment-friendly operation	Toxicity of existing dyes and development of non-toxic dyes yet to be studied. Chemical names of existing dyes are trade secrets	Not yet quantified since cost of substitute dyes not known	Marginal impact on effluent toxicity	The measure requires further basic studies on toxicity of existing dyes and development of less toxic dyes

(Contd...)

Waste minimisation measure	*Anticipated benefits*	*Technical requirements*	*Economic viability*	*Environmental impact*	*Remarks*
Use of the dye fixing agents	Reduced dye consmn. Reduced toxicity in effluent	Some of the indigenously available dye fixing agents are: SARSOLAN; TAMOL; MNOX	I = Nil S = Cost of dye fixing agents and possible savings yet to be quantified	Reduces toxicity in effluent	The measure is easily implementable. Attractive only for units having substantial dye reqmts.

I = Investment; S = savings; and P = payback period.

Note: The following abbreviations means: RM = Raw material; reqmt. = requirement, consmn. = consumption

Table 17.12. Waste minimisation measures in the paper-making section.

Waste minimisation measure	*Anticipated benefits*	*Technical requirements*	*Economic viability*	*Environmental impact*	*Remarks*
Fine-tuned dilution control at fan pump	Enables uniform pulp consistency Enables easy switchover from one grammage to other Reduces paper breakage	Equipment: Additional pipeline (of smaller diameter) with a valve to fine control the the water feed for dilution	I = Rs. 0.1–0.2 lakh S = Rs. 0.6–1 lakh P = < 1 year	No significant impact	Simple and easy to implement
Provision of high consistency pump in couch pit	Reduced fibre loss due to couch pit overflow during paper breakge Reduced consmn. of water in couch pit Reduced load on couch decker Reduced fibre loss in couch decker filtrate	Equipment: High consistency pump	I = Rs. 1–1.5 lakh S = Rs. 1–5 lakh P = < 1 year	Significant reduction in TS and COD	Saving and environmental impact depends on the extent of couch pit overflow
Controlled water pressure for edge cutting nozzle	Reduced paper breakage	Equipment: Separate water tank High pressure pump	I = Rs. 0.2–0.3 lakh S = Rs. 0.4–0.6 lakh P = < 1 year	Marginal reduction in TS and COD	Measure is easily implementable
Double felting to reduce press picking	Reduced paper breakage Higher fines allowable in pulp Increased production capacity due to higher machine speeds Reduced kerosene oil consmn. Reduced alumn consmn. at paper machine Reduced steam reqmt. in dryer because of better mechanical dewatering Reduction of VOC in waste-water	Equipment: Additional felt rolls	I = Rs. 3.5–4 lakh S = Rs. 4–6 lakh P = < 1 year	Reduced kerosene contamination in effluent	Will increase drying capacity of paper machine

(Contd...)

Waste minimisation measure	*Anticipated benefits*	*Technical requirements*	*Economic viability*	*Environmental impact*	*Remarks*
Recycling couch decker filtrate in pulp washing	Fibre recovery Reduced water consmn. Reduced pollution load	Equipment: Pipelines and pump	I = Rs. 0.5–1 lakh S = Rs 1.0–2 lakh P = < 1 year	Reduction in TS and COD load by 2–5%	The measure is easy to implement
Avoiding fan pump pit overflow by: Providing level control in fan pump pit or Recyling it to couch pit	Reduced fibre loss Reduced dilution reqmt in couch pit	Equipment: A small pipeline. The flow takes place by gravity Process: The level controller should actuate and control the fresh water supply	I = Negligible S = Difficult to quantity	Reduced in TS and COD load but difficult to quantify	The measure is very simple and easy to implement
Recycling wire pit water in showers	Reduced fresh water consmn. in showers Reduced effluent volume	Equipment: Multiplex filter unit High pressure pump Water collection sump	I = Rs. 4–6 lakh S = Rs. 1–1.5 lakh P = 4 years	Reduction in effluent volume	The measure is principally aimed at reducing fresh water. Additionally, the filter cake can be used for board-making
Removal of sand and waste-water	Prevention of sand and inerts from centricleaner Reduced wear and tear of pumps in ETP and hence reduced maintenance Prevention of choking of drains	Equipment: inert loading in ETP	I = Rs. 0.1–0.2 lakh Rifflers	Nil S = Nil	Frequent cleaning of drains is avoided
Recovery of fibre from centricleaner reject	Reduced fibre loss Reduced dependence on operators Reduced pollution load	Equipment: Fibre savers with high pressure pump or Hill screen with low pressure pump	I = Rs. 0.2–0.3 lakh S = Rs. 2–4 lakh P = < 1 year	Reduction in TS and COD load by 2%	An HP pump of the DAF system can be used to supply HP water
Provision of better nozzles cleaning showers	Reduced water consmn. Better cleaning efficiency resulting in better machine runability	Equipment: Fan flat type of other suitable nozzle in place of existing perforated pipe of simple nozzles High pressure pump	I = Rs. 1–1.5 lakh S = Rs. 0.5–0.7 lakh P = 2 years	Reduced effluent volume	Measure is easily implementable

		Technology: Nozzles and pump indigenously available			
Adjustment of paper width by edge cutting nozzles	Reduced paper trimming loss Reduced reprocessing of paper trimmings Marginal reduction in dryer steam consmn	Nil	I = Nil S = Would depend on market demand for paper of different widths	Nil	The width of paper is governed by market demand. The prevailing practice is to cut the paper to required size. Instead, the width should be reduced on deckel at the pulp web stage itself
Installation of broke pulper in paper machine section	Reduced reprocessing reqmt. Reduced filler and chemical consmn. Reduced handling reqmt. due to proximity of pulper to trimming generation point Reduced pollution load	Equipment: Broke pulper Pulp conveying pipes Pump Process: 10 sq m space is required in the paper machine section	I = Rs. 5–10 lakh S = Rs. 2–3 lakh P = 3–4 years	Reduced TS and COD load	The savings would be higher in the manufacture of width paper due to retention of fillers and other chemicals
Installation of additional press roll set	Increased mechanical dewatering Reduced steam consmn. in drying Increased speed of paper machine and hence increased capacity	Equipment: Press roll set Process: Space availability between existing press exising press roll and dryers is a must	I = Rs. 8–12 lakh S = Rs. 6–8 lakh P = < 2 years	Reduced air pollution due to lower steam reqmt.	The measure is usually applicable in cases where adequate space is already available. Shifting the dryer is quite cumbersome and expensive
Timely replacement of upper press roll to reduce press picking	Reduced paper breakage Reduce consmn. of kerosene and anti-picking agents Increase in capacity by 3–4% Reduced pollution load	Equipment: Chilled SS rolls and MS rolls are reported to give a better performance than conventional rolls	I* = Rs. 2 lakh S = Rs. 5–8 lakh P = < 1 year * based on average life of roll as six months	Reduced TS and COD load. Reduce air pollution due to less steam reqmt.	This measure requires close monitoring of the press roll to replace it as soon as press picking shows a rising trend

(Contd...)

Waste minimisation measure	*Anticipated benefits*	*Technical requirements*	*Economic viability*	*Environmental impact*	*Remarks*
Provision of high velocity hood in steam dryer	Increase in paper production capacity by 15% Marginal improvement in drying efficiency	Equipment: High velocity hood Process: The measure is recommended for final drying Techenology: Indigenously available	I = Rs. 17–20 lakh S = Rs. 18–20 lakh P = < 1 year	Marginal reduction in air pollution	The measure is cost intensive, however, the payback period is very attractive Major shutdown required to install the system
Save all for fibre recovery	Increased fibre recovery Reduced pollution load Reduced effluent volume Filtrate can substitute fresh water consmn. in in pulping section	Technology: Two types of saveall systems are avallable: Dissolved air flotation (DAF) sedimentation Equipment: Save all system pump Process: Additional energy consmn. Additional polyelectrolyte consmn. Manpower: Skilled manpower required for DAF	For DAF system I = Rs. 10–15 lakh S = Rs. 5–8 lakh P = > 2 years For sedimentation system I = Rs. 3–5 lakh S = Rs. 3–5 lakh P = 1 year	Reduction in TS and COD load by %	The DAF system is more energy- and chemical intensive but gives higher fibre recovery. It also requires meticulous care to run the system

I = Investment; S = savings; and P = payback period.

Note: The following abbreviations means: RM = Raw material; reqmt. = requirement, consmn. = consumption

Table 17.13. Waste minimisation measures in the utilities section.

Waste minimisation measure	*Anticipated benefits*	*Technical requirements*	*Economic viability*	*Environment impact*	*Remarks*
Use of soft water as boiler feed water	Reduced scaling of boiler tubes and thus lower tube failures Increased boiler efficiency and capacity Reduced blow-down heat loss Reduced boiler maintenance Reduced blow-down reqmt.	Equipment: Water softening agent Technology: Indigenously available Manpower: Skilled manpower is required	I = Rs. 1–2 lakh S = Rs. 2–3 lakh P = < 1 year	Reduced air pollution by 3–5%	The additional cost of regeneration chemicals and manpower reqmt is very low as compared to savings. The use of soft water has a beneficial effect on boiler life
Insulation of feed water tank and condensate recovery tank	Increased steam generating capacity Reduced fuel reqmt.	Equipment: Insulation material	I = Rs. 0.3–0.5 lakh S = Rs. 1–2 lakh P = < 1 year	Reduced air pollution	The measure is easily implementable
Proper insulation of steam pipelines	Reduced steam pressure and temperature drop Reduced heat loss from steam pipelines Better cooking due to availability of higher steam pressure	Equipment: Insulation material	I = Rs. 0.5–1 lakh S = Rs. 0.3–0.5 lakh P = 1–2 year pollution	Marginal reduction in air	The measure is easily implementable
Regular maintenance of DG set	Reduction in specific fuel consmn. by 5–10%	Process: Preventive maintenance	I = Rs. 1–2 lakh S = Rs. 2–3 lakh P = < 1 year	Reduced air pollution from DG set	Preventive maintenance plan has to be prepared and adhered to
Installation of maximum demand controller	Enables avoidance of penalties levied due to exceeding the contracted power demand	Equipment: MD controller Technology: Indigenously available	I = Rs. 1.5–3 lakh S = Difficult to quantify	No direct impact	The option is applicable to units with lower contract demand limits. The savings would depend on the number of times the MD exceeds contract demand
Provision of fuel (rice husk) feed control mechanism in boiler	Enables boiler operation at maximum capacity and efficiency by ensuring uniform fuel firing	Equipment: Feed controller Technology: Available indigenously	I = Rs. 1–2 lakh S = Difficult to quantify	Reduced air pollution. Reduced ash generation	Simple and easy to implement

(Contd...)

Waste minimisation measure	*Anticipated benefits*	*Technical requirements*	*Economic viability*	*Environment impact*	*Remarks*
Supply of make-up water in condensate tank	Reduces steam locking of condensate feed pump Reduces flash steam loss from condensate tank	Equipment: Pipings	I = Rs. 0.05–0.1 lakh S = Rs. 1–1.5 lakh P = < 1 year	Marginal impact	Simple and easy to implement
Combustion optimisation in boilers	Reduced reqmt. due to reduced stack and unburnt loss in ash	Manpower: Requires training of boiler operators to optimised combustion	I = Nil	Reduced air pollution	The measure requires improvement in operational practices
Use of fluidised bed boiler	Improvement in boiler efficiency by 10–15% over that of fixed grate boiler Better steam pressure control	Equipment: Fluidised bed boiler Technology: Indigenously available Manpower: Skilled operator required	I = Rs. 50–60 lakh S = Rs. 15–20 lakh P = 3–4 years	Reduced air pollution Reduced ash generation	The measure is more attractive for new mills and mills going in for expansion and requiring additional steam generation capacity
Insulation of condensate return line	Reduced heat loss Higher feed water temperature and hence faster load response	Equipment: Insulation material	I = Rs. 0.2–0.3 lakh S = Rs. 0.2–0.3 lakh P = 1 year	No significant impact	The measure is simple and easy to implement
Avoidance of condensate and steam leakages	Reduced heat loss Reduced make-up water	Nil	I = Nigligible S = Difficult to quantify	No significant impact	The measure requires timely repair and maintenance
Rationalisation of steam and condensate lines	Reduced temperature and pressure drop	Equipment: Pipelines	Difficult to quantity. Would vary from case to case	No significant impact	Special care needs to be taken to avoid unnecessary bends and submergence of condensate line in water

I = Investment; S = savings; and P = payback period.

Note: The following abbreviations means: RM = Raw material; reqmt. = requirement, consmn. = consumption

BARRIERS TO WASTE MINIMISATION AND ENABLING MEASURES

The most important barriers and enabling measures specific to the pulp and paper sector are:

1. Lack of commitment and involvement from management as well as employees, caused by insufficient care for housekeeping, limited recognition for initiatives from workers including limited job safety, fear of failure, etc.)
2. Lack of data (caused by poor record keeping, poorly developed management systems, lack of instrumentation and monitoring equipment, poor production planning, etc.)
3. Technical limitations (including limited access to technical information, limited availability of agro-based pulping know-how, outdated equipment, limited in-house technical and maintenance skills and facilities, etc.).
4. Economic limitations (low prices for water and agro residues, *ad hoc* investment policy, limited insight in production costs, poor production planning).

Enabling Measures

The following key enabling measures for entrepreneurs could then be elaborated:

1. Development of information systems (collection of baseline production data, materials and energy consumption data, economic data, reporting and feedback procedures).
2. Enhancing involvement and participation (involvement of workers, emphasis on non-production issues, reward schemes, delegation of decision powers, etc.)

The following section schematically represents some of the possible barriers, catalysts and enabling measures. For the sake of clarity and ease of understanding, the schematics have been classified in the following categories:

1. Organisational barriers (Table 17.14).
2. Systemic barriers (Table 17.15).
3. Attitudinal barriers (Table 17.16).
4. Economic barriers (Table 17.17).
5. Other barriers.

Table 17.14. Organisations barriers and enabling measures.

Barriers	*Enabling measures*
Non-involvement of workers in WM	Select WM team with leader who has authority to implement WM
Authority only with the owner	Include concerned supervisors and operators in WM team
High employee turnover	Recognise efforts of WM team
Payment on 'production basis'	Assign costs to various input

Table 17.15. Systemic barrier and enabling measures.

Barriers	*Enabling measures*
Lack of: Documentation and records	Conduct training session with WM team to clarify WM objectives, methodologies, improtance of planned production, data collection and analysis and problem-solving approach
Planning	Housekeeping drive shoud start with top-down appraoch
Analysis of data	Develop generic and specific manuals

(Contd...)

Barriers	Enabling measures
Management systems	Conduct sectorial and national workshops to increase awareness
Housekeeping	
In-house maintenance	
Non-availability of information on:	
Toxicity of material	
Safe substitutes	
Cleaner technologies	
WM experiences	

Table 17.16. Attitudinal barriers and enabling measures.

Barriers	*Enabling measures*
Fear of failure	Prioritisation of options by WM team
'Not me first' syndrome	Experiment with low-cost options
Indifference	Publicise early successes
Fast shifting priorities	Encourage experimentation
Trading rather than manufacturing approach	Calculate and control manufacturing costs
Suspicion of project objectives	Management review of WM progress on simple indicators on a regular basis
Lack of leadership	

Table 17.17. Economic barriers and enabling measures.

Barriers	*Enabling measures*
Internal:	Internal:
Shortage of working capital	Planned investments
Low return on WM investment	Assign costs to waste streams
Ad hoc investment criteria	
External:	External:
Low environmental costs	Devlop special WM financial schemes
Changing excess duties (short term policies)	Long-term industrial policy
Job work	Fiscal incentives for low resource-consuming WM practices
Seasonal nature of some industries	
Lack of grants and subsidies	

Table 17.18. Other barriers and enabling measures.

Barriers	*Enabling measures*
WM options not owned up by the company	All WM options should be generated at the shopfloor
Lack of regulations on environmental management systems, air emissions and solid waste	Form area-wise, voluntary WM groups

BIODEINKING: THE NOVEL, ECO-FRIENDLY WAY OF RECYCLING THE WASTE PAPERS

During photocopying and laser printing, inks get thermally fused in cellulosic fibres. This paper cannot be used in recycling of waste paper. Annually 7.0 million metric tonnes of waste paper is generated. This creates landfills and disposal problems in the biosphere. Enzymatic deinking can be the best eco-friendly alternative for hazardous chemical processes that are practiced in paper and pulp industries.

Enzymes are proven safe and easy to control. Enzymatic treatment to pulp increases brightness of the paper and lowers dirt counts without compromising strength of the paper. Lowered COD and BOD values in the effluents also help to save on effluent treatment costs. Indeed, enzymatic deinking can become a novel eco-friendly way to recycling the waste papers.

Paper manufacture is one of the largest industries of the world. According to a study published by the European Organisation for Economic Co-operation and Development the demand for paper is rising at a fast rate. The global market for paper was about 80 million metric tonnes per annum at the beginning of 1980s, that nearly doubled to reach 140 million metric tonnes in year 2000 and expected to reach 370 million metric tonnes by the year 2005.

In India more than 350 mills are engaged in production of 4.0 million tonnes per annum of paper and paperboards. The production is 6.7 million tonnes per annum by the end of year 2005. Thus, it is one of the fast growing business of the nation.

Industry of such a big size heavily depends on natural resources like timber as well as water and always been scrutinised on environmental aspects. It is estimated that for one tonne of paper production it requires 3.3 tonnes of trees and 0.4 tonnes of petroleum. Continuous supply of such a tremendous amount of wood is just impossible in the years to come. Secondly, what to do about waste paper that is clogging up landfills every year? About 7 million metric tonnes of waste paper is annually generated worldwide. Obviously recycling of waste paper is the best solution to attempt in order to save trees and energy. This in turn will reduce the effluents, pollutants and also the size of landfills. As a result the technologists are looking forward to develop the novel, eco-friendly way of recycling the waste papers. The trend in global paper industry is towards the greater use of the recycled waste paper.

The study by Freedonia group consultancy forecasted the secondary fibre is to account for 46 per cent of the paper industry's requirement by 2005. In India alone over 240 paper mills are based on recycling of paper and paper based products that accounts for 32 per cent of total production. Recycling of waste paper in to high grade writing paper requires removal of contaminants mainly the inks. In fact, it is considered as the major problem. Industry is familiar to the art of deinking for last 50 years but recycling of waste paper in to high grade writing paper is comparatively new trend.

The first deinking story started with a law professor *Dr. D. J. Claproth* in Germany in the year 1774. At that time fibres for paper making were short of supply in Europe. The professor had an idea to soak old books in water, stamp them for 12 hours, add a jug of lime to it and wash the slurry in hollander beater. This process of deinking pulp was of acceptable quality to manufacture new book paper.

Today, the process of removing ink and other contaminants from recycled firbre is termed as deinking. Generally, the ink contains a non-polar pigment or dye and an agent that carries the ink to paper and facilitates in binding. Such agents are typically water insoluble oils or polymers as in case of toners.

This makes them very difficult to remove. Washing and floatation are two most commonly used methods of deinking. Other methods like heat decolourisation, irradiation, organic solvent based deinking, magnetic deinking are still under experimental conditions. The choice of method depends on type of the ink to be removed and the desired quality of pulp.

Generally black newspaper inks are removed by washing while coloured magazines are deinked by floatation. Comparative studies of deinking technologies in different paper producing sectors of the world, namely North America, Europe and Pacific rim shows distinct differences in the deinking technologies. A generalised system for deinking of waste paper is shown in Fig. 17.2.

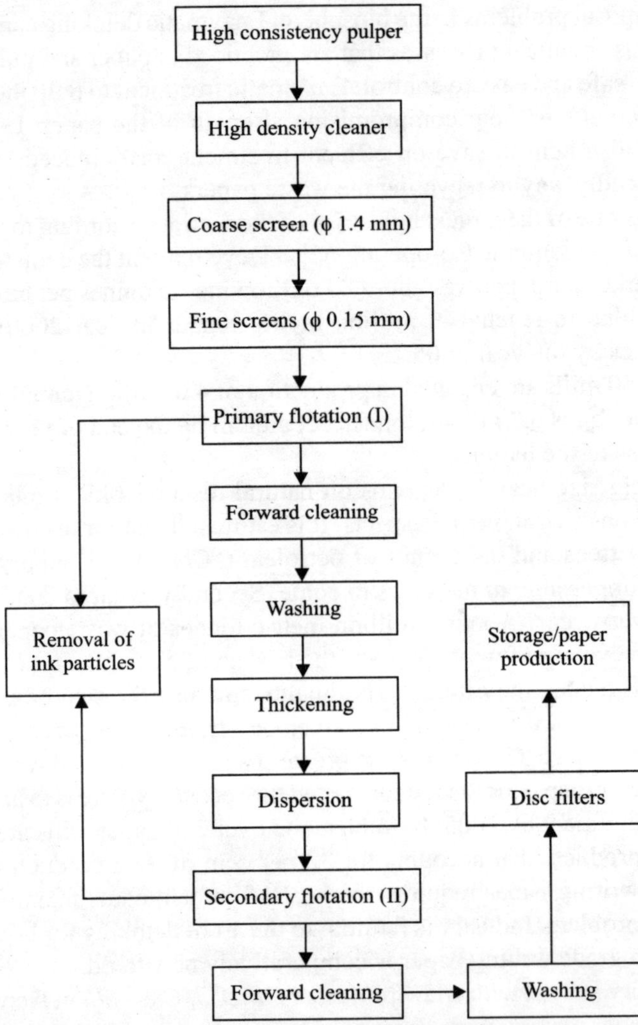

Fig. 17.2. Eco-friendly deinking process.

As the consumer demand for recycled paper is increasing, the manufacturers are also trying to introduce environmentally friendly production technologies. However, still many of these steps utilises chemicals like alkali, blending agents, reducing agents and surfactants. The demand for deinking and recycling chemicals is projected to reach 3079 million tonnes by the end of the year 2005 that was 1875 million tonnes in the year 1989. These chemicals have harmful impact on the environment. They increase chemical oxygen demands in waste-water generated in the overall process.

Fortunately modern biotechnology is offering some eco-friendly technologies for paper and pulp industries. Biotechnology can also help us in deinking process and to get rid of 'dirty' harmful chemicals (Table 17.19).

Table 17.19. Role of biotechnology in paper and pulp industry.

Process	*Mechanism*	*Micro-organism used*
Fibre handling	Pectin removal using pectinases	*Aspergillus* sp.
Biopulping	Using fungi for lignin breakdown (ligninase)	*Ceriporiopsis subvermispora*, white rot fungus *Lentus edodes*
Enzymatic pitch control	Sticky lipid removal by lipases	*Candida cylindracea*
Biobleaching	Xylan breakdown by xylanases	*Trichoderma* sp. *Trametous versicolour*
Sizing	Starch modification with amylases	*Bacillus stearothermophillus*
Refining	Removal fine particles with cellulases	*Trichoderma* sp.
Slime removal	Levan hydrolysis	–
Enzymatic deinking	Removal of ink particles from recycled fibre with alkaline active cellulases	*Humicola insolens*
Effluent treatment	Removal of lignosulphonates using microbial cultures	*Trametous versicolour, Aspergillus* sp., sulphate reducing bacteria

Deinking with Enzymes

Pulp is biobased material composed of cellulose, hemicellulose and lignin. It can be modified by microbial enzymes. The enzymes can accelerate reactions; they act on only specific substrates. The enzymes are safe and easy to control.

They can replace harsh chemicals and are biodegradable. Thus enzymes are more advantageous than chemicals. The enzymes used in deinking process are lipases, hemicellulases, amylases, esterases, and cellulases from different microbial sources.

Here the choice of enzymes to be used depends on the type of the ink to be removed and the type of pulp. For example if the furnished pulp stock includes paper with soybean oil based inks, as generally in case of coloured magazine prints then lipases are found useful in deinking of such pulp stock, whereas hemicellulases have proven useful in deinking of newsprint waste paper. Deinking with hemicellulases has benefit of improvised brightness of pulp due to its bio-bleaching effects. Amylases are also important in bleaching and brightening processes.

Deinking of Mixed Office Waste Paper with Cellulases

Any modern office produces waste paper that includes laser printed papers and photocopier papers. Toners and laser printing inks are synthetic nylon or polyester based polymers with embedded carbon black that do not disperse readily.

Toners are difficult to disperse than conventional oil based inks. Toners remain as large, flat rigid particles that separate poorly, from fibres during flotation and washing stages. Properties of deinked pulp are shown in Table 17.20.

Table 17.20. Important properties of deinked pulp.

Properties	*Test*	*Commercial importance*
Optical	Brightness, colour contrast	Appearance, colour printing
Cleanliness	Dirt specks, undetched ink	Appearance
Contaminants	Stickies	Running capacity of machines
Strength	Tensile index, tear index, burst index	Durability of paper
Surface	Smoothness, ISO-limits	Important in specialised applications like posters, calenders

Costly dewatering and dispersion steps followed by additional floatation and washing are required to improve the deinking of laser printed paper. Here is the real challenge for paper technologists because none of the conventional deinking methods can remove these inks from paper. Moreover, it is the most valuable grade paper available in market. In the enzymatic deinking mixed office waste paper cellulases play key role in releasing thermally fused toners from fibre surfaces.

Cellulase is a multi component enzyme system that hydrolyses lignocellulose. The enzyme complex typically consists of related cellulolytic activities including both endocellulase and exocellulase. In nature all of them act together for effective breakdown of cellulose polymer in to readily utilisable sugars like glucose, cellobiose and cellooligosaccharides.

Cellulase enzyme can be obtained from variety of microbial sources such as *Aspergillus niger, Trichoderma longibrachiatum (T. reesi), Penicillium* sp., *Fusarium* sp., *Humicola* sp., and *Bacillus* sp. Each of these micro-organisms can produced different kind of cellulase system that differs in their mode of action as well as properties like activity towards crystalline cellulose, activity and stability in acid or alkaline pH.

Different kinds of cellulases work best under different operational conditions. Acid cellulases that hydrolyse cellulose in acidic pH range of 3–6 can be used in acid sized (sulphite paper) paper. Alkaline cellulases that are effective between pH 8–10 are more useful since alkaline sized papers are more acceptable owing to their high fibre flexibility. Calcium carbonate is often used in the paper as a filler and brightener. It dissociates from fibres in alkaline environments.

This also leaves fibres with alkaline pH hence alkaline active and stale cellulases are necessary to process them for deinking. This automatically saves additional acid treatment required for acid cellulases to work. However, microbial alkaline cellulases are vary rare. A few *Bacillus* sp. produce cellulases that are active in alkaline pH. Unfortunately bacterial cellulases lack complete cellulase complex. Thermal stability of enzyme is an additional important pre-requisite for enzyme complex to be used for development of a commercial process. These all requirements led to shift research focus on screening of alkalophilic fungi that are capable of producing thermostable cellulases and are operative in pH range of 8–10. Commercially available enzyme preparations for deinking process are obtained on a large scale from micromycetes fungus *Humicola insolense*. The enzyme is complex produced by *H. insolense* active in a broad pH range of 5–10.

Deinking Process

The process of deinking with enzymes is similar to conventional chemical processes but it has the benefit of using lesser amounts of various chemicals. Fiberisation is generally the fist step in typical deinking operation in which the recycled pulp is collected in pulper. Efficient wetting of fibre surface in this stage needs the addition of a wetting agent such as sodium silicate.

This reduces surface tension on liquids for preventing ink particles from re-depositing on fibres. It is important to have maximum surface area for efficient deinking. Next step is sizing. In that specific chemicals are added to make the paper surface more hydrophobic and susceptible to enzyme action. This is followed by batch wise addition of the enzyme. Generally it is 10–15 per cent of the total volume of the pulp. The recycled fibres are mixed properly and pulping consistency is adjusted to 14 per cent. The process is generally carried out at pH 8.0 and at a temperature of 50°C for 30 minutes. Hereafter, deinking process involves series of floatation, washing and particles screening steps.

Mechanism of Enzymatic Deinking

In the floatation deinking process the pulp is diluted to 1 per cent consistency. Now the air bubbles are injected through the pulp. Ink particles that are more hydrophobic gets attached to bubbles that rise to the top creating forth layer. This layer is swept off by skimming action. Here the stability of air bubble is the most important factor. Generally ink particles are 40–50 μ in size.

Addition chemicals termed as 'collectors' are needed if the ink chemicals are smaller in size. All the ink particles can not be removed in a single run. So the process is repeated couple of times. The definite mechanism of deinking with enzyme is not yet understood but it is assumed that the enzyme reduces the hydrodynamic drag that in turn increases filtration and floatation rates. In floatation process, the air bubble rising to the top carries ink particles along with cellulose fibres. These fibres exert negative effects on floatation efficiencies. Cellulases simply 'shaves' fibres from the ink particles where fibres will settle down and ink particles will adhere to the air bubble in much better way experiencing much less drag from interaction with water. Thus, the ink particles can be carried to the top of the tank much easily. In short, the cellulase enzyme complex reduces the hydrodynamic drag produced due to the attachment of fibres to the floating air bubbles.

The enzyme complex can also reduce specific surface area of fibres, thereby lowering the interaction with contaminants. The enzyme activity enhances the draining resulting in enhancing filtration or floatation process. Once the floatation cycles are over, the pulp passes through different micro screens (0.15 mm) under pressure and the accepted fibre undergo subsequent washing steps.

This results in removal of stickies and other contaminants. Next job in deinking is to increase the brightness of fibres. Conventionally it can be done by sodium hypochlorite or hydrogen peroxide. But these chemicals are highly toxic. Microbial enzymes can have role to play in this process. Microbial hemicellulases, amylases have shown promising results in increasing brightness of the fibres.

Advantages of Enzymatic Deinking

The deinking with enzymes works best with chemical pulped fibre as compared to mechanically pulped fibre. This is because lot of lignin is associated with mechanically pulped fibres. In chemically pulped fibre, most of the lignin is removed in the black liquor. About 88 per cent of mixed office waste paper is composed of chemically pulped fibre.

As compared to water deinked pulps the enzyme releases upto 95 per cent of toner particles from recycled fibres. It is shown that properties of deinked pulp such as tensile index, burst index, tear index are comparable to pulps from convention deinking procedures. The enzymatic process preserves fibre integrity and also show improved pulp brightness. Effluent samples from the deinking plants indicated lower oxygen demands as well as less toxicity than that of traditional processes.

The enzymatic 'biodeinking' process requires lesser chemicals. This is a double benefit as it saves cost of waste treatment along with saving cost of chemicals required for the process. It also avoids landfill disposal costs. Enzyme deinked pulp shows high drainage rates resulting in faster machine speed and less energy cost. The technology fits well to current recycling technologies and it does not require additional investment for paper maker. This is considered as the most attractive benefits to the manufacturer.

The technology for paper and pulp manufacturing is changing rapidly. This industry is becoming a major prospective market for the enzyme industry. However, enzymatic deinking is relatively a new process. Some of the paper and pulp industries in India are aware of this technology. They have even developed some unique pilot plants to explore the eco-friendly enzymatic technology of the future.

The technologists from Forest Product Laboratory (Madison, WI, USA) have prepared some cost analysis of enzymatic deinking. It is approximately US $ 1.60 to 2.40 per 1000 kg of the recovered paper. In some cases, depending upon the quality of the waste paper, the expenses can be as high as US $ 18 per 1000 kg of the raw materials.

In comparison, conventional deinking, chemical cost is between US $ 3.30 to 8.00 when sodium hydroxide (1 per cent) is used alone or in combination with hydrogen peroxide. Considering many factors the cost of conventional deinking process and proposed enzymatic process will not differ much.

Expectations for better quality papers and awareness about environmental issues are going to be two important drives for paper industry in the years to come. Enzymatic deinking can fulfill both the demands, as it is the best eco-friendly way of recycling the waste paper.

18

Paint and Dye Industry

INTRODUCTION

Paints are the surface coatings with ingredients such as pigments, oils, resins and solvents. Paint is applied to any surface to fulfill two requirements, viz to give nice appearance to the surface and to protect the surface against weather, moisture, solar radiation, chemicals and mechanical damage. The solvent-based paints include pigmented non-tints, pigmented tints, lacquer thinners, unpigmented paints (clears), and stains. The water-based paints are emulsion paints (i.e., latexes).

Dyes are substances capable of colouring fabrics in such a manner that the colour cannot be removed by rubbing or washing. It should also be noted that every coloured substance is not a dye. For example, azobenzene is of orange red colour, but it is not a dye, because it is not capable of colouring the fibre. Thus, a substance may be called a dye, if it satisfies certain conditions as: (i) it must have a suitable colour; (ii) it must be capable of being fixed to the fabric directly or with the help of certain reagents, called mordants; and (iii) when fixed, it must not be fugitive, i.e., the colour must be fast to light and resistant to soap and water, and to a certain extent to dilute acids and alkalies.

The raw materials used for the manufacture of dyes are mainly aromatic compounds, such as benzene, toluene, naphthalene, anthracene, phenol, pyridine, carbazole, pyrene etc. These raw materials are known as primaries and exclusively obtained in the past from the distillation of coal tar. In recent years, petroleum and natural gas have become important sources of primaries such as benzene, toluene, naphthalene, xylene etc. Aromatic hydrocarbons in recent years have also been obtained from other sources such as water gas, methane and olefines by catalytic processes.

A large number of inorganic compounds have also been used in dye industry to serve various purposes. These materials include sulphuric acid, oleum, nitric acid, acetic acid, caustic soda, hydrochloric acid, chlorine, bromine, sodium chloride, sodium nitrite, sodium carbonate, sodium sulphate, sodium hydrosulphite, sodium sulphide, sodium dichromate, manganese dioxide, aluminium chloride etc.

A large number of important reactions such as nitration, oxidation, halogenation, sulphonation, reduction, condensation etc. have been used in the production of intermediates. These reactions give rise to the formation of substituted hydrocarbons which are functional in nature (They contain functional groups which undergo further chemical reactions to produce dye). Sulphonation, nitration, animation by reduction, halogenation, dehydration, Friedel-Craft's reaction etc. are the most important and common unit processes which are involved in the synthesis of dye. Aliphatic compounds, aromatic compounds and heterocyclic compounds are the three main types of intermediates needed for the manufacture of dyes. The most common aliphatic compounds used as dye intermediates are alcohols, carboxylic acids, esters, chloro derivatives, aldehydes, ketones, amines etc. They are mainly used as solvents or as reagents or reactants.

PAINT INDUSTRY

The paint manufacturing industry comprises of establishments engaged in the manufacture of paints, varnishes, lacquers, enamels, paint and varnish removers, and allied paint products. The industry generates large quantities of both hazardous and non-hazardous wastes. Therefore, reducing waste is of high priority in this sector. Paint, coating, and ink manufacturers represent approximately 44 per cent of the solvent market. Solvent wastes (or solvent containing wastes) disposed of by the paint manufacturing industry have the highest volume of solvent wastes generated in most states.

Process Description

The paint manufacturing industry's two main products are solvent-based paints and water-based paints. In a typical plant, both types of paints are produced. The steps involved in manufacturing paint are illustrated in the block diagram in Fig. 18.1.

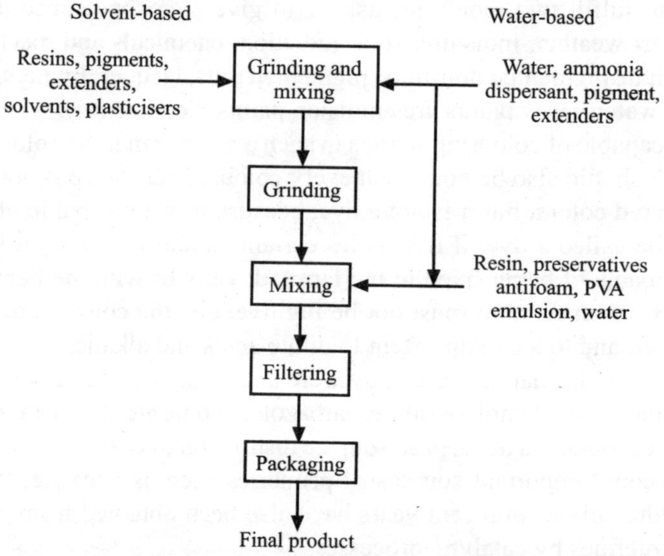

Fig. 18.1. Block diagram of paint manufacturing process.

The production of solvent-based paint begins with mixing of resins, dry pigment and pigment extenders, along with solvents and plasticisers in a high-speed mechanical mixer. After the mixing operation, the batch is transferred to a mill for additional grinding and mixing. The type of mill employed is a function of the types of pigments being handled. Examples of the types of mills used are roll mills, sand mills, pin mills (e.g. Alpine mill) and hammer mills. Then the paint base or concentrate is transferred to an agitated tank where tints, thinners and additional resins are added and mixed.

After reaching the proper consistency (viscosity and/or concentration), the batch is filtered via a cartridge filter to remove any non-dispersed pigment and the paint is transferred to a loading hopper. From the hopper, the paint is poured into cans, labeled and moved for storage or to the warehouse.

The water-based paint manufacturing process is similar to the solvent based process, except for the use of water in place of solvent and the sequence of addition of materials. Manufacturing of water-based paints begins by mixing water, ammonia and a dispersant in a high speed mixer. Dry pigment and pigment extenders are also added and mixed. After mixing the batch is milled and transferred to an agitated tank where resin, plasticisers, preservatives, anti-foaming agents, polyvinyl acetate emulsion and water as thinner are added. After the mixing operation, the water-based paint undergoes the same procedure as used in the solvent-based paint manufacturing process. At many paint manufacturing facilities, all dispersion operations are accomplished in a single high-speed mixer rather than through grinding and mixing operations in a mill and then in an agitated tank.

Waste Description

Paint industry uses varied raw material such as resins, solvents, drying oils, pigments and extenders. The major waste generated by the paint manufacturing industry are empty raw material packages containing trace elements, equipment cleaning wastes and spills. Empty raw material packages are generated during unloading of materials to high speed mixers or mixing tanks. Water solvents are generated from equipment cleaning. Even after distillation of waste solvents for reuse, a residual paint sludge remains. The paint sludge contains solvents and residual toxic metals such as mercury, lead and chromium. Waste rinse water is generated from equipment cleaning with water and for caustic solutions. Wastes containing undispersed pigments are contained in waste filter cartridges.

The equipment cleaning wastes can be minimised by employing more efficient cleaning methods, like reduction in the frequency of equipment cleaning, use of rubber wipers to reduce the amount of paint left on the walls of a mix tank, use of teflon-lined tanks to reduce adhesion and use of plastic or foam to clean pipes to improve drainage.

Pigment waste from bags and packages may be reduced by using water soluble bags. Paste pigments, which are wetted or mixed with resins, may be used in place of dry powder pigments that cause particulate or dust emissions. Bag filters or metal mesh filters may be used instead of cartridge filters. The most effective way of reducing wastes associated with bags and packages is to segregate hazardous materials from non-hazardous materials. For example, empty bags and packages containing residual amounts of hazardous materials should be stored in plastic bags to eliminate dusting, which can cause contamination of non-hazardous materials. By installing a separate dedicated bag house for each production step, the collected pigment dust or resin dust can be recycled to the process step.

Generation of waste-water at source should also be practiced. This is to reduce the effluent load rather than finding methods to treat it. Good housekeeping can reduce generation of both waste-water and solid-wastes and spills and rejects can be reduced by increased automation. Obsolete products and rejects can also be blended into new batches of paint.

Pollution Prevention Options

The wastes generated by the paint manufacturing industry include equipment cleaning wastes, spills and off-specification paints, leftover pigment in packages, pigment dust from air pollution control equipment (usually a baghouse), waste filter cartridges and obsolete products and rejects. Equipment cleaning generates most of the waste associated with paint manufacturing.

Equipment cleaning wastes can be minimised by employing more efficient cleaning methods. Other process modifications may be used to minimise wastes from off-specification paints and obsolete products and rejects. Better housekeeping and operating practices should be applied to all waste streams.

Process Modifications

Because equipment cleaning is the largest source of waste in the paint manufacturing industry, reduction in the frequency of equipment cleaning (e.g. mixing tanks) is the most effective approach to pollution prevention. Methods that can reduce the frequency of cleaning include:

1. Use of rubber wipers to reduce the amount of paint left on the walls of a mix tank. This operation can be done either manually or mechanically. Many new mixers are equipped with automatic wall scrapers
2. Use of teflon-lined tanks to reduce adhesion and to improve drainage.
3. Use of a plastic or foam 'pig' to clean pipes. Plastic or foam pigs (slugs) can be used to clean paint from pipes. The pig is forced through the pipe from the mixing tank to the product hopper. This increases the yield and reduces the frequency of pipe cleaning required. Inert gas is typically used to propel the pig and the equipment (launcher and catcher) must be designed carefully to prevent spills, sprays and potential operational hazards.

Equipment used for producing water-based paints is cleaned either with water or a caustic solution. Equipment used for producing solvent-based paints is cleaned with solvents or a caustic solution. However, caustic solutions are more frequently used for water-based paint equipment than for solvent-based paint equipment. Therefore, the amount of waste rinse water generated is a function of the ratio of the water-based paints manufactured to the solvent-based paints in a given facility. The amount of waste-water or waste caustic solution generated may be minimised by employing the following methods:

1. Use of high-pressure spray heads. After the tank walls are scraped, high-pressure spray hoses can be used in place of regular hoses to clean water-based paint tanks. High-pressure wash systems can reduce water use by 80 to 90 per cent. For tanks used in making solvent-based paints, a built-in high-pressure cleaning system may be employed.
2. Use of a countercurrent rinsing sequence. For facilities with additional space, a countercurrent rinsing operation may be employed to increase the rinsing efficiency and to minimise the amount of waste-water generated.
3. Encouragement of dry cleaning methods. Equipment may be cleaned by hand with rags rather than by using hoses.

Pigment waste from bags and packages may be reduced by using water soluble bags. In water-based paint manufacturing, bags containing toxic pigments can be dissolved or mixed with the paint. This method is frequently used for mercury compounds and other paint fungicides. However, soluble bags cannot be used for producing high quality, smooth finish paints since the bag material can affect quality. Dissolved bag material can also increase the load on the filters.

Paste pigments, which are wetted or mixed with resins, may be used in place of dry powder pigments that cause particulate or dust emissions. Bag filters or metal mesh filters may be used instead of cartridge filters. Metal mesh filters are available in micron sizes and can be cleaned and reused indefinitely.

Spills and rejects can be reduced by increased automation. Obsolete products (resulting from changing customer demands, newer and better products, and expired shelf life) and rejects can also be blended into new batches of paint.

Resource Recovery and Recycling

Waste solvents generated by equipment cleaning can be collected and used in the next compatible batch as part of the formulation. For example, wash solvent from each solvent-based paint batch can be collected separately and stored. When the same type of paint is being produced again the waste solvent from the previous batch is used in place of virgin solvent. Waste solvents can also be collected and recycled by distillation, either on-site or off-site.

WEALTH FROM WASTES AT TATA PIGMENTS

Waste pickle liquor, which is an objectionable waste, cannot be directly discharged into the river without proper treatment. It is usually treated with lime and the resultant sludge is used for filling low-lying areas. Apart from being a very filthy operation, it involves substantial amount of recurring expenditures. Tata Pigments Ltd. has been able to convert waste pickle liquor generated in the plant of Tata Iron and Steel Co. Ltd. (TISCO) into a valuable product, viz., synthetic iron oxide (SIO) pigment, and thus generate wealth from waste and simultaneously take care of environmental pollution. Besides the other usages, the pigment produced is used for paint manufacturing.

Manufacturing Process

This process consists of two major steps, the first one is the production of SIO (yellow) through precipitation process and the next one is the conversion of yellow to red iron oxide through calcination. Through extensive research carried out in the laboratory and in the pilot plant, the process parameters were optimised to achieve the desired quality of products with maximum yield.

The first step in this process is the formations of the small yellow iron oxide nuclei or seed crystal. This is accomplished by reacting an alkali like sodium hydroxide with a solution of ferrous sulphate, i.e., treated waste pickle liquor in the presence of air under conditions that will produce extremely fine iron nuclei. The seed crystals, thus formed are transferred to a reactor containing iron scrap and dilute ferrous sulphate solution. With oxygen of the air, the ferrous sulphate is oxidised to hydrated iron oxide which precipitates on the seed crystals and thus make them grow in size. The oxidation of ferrous sulphate also produces an equivalent amount of sulphuric acid. The acid again reacts with the metallic iron scrap to form more ferrous suphate. This chain reaction can be stopped only by separating the metallic scrap from the other reactants. The operating parameters are controlled so as to achieve the desired growth in the size and shape of the yellow coloured hydrated ferric oxide crystals, as this reflects largely on the quality of the final product. The slurry discharged from the reactor is subjected to various unit operations like filtration, washing, drying and pulverisation to produce SIO (yellow) pigment.

Red pigment is made by calcining this yellow pigment at elevated temperatures. The development of different shades of red depends upon the temperature as well as the duration of calcination.

Unlike the dry process, the wet process provides an opportunity to use waste pickle liquor directly in the system instead of iron sulphate in crystal form. Moreover, there is no nuisance of air pollution in the wet process. The acidic filtrate and wash water generated in the wet process could be easily treated by conventional methods to take care of pollution. So, they are sent to the effluent treatment plant (ETP) for neutralising with lime under controlled conditions, prior to disposal.

Quality Upgradation

With a view to compete with the international synthetic iron oxide pigments, Tata Pigments Ltd. decided to import technology from USA. This technology is based on direct precipitation of red pigments without passing through yellow hydrated ferric oxide. The precipitated iron oxide is produced in an aqueous medium. The reaction consists of nucleation in which a seed nucleus is formed by the reaction of an iron salt and an alkali. The seed nuclei are then grown in the presence of a ferrous salt solution and scrap steel. The iron dissolves and is re-precipitated on the nuclei causing them to grow in size. Particle size controls the shade and tinting strength. After the attainment of the desired shade the precipitated red iron oxide pigment is filtered, washed, dried and pulverised.

The technical viability of the imported technology was tried out successfully on a pilot plant scale and measures are being taken for commercialisation. A plant having a capacity of 2200 TPA is expected to be commissioned shortly. It is envisaged that with the implementation of new technology there would be a large saving in thermal energy as some of the energy intensive operations like calcination, filtration and drying get eliminated. Moreover, the overall cycle time is expected to reduce by about 40 per cent. As a result, a drastic reduction in the cost of production and enhancement in productivity are foreseen. On the quality front, it is conformed through the pilot plant scale trials that the product has unique brilliancy and tintorial strength and would provide further opportunities for export.

DYE AND DYE INTERMEDIATES

Basic Operations in Dyeing

The basic operation of the dyeing process involves the following important steps : (i) Preparation of the Fibre; (ii) Preparation of dye bath. In addition to dye, some other ingredients such as wetting agents, carriers, salts, retarders etc. have also been added to the dye bath to improve certain properties. For example, carriers improve the rate of dyeing of hydrophobic fibres like polyester fibres, because they act as swelling agents. The colourless retarders slow down the dyeing process to the desired level by competing with dye for the reactive sites on the fibre. When wool is dyed with acid dyes, acetic acid is used as a retarder.

Manufacturing process

The manufacturing process generally adopted involves conversion of simple organic products like benzene, toluene, xylene, naphthalene, anthracene etc. into a vast number of complex chemical intermediates and dye through several steps of operation like: (i) chemical conversion; (ii) sulphonation; (iii) neutralisation; (iv) fusion; (v) chlorination; (vi) nitration; (vii) reduction; (viii) animation; (ix) acetylation; (x) hydrolysis; and (xi) carboxylation. Each process requires different operations like charging, reflux, distillation, filtration, washing, drying, grinding etc.

The intermediates are reacted in several processes like hydrochlorination, sulphation, diazotisation, coupling, condensation, phosgenation, thionation etc. in the manufacture of dyes. The process requires large number of chemical compounds but often in limited quantity. The raw material is fed into the reactor where reactions are carried out ordinarily at atmospheric pressure. Usually the reactions are exothermic for which adequate temperature control is maintained to avoid side reactions. Generally batch process is followed for production of dyes.

Temperature control is accomplished primarily by addition of ice to the reaction tank. When the reaction is complete the dye particles settle out from the mixture. The liquid of the reactor is then sent to a plate and frame filter press where the dye particles are separated from the mother liquor which is discharged as waste-water.

The filter cake is first washed with compressed air while still in the press. The moist cake is discharged into shallow trays which are placed in circulating air dryer. The moisture is removed at temperature between 50 and 120°C. The dried dye is ground and mixed with diluent such as salt to improve the colour strength.

Waste Generation

In the *dye industry* waste is generated due to synthesis of dye and dye intermediates; steam generation and cooling system; washing and rinsing of reaction kettles, filter press, floors, etc.; domestic and other miscellaneous activities.

The water consumption pattern varies from one industry to another. In the same industry, the rate of water consumption often changes due to frequent changes of the feed material synthesis reaction and desired product. The change of product pattern needs cleaning and washing, which consumes a substantial quantity of water.

The air pollutants are generated from the reaction vessel. These emissions are scrubbed by water. The other emissions are of fugitive nature and should be controlled by proper modification of plant and maintenance of equipment.

The solid wastes in the dye intermediate industry are generated from filter press from the process house, physico-chemical waste-water treatment plant and the biological waste-water treatment plant.

The treatment of wastes involves:

1. Inplant control: It is essential to have proper inplant control measures before going for waste-water treatment.
2. Reduction of waste: The volume of waste can be reduced by proper control of fresh water consumption. The pollution load can also be reduced by recovery of chemicals and solvents as far as practicable.
3. Process modification: The production process equipment should be modified so as to generate less wastes.

The water usage in the industry is mainly for the following purposes:

1. Synthesis of the dyes and dye intermediates.
2. Steam generation and cooling system.
3. Washing and rinsing of reaction kettles, filter press, floors, etc.
4. Domestic and other miscellaneous activities.

The water consumption pattern varies widely from one industry to another. In the same industry the rate of water consumption often changes due to frequent changes of the feed material synthesis reaction and desired products. The change of product pattern needs cleaning and washing which consumes a substantial quantity of water. Thus water requirement of a dye and dye intermediate industry depends on the following factors:

1. Type of dye produced.
2. Number of products.
3. Gross production.
4. Pattern of working of factory i.e., continuous or in one shift only.
5. Frequency of change of product pattern etc.

The process water consumption is highest, next to it is the cool and boiler make-up water requirement. The generation of waste-water follows the trend of consumption of water. The waste-water quantity varieties according to the number of batches of products manufactured in a day, week or month, the duration of synthesis of the dye in the reactor vessel and the duration of the washing and rinsing operations.

The frequency of discharge may be intermittent in the small dye units, operating one or two batches per day, but in the case of the medium and larger units, which operate a number of reactors simultaneously, waste-water discharge is continuous.

The waste-water generation sources are as follows:

1. Process waste-water including left out mother liquor.
2. Washing and rinsing wastes.
3. Sanitary and other miscellaneous waste-water.

Waste-water characteristics

The process waste-water is mainly the mother liquor left over after the product is isolated and separated by filter press. This waste-water is of smaller volume and highly concentrated in terms of pollutants. The vessel washings also contain similar type of pollutants but with lower concentration. It has been identified that the waste-waters of the industries have the following characteristics:

1. High levels of BOD and COD.
2. High acidity.
3. High TDS.
4. Deep colour of different shades.
5. High levels of chlorides and sulphates.
6. Presence of phenolic compounds.
7. Presence of heavy metals, e.g., copper, cadmium, chromium, lead, manganese, mercury, nickel and zinc.
8. Presence of oil and grease.

The dye industry waste-waters, if derived from naphthalene and anthracene bases are resistant to biodegradation. The colour removal is also not adequate by the conventional chemical and biological treatment.

Treatment of Wastes

The data indicate that the waste-waters are amendable and responsive to biological treatment. However, removal of colour from the waste-waters are difficult to achieve. In general the small and medium-scale industries have no organised treatment facility.

The large-scale dye and dye intermediate production units should provide appropriate waste-water treatment facilities. The small and medium-scale dye producing units may opt for combined effluent treatment plant (CETP). The unorganised scattered units may be resettled in organised industrial estates and CETP may be installed to take care of the waste-waters.

Reduction of waste

The volume of waste-water can be reduced by proper control of fresh water consumption. The cooling water blow-down may be reduced by raising the concentration factor. Timely maintenance of the units may be done to prevent leakage, spillage etc. Minimal usage of water for washing and rinsing may be practised. The last wash water may be recycled for first washing. Application of counter-current washing may also reduce waste-water generation. Dry cleaning of the floor is preferred. When floor washing is absolutely necessary, the treated waste-waters may be used. The pollution load can be reduced by recovery of chemicals and solvents as far as practicable. Spills, leakages, overflows etc. may not be allowed to join the waste-water stream.

Segregation

Storm waters need segregation to reduce the volume of waste-water. Similarly the uncontaminated and less contaminated waste-water streams like cooling water blow-down, boiler blow-down, condensate etc. are to be segregated and should not be permitted to join the process waste-water stream. The highly contaminated and coloured mother liquors should be segregated and collected separately. Sometimes strong wastes of the reaction vessels are required to be discharged. These unforeseen strong wastes are to be collected in separate holding tanks and drawn into the waste-water treatment plant at a regulated rate.

Process modification

The production process equipments should be modified so as to generate less wastes. The raw materials used in the synthesis may be substituted by choice of more readily biodegradable chemicals.

Effluent control

The waste-water streams originating as mother liquor and strong waste-waters as mentioned earlier should be segregated and collected in two separate collection tanks. The rest of the process waste-waters may be diverted to the equalisation pond. Oil and grease and the floating matters present in the waste-water streams are to be arrested by putting up appropriate control measures at the boundary limit of each plant. The waste-water collected in equalisation pond will be monitored to ascertain the concentration of pollutants. Based on the data obtained, the waste-waters of holdings, ponds are to be diverted to the equalisation pond at regulated rates so that the concentration of pollutants are acceptable to the biological treatment system.

The equalisation of waste-waters is followed by pH adjustment and clarification in primary clarifier. Clarifier may be of sludge blanket type. Ferrous sulphate and lime may be used for precipitation of heavy metals. The overflow of the clarifier after pH adjustment will flow to the extended aeration type biological treatment unit.

The effluent of biological treatment unit will be clarified in convenient type clarifier using coagulant. Figure 18.2. gives the scheme of the proposed integrated treatment of waste-water.

The reduction of colour may be achieved by segregating and controlled discharge of the mother liquor, which also contains most concentrated form of the chemicals left after dye synthesis. It is envisaged that substantial colour reduction of the waste streams may be achieved by the three-stage treatment system as proposed above. However, if the desired limit of colour (400 Hazen unit) is not achieved by the above treatment then one or both of the following steps are to be adopted:

1. Evaporation of the segregated mother liquor and washings of vessels by solar evaporation or by indirect heating using steam.
2. Powered activated carbon adsorption of the treated waste-water.

Identification and characterisation of major source of solid waste

There are two major sources of solid effluents from H-acid plant, viz., gypsum sludge from neutralisation and iron sludge from reduction. These solid effluents also have retained organics to the extent of 1.08 per cent and 3.7 per cent (w/w), respectively. It is imperative to leach them to levels of 0.05 per cent and 0.1 per cent, respectively, before their use as land filling agents. However, most industries do not conform to these limits for retained organics before dumping, thus giving rise to soil and water pollution.

Water consumption and waste-water generation

The present H-acid process followed by Indian manufacturers requires water of about 19 tonnes in the process and one tonne for washing per 1 tonne production of H-acid. Out of this water, about 9 tonne is evaporated in the 'concentration' step. The steam generated can be used for further concentration, if multiple effect evaporation system is used. This reduces the steam requirement of the process. All the condensates can be recycled to the boiler, so that no make-up water is required for utility plants. About 11 tonnes of water is in the liquid effluent. Most Indian manufacturers give only primary treatment of neutralisation to the liquid effluent. Effluent after the primary treatment contains large quantities of inorganic and organic compounds, and this water cannot be reused.

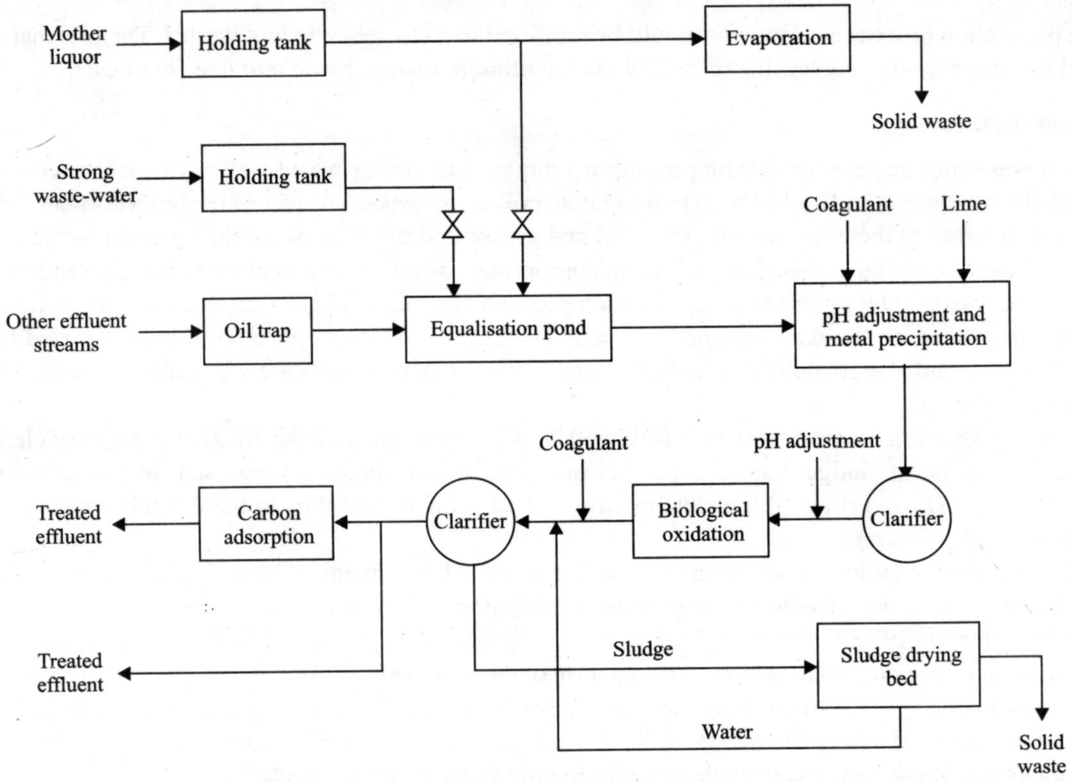

Fig. 18.2. Dye manufacturing effluent treatment scheme.

It is possible to reduce the requirement of water by modifying the process. The process of G-acid requires about 5 tonnes of water–4.2 tonnes are required in the process and about 800 litres for scrubbing and washing. There are no effluents, except spent scrubber solution and washes. This is a small quantity and can be treated with effluents of other plants.

Recycling and Reuse of Wastes

1. Ammonia issuing out of alkali fusion reactors may be suitably scrubbed into water.
2. This resultant ammonia/ammoniacal solution may be utilised to neutralise SO_x and NO_x vapours.
3. The solid waste of gypsum may be given a thorough washing with alcoholic solvents to considerably reduce its organic content.
4. The treated gypsum may be used as a building material in place of cement.
5. The iron-sludge may be converted into value-added ferrous and ferric salts by reacting it with spent acid.

6. The iron-sludge may be converted into a valuable inorganic red pigment consisting mainly of iron oxide.
7. Alternatively, the process can be modified to eliminate solid wastes. This will involve isolation of nitromass, use of spent acid after isolation for neutralising fusion mass, catalytic hydrogenation of nitromass and recycle of catalyst. Methanol can be used as a solvent in hydrogenation and fusion. Methanol-water mixture can be recovered and reused.
8. The bio-degradability of the waste-water effluent can be considerably enhanced by separating the organics from it by extraction with a tertiary alkylamine dissolved in a water-immiscible solvent.
9. The naphthalenesulphonic acids thus obtained in a concentrated form can be further concentrated by drying techniques such as spray or vacuum drying.

19

Fertiliser Industry

INTRODUCTION

Fertilisers produced in India can be classified as nitrogenous fertilisers, phosphatic fertilisers and complex fertilisers. Plants may be producing only nitrogenous fertilisers, like urea, ammonium sulphate, ammonium nitrate and ammonium chloride, or only phosphatic fertilisers like super phosphates; there are plants where complex fertilisers containing both nitrogen and phosphates, like ammonium phosphate and ammonium sulphate phosphate, are produced. Also, some fertiliser units are only involved in mining and undertake no other activity.

MANUFACTURING PROCESS

Ammonia is the principal intermediate in the manufacture of all nitrogenous fertilisers. So, except when the by-product ammonia is available from a coke-oven, raw materials for nitrogenous fertiliser production are carbonaceous material, which are required for making ammonia. Thus, all nitrogenous fertiliser plants essentially have an ammonia production unit and a reactor where the synthetic ammonia is reacted with other chemicals to produce the final product. The plant may have auxiliary units to produce reacting chemicals also.

Steps in Manufacture

Basic process steps in the manufacture of urea from carbonaceous raw materials like naphtha are:

1. Reaction of the carbonaceous material with steam and air to form a mixture of hydrogen and carbon monoxide, known as synthesis gas.
2. Reaction of carbon monoxide with steam over a catalyst to form more hydrogen and carbon dioxide.
3. Separation and purification of carbon dioxide.
4. Removal of residual carbon monoxide from gas mixture.
5. Synthesis of ammonia by reacting hydrogen and nitrogen over a catalyst (Nitrogen is supplied as air in an earlier step).
6. Synthesis of urea by treating ammonia with carbon dioxide in a reactor at higher temperature and pressure.

Plants using by product ammonia from other manufacturing plants (like coke oven) have to produce carbon dioxide separately for the production of urea.

Ammonium sulphate may be produced by reacting anhydrous ammonia with sulphuric acid, usually obtained as a by-product from other manufacturing plants. Ammonium sulphate may also be manufactured from gypsum or from calcium sulphate sludge, obtained from the phosphatic fertiliser plant, using ammonia and carbon dioxide obtained from ammonia plant. In this process, calcium sulphate is reacted with ammonium carbonate solution to produce ammonium sulphate.

Ammonium nitrate is produced when ammonia is reacted with nitric acid. Normally, the required quantity of nitric acid is produced in the same plant, by oxidising ammonia.

Super phosphate is produced by merely mixing the phosphate ore (commonly known as phosphatic rock) with sulphuric acid to convert the phosphate to 'monocalcium phosphate'. The by-product calcium sulphate of this process may be used in the manufacture of ammonium sulphate.

Ammonium phosphate is made by treating phosphoric acid with ammonia; the phosphoric acid production process involves the following steps:

1. Dissolving phosphate rock in enough sulphuric acid.
2. Holding the mixture until the calcium sulphate crystals grow to adequate size.
3. Separation of the phosphoric acid and calcium sulphate by filtration.
4. Concentration of acid to the desired level.

WASTE IN FERTILISER PLANTS

In *Fertiliser* plants, a variety of wastes are discharged as water pollutants in the form of processing chemicals like sulphuric acid; process intermediates like ammonia, phosphoric acid; final products like urea, ammonium sulphate, ammonium phosphate, etc.

In addition, oil bearing waste from compressor houses of ammonia and urea plants, some portion of the cooling water and the wash water from the scrubbing towers for the purification of gases, also come as waste.

Wash water from, the scrubbing towers may contain toxic substances like arsenic, potassium carbonate, etc., in a nitrogenous fertiliser plant, while that in a phosphatic fertiliser plant may contain a mixture of carbonic acid, hydrofluoric acid and fluorosilicic acid. Both alkaline and acidic wastes are also expected from the boiler feed water treatment plant, the wastes being generated during the regeneration of anion and cation exchanger units.

Additional pollutants like phenol and cyanide will be introduced in the list of pollutants in the fertiliser plant whereas ammonia is derived from the waste ammoniacal liquor of coke ovens. All the components of waste from the fertiliser plants induces adverse effects in water streams. Acids and alkalies can destroy normal aquatic life while arsenic fluorides and ammonium salts are found to be toxic to fish.

Major pollutants in the fertiliser waste-water, for which the treatment is necessary include oil, arsenic, ammonia, urea, phosphate and fluoride. Oil is removed in a gravity separator. Arsenic containing waste is segregated and after its concentration, the solid waste is disposed of in a safe place. Phosphate and fluoride bearing wastes are also segregated and chemically coagulated by lime; clarified effluent which still contains some amount of phosphate and fluoride is diluted by mixing with other wastes.

Waste-water from Fertiliser Plants

A variety of waste streams are discharged from fertiliser plants in the form of:

1. Processing chemicals like sulphuric acid.
2. Process intermediates like ammonia, phosphoric acid, etc.

3. Final products like urea, ammonium sulphate, ammonium phosphate, etc.

Effects of Wastes on Receiving Streams

All the components of waste from the fertiliser plants induce adverse effects in streams. Acids and alkalies can destroy normal aquatic life. Arsenic, fluorides, and ammonium salts are found to be toxic to fish. Amines are not only toxic to fish but also exert a high oxygen and chlorine demand. Presence of different types of salts renders the stream unfit for use as a source of drinking water in the downstream side. Nitrogen and other nutrient content of the waste encourages growth of aquatic plants in the stream.

Treatment of Fertiliser Waste-water

Major pollutants in fertiliser waste-water for which treatment is necessary include oil, arsenic, ammonia, urea, phosphate and fluoride. Oil is removed in a gravity separator. Arsenic containing waste is segregated and after its concentration, the solid waste is disposed of in a safe place. Phosphate and fluoride bearing wastes are also segregated and chemically coagulated by lime; clarified effluent, which still contains some amount of phosphate and fluoride, is diluted by mixing with other wastes. Several alternatives are there for the treatment of ammonia bearing wastes, including:

1. Steam stripping.
2. Air stripping in towers.
3. Lagooning after pH adjustment.
4. Biological nitrification and denitrification.

For all practical purposes, 'steam stripping' for ammonia removal from fertiliser wastes has been found to be uneconomical. Removal of ammonia gas from the solution in an air stripping tower, packed with red wood stakes, is found to be a very efficient method.

Very encouraging results are obtained from some laboratory and pilot plant studies conducted by NEERI in the removal of ammonia by simply lagooning the waste. It was found that considerable reduction in the ammonia content can be accomplished just by retaining the ammoniacal waste in an earthen tank, about 1 m deep, for a day or two, after pre-treatment of the waste by lime to increase the pH to 11.0. However, no reduction in urea content was observed within this period in wastes containing urea; thus waste containing both urea and ammonia required to be retained in the lagoon for a longer period, to allow urea to decompose to ammonia first.

Biological nitrification involves oxidation of ammonia to nitrate, via nitrite under aerobic conditions; this is followed by denitrification of the nitrified effluent under anaerobic condition, in which gaseous N_2 and N_2O are the end products and are released into the atmosphere. The denitrification requires addition of some quantity of carbonaceous matter in the reactor.

In all the ammonia removal methods described above, urea remains untouched. If urea removal is required, wastes containing urea must be retained for a sufficiently long time in an earthen lagoon to allow it to decompose first to ammonia. Figure 19.1. shows a flow diagram for the effluent treatment of a complex fertiliser plant. Pollution control aspects related to single super phosphate (SSP) fertiliser, straight nitrogeneous fertiliser and complex fertilisers are discussed here.

SOLID WASTE

Solid waste generated in the fertiliser plant would depend upon the process, raw material etc. and may result directly from a process or from the treatment of other effluents. They may be formed continuously or may be formed intermittently. The solid waste generated can be disposed of by any of the following modes of disposal.

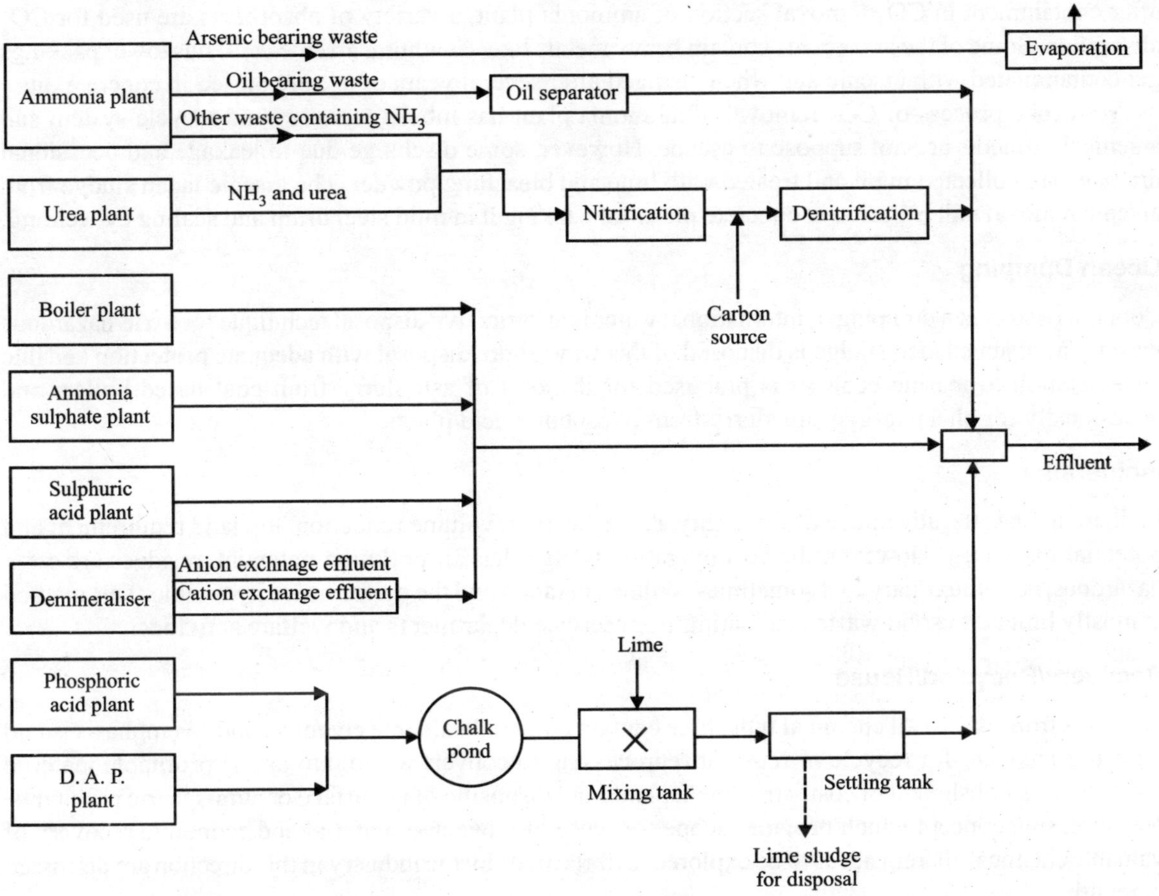

Fig. 19.1. Flow diagram for effluent treatment of a complex fertiliser plant.

Open Land Disposal

This mode of disposal has the advantage of ease of operation and low cost. However, it can be used for only non-toxic waste. Chalk sludge/waste generated during the manufacture of ammonium sulphate and phospho-gypsum by-product of phosphoric acid plant are disposed openly on land.

Land Filling

The chromate sludge from chromate removal unit is disposed of by permanent containment or by land filling. The relatively non-toxic sulphur sludge, from sulphuric acid plant, sludge generated by neutralising phosphoric acid plant effluent by lime, clariflocculator sludge from water treatment plant are disposed of by land filling. However, the land requirement is high and special safety measures like complete insulation of the deposited material from ground water or placing the land fill site over non-water bearing sediment needs to be taken.

Burying

Burying is practised for small quantities of seasonal waste contaminated with toxic substance, that too after containment in CO_2 removal section of ammonia plant, a variety of absorbents are used for CO_2 absorption. Some of the absorbents contain heavy metals like vanadium, arsenic etc. Thus tower packings get contaminated with arsenic and when changed after several years of use are buried in concrete pits.

Vetrocoke process of CO_2 removal of ammonia plant has inbuilt recovery and recycle system and arsenical effluents are not suppose to escape. However, some discharge due to leakage and occasional drainage are collected in pit and treated with lime and bleaching powder. The arsenic laden sludge from arsenic removal unit is buried in concrete pits after packing it in mild steel drum and scaling by welding.

Ocean Dumping

Containerised ocean dumping in international water is an attractive disposal technique for toxic/hazardous waste. The arsenic laden sludge is disposed of this way. Shore disposal with adequate protection and due consideration to marine ecology is practised for disposal of ash slurry from coal based boilers and occasionally for phospho-gypsum slurry from phosphoric acid plants.

Incineration

Incineration has the advantage of flexibility, detoxification, volume reduction, lowland requirement and potential for energy. However, the cost operation is high, has air pollution potential, residues are often hazardous, needs auxilliary fuel sometimes, skilled operation and the performance is variable. This method is mostly limited to solid waste originating from service departments and welfare activities.

Recovery/Recycled/Reuse

Effluents from almost all chemical industries contain potential valuable chemical and so emphasis is laid on waste recovery for recycle or reuse or reprocessing to convert waste into useful/profitable material rather than establishment of costly treatment plants and disposing of as surface or atmospheric pollutants. Newer design concept which provides scope for reuse of unwanted material and economic recovery of valuable chemical therein needs to be explored. Efforts made by the industry in this direction are discussed herewith.

In ammonia plants based on partial oxidation during gasification of oil about 2 per cent carbon is kept unburnt and thus a large quantity of carbon is produced everyday. In the conventional plant 75 per cent of the carbon is recovered. But the unrecovered carbon causes severe pollution problem. In the latest technology, a total recycle process is incorporated by which the carbon is fully recovered and recycled. A small portion of carbon slurry is bled off. Possibility of converting the carbon from the bled off into fuel by mixing it with coal rejects from the steam generation plant is being explored for all plants where full recovery of carbon is not possible.

Chalk sludge generated during the manufacture of ammonium sulphate is suitably mixed with ammonium nitrate to form CAN (25 per cent). It can also be utilised for cement production.

Phospho-gypsum is utilised of course to a limited extent only for the manufacture of ammonium sulphate because ammonium sulphate is low nutrient fertiliser and finds restricted use. Phosphogypsum is used as soil conditioner in alkaline soil or making building material like bricks, plastic board, wall plaster etc. But only a small quantity is utilised in this manner and the remainder is disposed off on land. A proven technology exists for the conversion of phospho-gypsum into CaO and SO_2. Calcium oxide can be used in the manufacture of cement and sulphur di-oxide for sulphuric acid production. This would totally eliminate solid waste besides recovery and recycle of sulphur into useful product. Despite all the advantage this process has not been found to be commercially viable in our country although one plant is operating successfully in Austria even today. Fly ash generated in steam generation plant is recovered by electrostatic precipitaters and is used for ravine filling when low lying land is available for production of cement, building bricks, as soil conditioner etc.

20

Pharmaceutical Industry

INTRODUCTION

Pharmaceutical plants discharge highly alkaline and toxic liquid wastes. If crude waste from an antibiotic waste is discharged into a stream, it not only imparts an objectionable odour to it, but also adversely affects its biological population. This waste should not be allowed to be discharged into a municipal sewer, unless the sewage treatment plant is properly designed to handle a widely varying and concentrated waste from such a plant. Penicillin waste is found to have a disturbing effect on the process occurring within a sludge digestion tank.

Synthetic drugs plants utilise large number of both organic and inorganic chemicals, and usually produce a variety of drugs in different sections. The volume and composition of the liquid wastes not only vary from plant to plant, but also from section to section in a plant, producing different types of drugs from different raw materials and using varieties of processes.

Wastes containing toxic elements like cyanides and heavy metals, if discharged without any treatment, are harmful to aquatic life in the streams. These toxic elements interfere with the biological sewage treatment units very badly.

Due to the great pollution potential and the diversified characteristics of the waste from different sections of a plant, the planning for the treatment of synthetic drug wastes should be preceeded by a careful study of each waste. Segregation and equalisation very often improve the overall treatment efficiency and reduce the cost of treatment. If the wastes have high COD:BOD ratios, benchscale laboratory biological treatment studies with acclimatised seed sludge are necessary for proper planning.

The acidic wastes, wastes containing toxic elements like cyanide, and those containing offensive-odour-producing compounds are usually segregated and treated separately. Acidic wastes may be neutralised with lime. The odour-producing compounds are usually destroyed by chlorination; compounds which are resistant to chlorine may be destroyed by heating in a furnace.

MANUFACTURING PROCESS

The drug manufacturing and processing industry can be categorised into five areas:

1. Fermentation.
2. Chemical synthesis.
3. Natural extraction.

4. Formulation and packaging.
5. Research and development.

Process Description

Fermentation

Batch fermentation processes are typically used for producing steroids and antibiotics. The fermentation process consists of three steps:

1. Inoculum and seed preparation.
2. Fermentation.
3. Product recovery or isolation.

An example of a fermentation process is illustrated in Fig. 20.1.

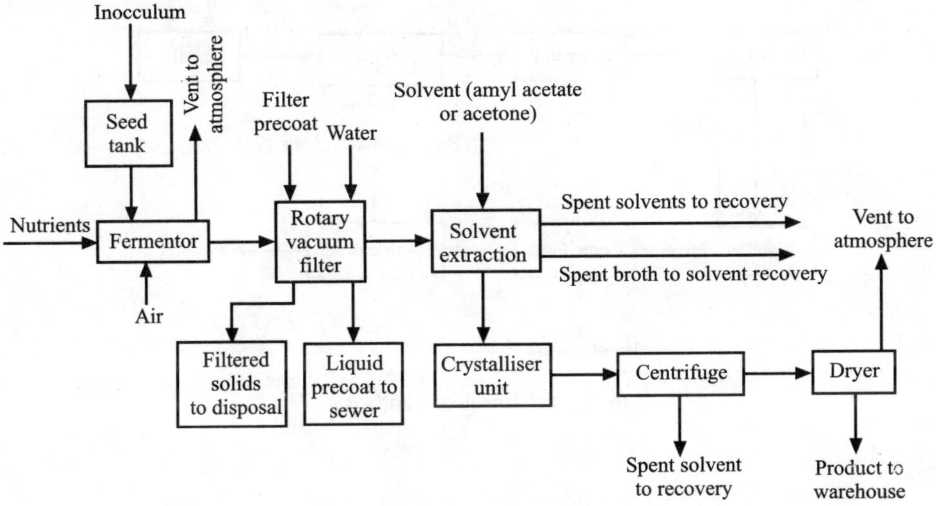

Fig. 20.1 Fermentation process.

Chemical synthesis

The majority of drugs today are produced by chemical synthesis. Typically, one or more batch reactors (depending on the number of chemical steps) are utilised in series to make the final product. The product recovery (or isolation) steps may involve separation and/or purification.

The intermediates can also be isolated in order to avoid carrying impurities to the next chemical reaction (impurities may interfere with the next reaction, reducing yield). The types of chemical reactions and the isolation processes used are quite diverse, depending on the chemical structure of the desired final product. An example of a chemical synthesis is illustrated through a process flow diagram of a product (Fig. 20.2).

Natural extraction

Drugs manufactured by natural extraction include allergy relief medicines, insulin and morphine. These drugs are extracted from natural sources (e.g. roots, leaves, animal glands, fungi, etc.). Typically, these drugs are extracted from natural sources because chemical synthesis is difficult owing to their large molecular sizes. During each extraction (or process) step, the volume of the intermediates diminishes greatly and the final purification can occur on volumes less than 1 per cent of the initial volume. Therefore, conventional batch or continuous processes are not usually suitable for natural extraction.

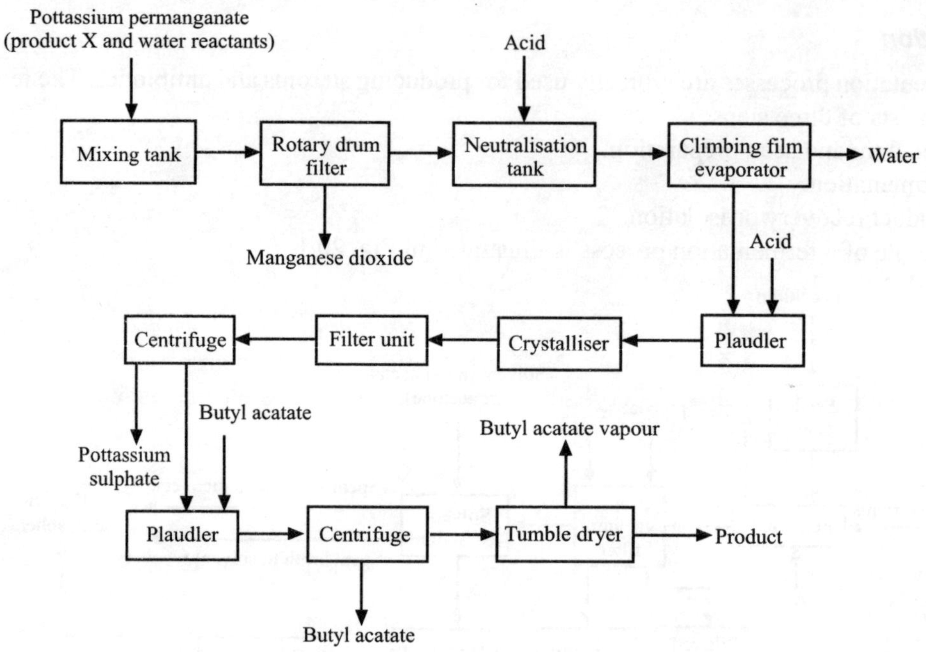

Fig. 20.2. Chemical synthesis.

Formulation

The active ingredient produced by either fermentation, chemical synthesis, or natural extraction is then formulated to tablets, capsules, liquids, creams and ointments. The active ingredient is blended to the right dosage by adding fillers or binders. Fillers or binders used for tablet formulation are starch and sugar. The blend is then compressed via wet granulation, direct compress, or slugging.

In wet granulation, the active ingredient and filler in powder form are blended and wetted with a binder solution to form coarse granules. These granules are dried, mixed with lubricants (e.g. magnesium stearate) and compressed into tablets.

After tablets, capsules are the most widely-used oral dosage form solid drugs. Hard gelatin capsules are prepared by depositing a gelatin film on the capsule (using a solution of gelatin and cooling). The capsules are dried, trimmed and the separate sections of the capsule are filled and joined.

Liquids are prepared for injection and oral use in aqueous solutions, syrups, suspensions and tinctures. The liquids are prepared by mixing the active ingredient with a solvent and preservatives (to prevent mould and bacterial growth), and filtering at high pressure. Suspensions and emulsions are prepared by using colloid mills and homogenisers, respectively.

Creams and ointments are prepared for topical use, usually by blending the active ingredient with petroleum derivative petrolatum and passing the mixture through a colloid or roller mill.

WASTE FROM DRUG INDUSTRY

Fermentation

Fermentation processes generate large volumes of spent fermentation medium. Filtration processes, which remove the solid remains of micro-organisms (mycelia), yield large quantities of solids as spent filter cake. The filter cake contains mycelia, filter aids and small amounts of residual product. After product recovery, the spent filtrate is discharged as waste-water. Other sources of waste-water are equipment-cleaning operations and fermenter vent gas scrubbing. Waste-waters from fermentation processes are typically high in BOD (biological oxygen demand), COD (chemical oxygen demand) and TSS (total suspended solids), with a pH range of 4 to 8.

Chemical Synthesis

Wastes from chemical synthesis are complex and diverse as a result of the variety of reactions and operations employed. Chemical synthesis processes generate acids, bases, solvents, cyanides and metal wastes among others. Solid wastes are generated as waste filter cake, which may include various inorganic salts (e.g. sodium sulphate, ammonium sulphate, etc.), organic by-products and metal complexes. Typically, spent solvents are recycled on-site within the process (via extraction and distillation). Still bottom wastes are generated as a result of recovery and recycling operations. Chemical synthesis processes also generate aqueous waste streams, which include mother liquors from crystallisation operations, aqueous wash layers from extractions, contaminated scrubber water and equipment-cleaning wastes. Waste-waters from chemical synthesis processes typically have high BOD, COD and TSS, with the pH ranging from 1 to 11.

Natural Extraction

Wastes resulting from natural extraction processes include raw materials such as leaves and roots, spent solvents and waste-water. Waste-waters from natural extraction processes are typically low in BOD, COD and TSS levels, with the pH ranging from 6 to 8.

Formulation

Wastes generated in formulation processes include wastes resulting from cleaning, sterilisation, spills and rejected products. During blending and/or tableting operations, dusts are generated, which are typically recycled back into the formulation process. During tablet coating and capsule manufacturing, there are air emissions from solvent handling.

The primary source of waste-water is equipment cleanouts. Waste-waters typically contain inorganic salts, sugars, syrups and residual amounts of the active ingredient. Waste-waters are usually low in BOD, COD and TSS, with neutral pH.

POLLUTION PREVENTION OPTIONS

Pollution prevention options in the drug manufacturing and processing industry are source reduction through process modifications, recycling and resource recovery, and good housekeeping and operating practices. Process modifications include product reformulation, material substitution and process step modifications. The following sections describe some of the pollution prevention options available to the drug manufacturing and processing industry.

Process Modifications

Product reformulation involves the substitution of one or more raw materials, which may reduce the volume and/or toxicity of wastes generated. However, product reformulation is difficult for the pharmaceutical industry because of the time required for product redevelopment and testing. Testing is required to ensure that the reformulation has the same medicinal effect as the original drug. The time and the testing required for FDA approval of the reformulated drug can be long and drawn out. Therefore, product reformulation is usually not an economic or practical option.

Material substitution is the replacement of a hazardous material in manufacturing or processing with a non-hazardous material. Material substitution may also reduce the volume and toxicity of the wastes generated, but without changing the product itself. For example, organic solvents used during tablet-coating processes can be substituted with water-based solvents; reducing the volatile organic compound emissions into the atmosphere. Another example is the use of less hazardous organic solvents (e.g., ethyl acetate, isopropyl acetate, toluene, etc.) in the place of more toxic (more rigidly controlled by CPCB) solvents, such as methylene chloride and benzene. In fact, many drug companies have specific guidelines prohibiting the use of certain solvents (e.g., methlyene chloride and benzene) even at the research and development stage.

Another approach in material substitution includes the use of aqueous cleaning solutions (e.g., soap and water) in place of solvent-based solutions. The current trend in the industry is to replace chlorinated solvents with non-chlorinated solvents.

Generation of wastes can be reduced through process modification or modernisation. Process modifications involve changes in operating conditions and/or equipment. Process modernisation includes installing updated control systems or increasing control levels. Another form of modernisation is increasing the degree of process automation.

Incomplete chemical reactions can result in the increased formation of by-products, thus increasing the amount of wastes generated. Possible causes of incomplete reaction are inadequate feed flow control and/or reactor temperature control. Installing updated control systems can improve reactor/reaction efficiency and reduce by-product formation. An increased degree of automation can also reduce wastes generated by preventing operator errors. Other process considerations for waste minimisation are:

1. Install internal recycling systems for solvents and cooling water.
2. Select new or improved catalysts to improve reaction conversion.
3. Consider using a continuous process in place of batch process.
4. Modify tank and vessel dimensions to minimise wetting loss (e.g. use of conical vessels).
5. Optimise crystallisation conditions to reduce the amount of product loss to the mother liquor and cake wash.
6. Optimise crystallisation conditions to obtain the desired crystal size, thus minimising the need for recrystallisation (crystal size too small to filter) or milling (crystal size too large).
7. Consider new reactor-filter-dryer systems that can handle more than one operation (e.g., Rosenmund filter-dryer).
8. Utilise available computer programmes/expert systems that can evaluate the entire process for efficiency and characterise the waste streams, thus determining the major source of waste generation and optimising the process.

One example of pollution prevention via process modification is the use of supported liquid membranes (SLMs) for liquid solution separation during reaction, work-up, and/or isolation sequences. The SLMs are thin layers of pure liquid or liquid solutions immobilised in microporous inert supports. They have the extraordinary capacity of removing solutes from a feed solution to a strip solution. For example, an aqueous solution containing a solute is fed into an SLM and the solute is extracted into a contained or supported organic solvent, which in turn, is stripped to the aqueous strip solution. The organic solvent is continuously renewed and this specific feature allows significant reduction in solvent usage.

Other advantages of SLMs are the high separation factor per stage and low capital/operating costs. The major disadvantage is the instability of the SLMs, which prevents long-term operation. In an attempt to overcome SLM's shortcomings, the hollow fibre contained liquid membrane (HFCLM) technique was developed. HFCLM uses microporous hollow fibres to contain liquid membranes in a permeator (Fig. 20.3). In theory, the solvent contained in the membrane can be used indefinitely, thus reducing the amount of solvent used for extractions. However, depending on the type of process, the solvent may have to be replaced as the performance starts to degrade because of the accumulation of impurities.

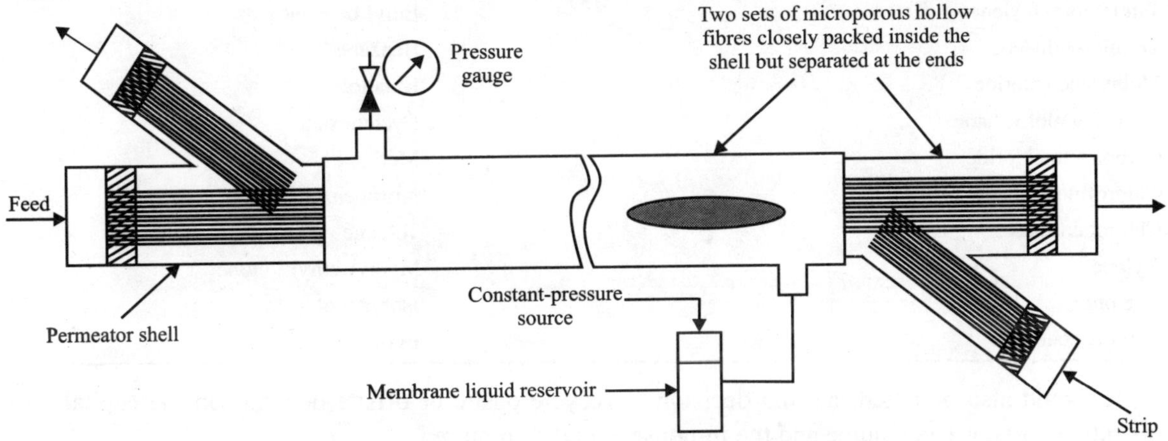

Fig. 20.3. Hollow fibre contained liquid membranes.

Although significant waste reduction can be achieved through process modifications, there are problems associated with this option. Extensive process changes can be very costly, owing to the installation of new equipment and downtime. New processes must also be tested and certified by the appropriate agency to ensure that the resulting product is acceptable. An important parameter to consider is the impurity profile of the drug manufactured by the new process, because an identical or better impurity profile (i.e., fewer impurities) is essential for the validation of clinical testing. In addition, the existing management may be reluctant to change a working (successful) process. Such resistance is particularly likely against adopting continuous processes in place of existing batch processes.

Equipment cleaning efficiency can be improved significantly by employing high-pressure spray nozzles (as a preliminary cleaning method) to clean reactor walls, rather than conventional solvent or caustic solution boil-outs. The use of high-pressure nozzles can also reduce solvent usage appreciably.

Resource Recovery and Recycling

Recycling and resource recovery include direct reuse of the waste generated-reclaimed for a separate use or direct reuse within the process via separation of impurities. However, the high degree of quality control required in the pharmaceutical industry limits the options available for reuse within the process.

Fermentation wastes may be used as supplements to animal feeds, soil conditioners and fertilisers. For example, mycelium filter cake and other by-products of the production of antibiotics by fermentation are widely sold as protein supplements to animal feed, as long as the antibiotic activity in the filter cake does not exceed 2 grams per tonne of cake and less than 3 pounds of cake is used per tonne of feed.

Inorganic salts such as ammonium sulphate and sodium sulphate (typical by-products of some chemical synthesis processes) can be reclaimed. Ammonium sulphate can be used as raw material for fertiliser and sodium sulphate can be dried for sale to the glass industry. Recycling is a particularly good option for spent solvents. Spent solvents can be used for equipment cleaning. They can also be purified for reuse in the process for reaction mediums, extraction mediums, and coating mediums. Commonly recycled solvents in the drug manufacturing and processing industry are summarised in Table 20.1.

Table 20.1. Commonly recycled solvents in the pharmaceutical industry.

Tetrachloroethylene	Ethyl benzene
Trichloroethylene	Benzene
Methylene chloride	Butanol
1,1,1-Trichloroethane	Cyclohexane
Carbon tetrachloride	Methanol
Chlorofluoroethanes	Nitrobenzene
Chlorobenzene	Toluene
Xylene	Methyl ethyl ketone
Acetone	Isobutanol
Ethyl acetate	Pyridine

It should also be noted that the decision to recycle onsite or offsite depends on the capital cost, operating costs, waste volume and the in-house expertise required.

The following steps can improve solvent waste recyclability:

1. Segregate chlorinated solvents from non-chlorinated solvents, aliphatic from aromatic solvents, and aqueous from flammable solvents.
2. Minimise solids concentration in solvent wastes.
3. Label all solvent wastes properly, including the composition and generation source.

REMOVAL AND RECOVERY OF ALUMINIUM SALTS FROM PHARMACEUTICAL EFFLUENTS

The effluents generated from Friedal Craft's reaction process in pharmaceutical industries contain 4 to 15 per cent aluminium and if they are not separated and mixed with main effluents the aluminium salts cause serious problems in the biological waste treatment due to osmotic imbalances, in addition to environmental pollution problems when the final effluents are discharged. An economically viable ion exchange process for the removal and recovery of $AlCl_3$ from pharmaceutical effluents was developed under optimised conditions. The design parameters for the fixed bed batch adsorber like the mass transfer co-efficients, total exchange capacities, M/L values for the removal of $AlCl_3$ from pharmaceutical effluents were determined. The column reaction parameters like the breakthrough constants, the bed depth service time were also calculated. The optimum conditions for recovery of high pure $AlCl_3$ suitable for re-use in pharmaceutical industries were determined and the cost analysis for the recovery process was made.

Anhydrous aluminium chloride is used as a catalyst in Friedal Craft's reaction in pharmaceutical industries for a wide variety of purposes in chemical industry for carrying out reactions like dehydrogenation, decarboxylation, oxidation, alkylation, acylation, desulphonation, amylation and polymerisation. Aluminium chloride is used particularly in industries which manufacture drugs like ibuprofen, ciprofloxin and R. floxcin etc. in large quantities in tens of tonnes per day.

The pharmaceutical industries in and around Hyderabad itself use nearly 80 tonnes of $AlCl_3$ per day which is being dumped into the effluent due to lack of economically viable recovery process for $AlCl_3$ in pure form from this effluent.

Segregation of Friedal Craft's reaction process effluents and removal and recovery of aluminium salts from these effluents are likely to solve to a certain extent these problems and further the recovered $AlCl_3$ can be re-used by the same industry. In our laboratory scale studies it was observed that the Friedal Craft's reaction process effluents (2 to 4 pH) can be directly treated by ion-exchange process using amberlite IR 120 cation exchange resin which can be used to remove and recover $AlCl_3$ quantitatively by operating the process under experimentally determined optimum conditions. $AlCl_3$ can be recovered in sufficiently pure form suitable for re-use by the same industry with high efficiency. The economic viability of the process has also been evaluated by comparing with other processes and the cost of $AlCl_3$ recovered.

Experimental

Batch experiments were performed by shaking fixed amount of resin with 50 ml of aqueous solution of the salt of desired concentration, temperature and pH in different glass bottles at 30°C in a shaking water bath with an invariable speed of 60 rpm. At the end of pre-selected intervals, the bottles were removed, the contents centrifuged and the supernatant analysed for the $AlCl_3$ titrimetrically. Data for isotherm study were obtained by allowing the resin to remain in compound with solution for three more hours after attainment of the equilibrium to ensure complete saturation.

Column experiments were also conducted with a variety of bed height, flow rates and solution concentrations. The exchange capacity of the resin was studied by observing the amount of aluminium retained by column.

Results and Discussion

The magnitude of $AlCl_3$ removal by amberlite IR 120 resin as a function of contact time and concentration was systematically investigated. The extent of uptake of $AlCl_3$ by resin increases with an increase in contact time and concentration and attains equilibrium just in 30 minutes (Fig. 20.4). It was observed that with increase in temperature from 30°C the absorption increases upto 60°C with the initial concentrations in the range of 200–2000 ppm and the rate of removal of $AlCl_3$ also increases felicitating attainment of equilibrium in 10 minutes.

It was observed that maximum sorption of aluminium ion can be obtained in the pH range of 3 to 4 which may be due to the formation of cationic species $Al(H_2O)_6^{3+}$ to larger extent in this pH range where quantitative exchange with protons of the cationic exchange resin occurs. At higher pH values the water molecules of $Al(H_2O)_6^{3+}$ will lose protons yielding gelatinous $Al(H_2O)_4^{3+}(OH)_2^{2-}$ or $Al(H_2O)^{3+}(OH)_3^{3-}$ which does not exchange with the protons of the cationic exchanger. Equilibrium isotherm absorption data were obtained with the equilibrium time at different pH ranges to determine the maximum sorption capacity of the resin.

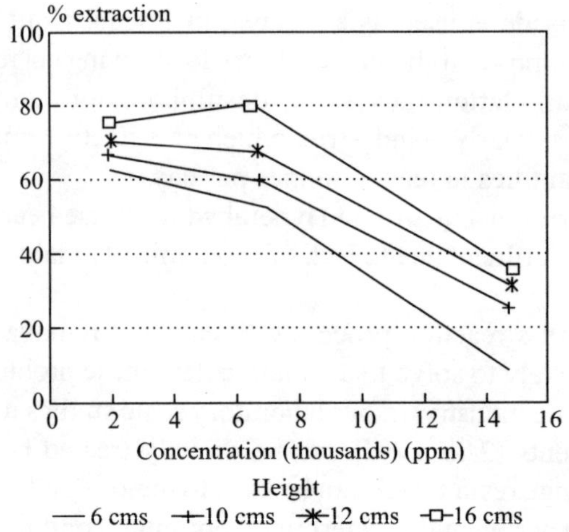

Fig. 20.4. Effect of height at various concentration.

Column studies

Column tests were conducted with resin bed volumes of 6.9 cm^3 (5.0 g) and with bed volume of 27.0 cm^3 (15.0 g). Flow rates were generally kept in the range of 0.416 l/m^2–s to 2.08 l/m^2–s.

The column studies indicated that 2 to 4 pH range was effective for the recovery of aluminium from different initial concentrations of the influent supporting the results obtained in batch studies. A flow rate of 0.416 l/m^2–s was found to give maximum recovery of aluminium ions. In the event of dealing with large volumes of solutions 0.832 l/m^2–s flow rate is also recommended to speed up the process operating without sacrificing much on the adsorption efficiency. The breakthrough capacity and total capacity were also monitored for different conditions. Aluminium which was absorbed on the resin was recovered by eluting the column with 2 M hydrochloric acid. The recovery of aluminium was almost complete.

The efficiency of adsorption of resin for aluminium at different influent concentration from 3 gms to 15 gms keeping the final volume constant was examined systematically. As the concentration of Al increases in the aqueous phase, the quantity of absorption of Al per gram of resin also gradually increased while the percentage extraction gradually decreased. The resin medium was found to be suitable for treating the effluent even at high concentration without affecting the resin matrix and sacrificing the adsorption efficiency. Similarly, when the quantity of Al kept constant and the quantity of resin adsorbent increased, the quantity of Al absorbed in the resin column was also increased.

Theoretical plates

The column efficiency is dependent upon the shape of the breakthrough curve. The more efficient the column the steeper the breakthrough curve. A quantitative measure of the column efficiency is the number of theoretical plates N. It was seen that the number of theoretical plates found at 15 cms height and 12 cms height column is used as the height for conducting all the experiments.

Distribution co-efficients

The rate at which two constituents separate in the column is determined by the ratio of the two corresponding distribution co-efficients. The data for the weight distribution co-efficients and bed distribution co-efficients are given in Table 20.2.

Table 20.2. Distribution coefficient at different flowrates.

Conditions: pH = 3.0; temperature = 30 ± 1°C; concentration = 150 mg/ml; and height = 12 cms.

Flow rate (ml)	K_d	λ	λ'
1	1.52	2.08	1.84
2	1.41	1.84	1.70
5	1.25	1.04	0.93
8	0.905	0.995	0.61
12	0.639	0.545	0.31

K_d = distribution co-efficient; λ= weight distribution co-efficient; λ'= bed distribution co-efficient.

Interference of various metal ions in the adsorption of aluminium

The selectivity of the absorption/extraction of the aluminium ions by IR 120 resin was studied and the results presented in Table 20.3. Quantitative recovery of aluminium can be achieved even in the presence of large concentration of calcium α = 1.01 and magnesium α = 0.787 by this process under optimum conditions. However, iron and be separated to a certain extent.

Table 20.3. Separation factor α with metal ions.

Metal	K_d	α
Al	1.52	–
Mg	0.787	$\alpha_{Mg}^{Al} = 1.93$
Ca	1.01	$\alpha_{Ca}^{Al} = 1.50$
Fe	1.93	$\alpha_{Fe}^{Al} = 0.787$

Note: Conditions: pH = 4; amount of solution taken = 100 ml; temperature = 30° ± 1°C; amount of resin = 1 gm; concentration of Al solution = 15 mg/ml; and concentration of metal ions = 5 mg/ml.

Application

Recovery of $AlCl_3$ from pharmaceutical industries

A typical effluent from pharmaceutical industry was chosen (pH 2.0 $AlCl_3$ 4 per cent) which manufactures ciprofloxcin and utilises aluminium chloride two tonnes per day. The efficiency of the recovery of $AlCl_3$ from this pharmaceutical effluent at optimum conditions was studies both by batch (Table 20.4).

Table 20.4. Batch studies for the pharmaceutical effluent.

Conditions: pH = 3.0; temperature = 30°C; and concentration of solution = 4% in 100 ml.

Dosage of resin gms	*Adsorbed %*	Q_e *mg/g*	C_e *% ppm*	l/Q_c	l/C_c 10^{-5}	$\log Q_c$	$\log C_c$
5	4	88.9	11010	0.0112	9.08	1.94	4.04
10	10	111.3	18090	8.98×10^{-3}	5.5	2.04	4.25
15	20	155.8	21518	8.14×10^{-3}	4.64	2.19	4.33

(Contd...)

Dosage of resin gms	*Adsorbed %*	Q_e *mg/g*	C_e *% ppm*	I/Q_c	I/C_c 10^{-5}	*log* Q_c	*log* C_c
20	30	169.9	25766	5.88×10^{-3}	3.88	2.23	4.41
30	50	185.5	20809	5.39×10^{-3}	3.24	2.26	4.48
40	70	194.7	33610	5.13×10^{-3}	2.97	2.28	4.52
50	95	211.4	35019	4.3×10^{-3}	2.00	2.32	4.54

Q_e = Amount absorbed in mg/gm; and C_e = amount remaining in ppm.

Design aspects from batch and column studies

The M/L values were plotted for the batch reactor (Fig. 20.5) and the β_1 is the mass transfer co-efficient and was found to be 7.06×10^{-7} cm s^{-1} and the $\beta_1 \times S$ was found to be 0.0295 min^{-1} where S it the outer surface of resin per unit volume of particle free slurry. For column design parameters the zone formation area was calculated as 0.015 m., the fractional capacity as 0.45, the formation time as 59.99 minutes, the percentage saturation as 98.5 per cent and theoretical break-point as 0.298 and experimental break-point as 0.33.

The removal of aluminium chloride using series of columns was carried out. In this case the removal of 15 per cent aluminium chloride from the solvent using multilevel columns in series it was found that the adsorption efficiency of almost 100 per cent can be achieved at the end of the 5th column. These results in (Table 20.5 and Fig. 20.6) indicate that the resin can be used in series column very efficiently for the removal and recovery of Al at various concentrations also. This technique is specially more useful when the alternative treatment methods like solvent extraction fail in terms of concentration gradient and where there is no suitable biological treatment.

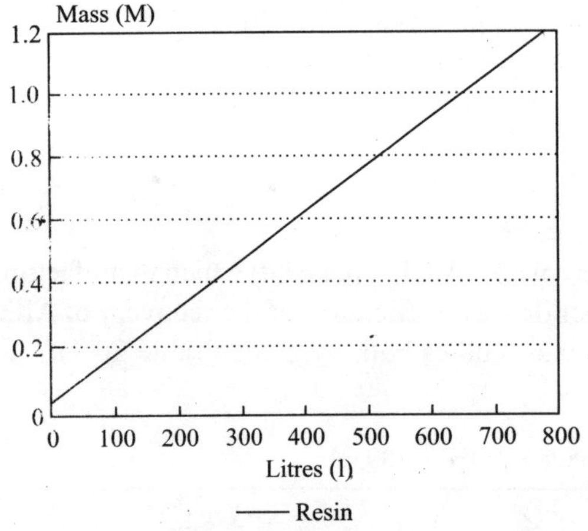

Fig. 20.5. M/L values.

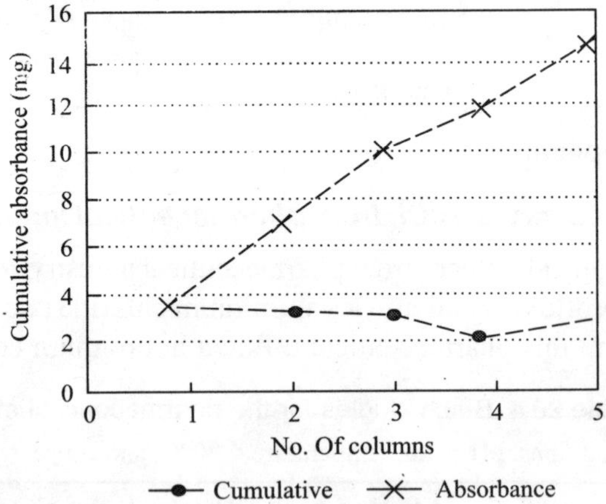

Fig. 20.6. Cumulative absorbance series column studies.

The resin can be re-used without any change in the adsorption capacity upto 100 cycles. The major advantage of this process is the desorption and recovery of the adsorbed Al in high pure state with the solvent HCl which also includes regeneration step and the stability of the resin even after several recoveries. Since recovery of aluminium is one of the main aims of the present work the desorption studies were also carried out using different molarities of HCl. It was observed that with 2 M HCl the aluminium can be quantitatively desorbed and the same HCl can be used for the desorption of further columns. The re-crystallisation of aluminium chloride using various solvents was studied. The results show that the re-crystallisation of $AlCl_3$ in more efficient when acetone is used as a solvent rather than when methanol is used as the percentage of recrystallisation is low when methanol is used. However HCl can also be distilled and recovered and both $AlCl_3$ and HCl can both be recovered by trapping the heat energy wasted in several processing in the industry using suitable heat exchangers.

Table 20.5. Column in series showing the cumulative adsorption.

Column no.	*Aluminium influent*	*Quantity (mg) effluent*	*Quantity adsorbed (mg)*	*Cumulative quantity absorbed (mg)*
1	15000	11480	3520	3520
2	11480	8100	3380	6900
5	8100	4800	3300	10200
8	4800	2800	2000	12200
12	2800	200	2600	14800

Note: Conditions: pH = 3.0; temperature = 30 ± 1°C; concentration = 150 mg/ml; height = 12 cms; and weight = 20 gms; flow rate = 1 ml/min; and volume of solution.

Economics of Recovery Process

To examine the economic feasibility of the processes the cost of chemicals used the prepared columns, and cost of $AlCl_3$ recovered are evaluated and compared with chemical coagulation (Table 20.6). To recover 1 kg of aluminium chloride from 25 litres of industrial effluent the cost of input chemicals comes to Rs. 80 which is arrived at taking into consideration that the resin can be used for at least 100 cycles and the recovery of HCl and acetone after use. However, no costs were included for pumping because the flow chart involving gravitational flow is considered.

Table 20.6. Comparative cost of removal/recovery $AlCl_3$ by various methods.

Method	*Removal efficiency*	*Cost of recovery*	*Recovery of $AlCl_3$*	*Remarks*
Lime coagulation	98%	Rs. 70	Cheap but process tedious	Solid residue remains as waste
Ion exchange process	99.3%	Rs. 80	Simple, $AlCl_3$ obtained in pure form	Effluent directly treated. No solid generated

21

Re-Refining of Used Lubrication Oil

INTRODUCTION

Lubrication oil is a petroleum based product used in all the industries such as, chemical, mechanical, textile, metallurgical, mining, sugar, pulp and paper, etc. After service life the oil is removed and disposed. It is a renewable resource, which can be reclaimed and regenerated for the re-use. Considering the ever-growing demand for petroleum products and dependence on import for crude oil, India can ill-afford to waste it. Various processes have been developed in India and abroad, for the reclamation of waste oil. Among the process, vacuum, molecular distillation technique is pollution free and cost effective.

Lube base oils are the most valuable constituents of crude oil. Virgin lube oils are recovered from lube crudes by solvent extraction followed by wax removal. Average crude oils have 3–8 per cent lube content. Whereas used lubrication oils have 60–75 per cent recoverable lube content, which if burned or dumped would mean the loss of a valuable natural resource. The lube oil is largely a mixture of hydrocarbons, paraffin, naphthalene, and aromatics and at times olefins. During service life, the oil deteriorates mainly due to two reasons namely contamination and oxidation or chemical breaks down. The main extraneous contaminants in the used oil are:

1. Air borne matter.
2. Combustion products.
3. Unburnt fuel.
4. Products of corrosion.
5. Metallic particles.

Oxidisation results in development of many intermediate products. This include acids that may attack metal parts, and that remain in soluble form in the oil at first but become insoluble later, contributing to sludge formation. Other oxidation products are lacquer and dark solid carbonaceous matter, both of which tend to be formed where thin oil films are subjected to higher temperature. However, there is only very little change in chemical composition of the oil. Major part of the mineral hydrocarbon remains in good condition. Through techniques of reclamation and re-refining the oil can be regenerated for further use.

TECHNICAL ASPECTS

Product Specifications

The product is re-refined lube oil. The BIS number of the product is BIS–9048. The lube oil requirements of the consumers vary according to their requirement.

The classification based on viscosity for automobile engine oils that are SAE 10, SAE 20, SAE 30, SAE 40 and SAE 50. The classification based on performance level is SA/SB/SC/SD/SE/SF/SG for petrol engines and CA/CB/CC/CD for commercial diesel engines.

Manufacturing Process

While reclamation techniques involve physical methods like setting, decantation, heating, dehydration, filtration and centrifuging, re-refining covers methods of purification of used oil through distillation, hydro treatment, chemicals etc. No single method however, is adequate enough for making the oil fit for further use and therefore a combination of methods are employed.

Various methods of reclamation and re-refining are grouped as under:

Reclamation techniques

1. Mechanical filtration.
2. Vacuum dehydration.
3. Centrifugal separation of impurities.
4. Magnetic filtration.
5. Chemical treatment with acids, alkalies, solvents and hydro-finishing.

Re-refining techniques

1. Vacuum molecular distillation.
2. Acid and clay treatment process (developed by Indian Institute of Petroleum (IIP), Dehradun, Defence Research Laboratory, Kanpur, Regional Research Laboratory, Jorhat, etc.)
3. Re-cyclon process.
4. BERC process.
5. KTI process.
6. REVAC process.
7. RTI process.
8. Bernd Meinken process.
9. Lubrex re-refining process.
10. Balmer Lawrie process.

Of the above processes, vacuum molecular distillation technique developed by Chem Vac Engineers, Mumbai is considered to be more suitable in view of its pollution free nature and cost effectiveness. The process flow chart of the above process is given in Fig. 21.1. The processes are outlined hereunder:

Vacuum Molecular Distillatian Process

In this process the following steps are followed:

1. The waste oil is pre-treated with two types of polymers, which will flocculate the carbon sludge substantially and separated by centrifuging/filtering.
2. The waste oil from the previous step is passed through phase separators, coalescer filters to remove free and emulsified water.
3. The waste oil is then passed through a thin film evaporator, under medium vacuum and temperature, to remove, diesel, dissolved water if any and some lighter fractions.
4. The undistilled waste oil is passed through a molecular distillation plant, under higher vacuum and temperature, to distill off lighter oil (approximately equal to SAE 20 in viscosity).

5. Subsequently, the oil is subjected to high vacuum, high temperature in an induction heated molecular distillation plant (above 340°C and at a pressure of 0.001 mm of Hg). The distilled oil will have a viscosity of about SAE 40 at ambient temperature and pressure.
6. The oil is bleached with active clay to get very light coloured base oils, to be blended with additives to get finished oils for the automotive and industrial applications.
7. The final residue will be semi-solid at ambient temperature and can be processed for the manufacture of asphalt, or briquetts, or to make greases for slow moving machineries.
8. The used clay can be reclaimed, re-activated 3–4 times, by using solvent extraction and high temperature activation process.

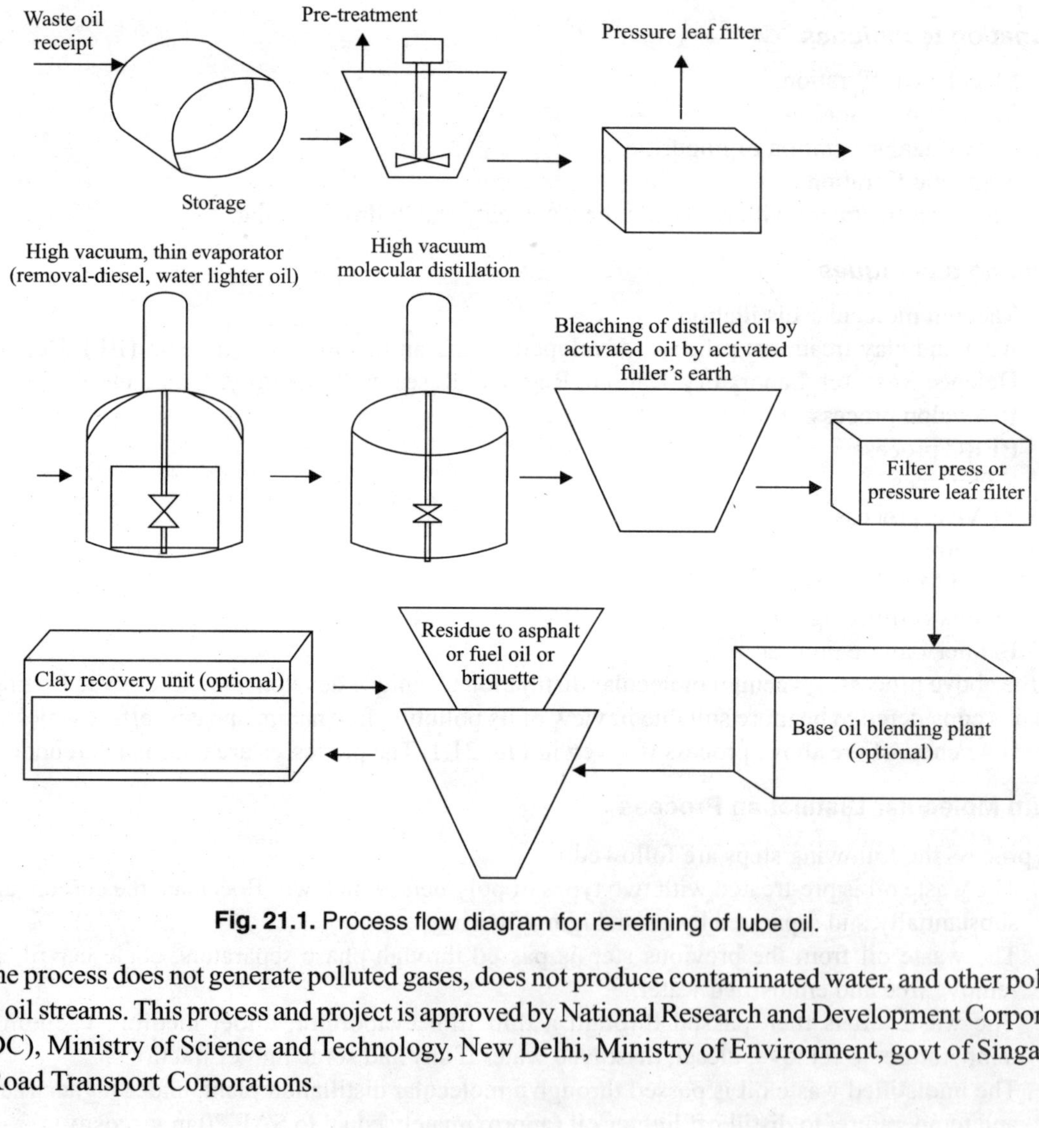

Fig. 21.1. Process flow diagram for re-refining of lube oil.

The process does not generate polluted gases, does not produce contaminated water, and other polluted waste oil streams. This process and project is approved by National Research and Development Corporation (NRDC), Ministry of Science and Technology, New Delhi, Ministry of Environment, govt of Singapore, and Road Transport Corporations.

Salient features of the process

1. The process is acid free and sludge free. Instead of by-product liability, it generates bottom residue, which is salable as finished products, like as asphalt/bitumen modifier. It has passed Environmental Protection Agency, USA regulations, toxicity characteristics, leach tests etc. This means that, none of the harmful chemicals and metals that can be originally found in the used lube oil will against enter into the environment.
2. The process eliminates the use of sulphuric acid and generates no acid sludge, which has disposal problems due to the tough anti-pollution regulations. Since the acid treatment is absent, the process greatly improves the plant economics.
3. The process minimises the quantity of bleaching clay needed, to about 3–5 per cent of the distilled products. This is substantially lower as compared to 18–20 per cent used in the acid clay process. The clay consumption in the process is minimum due to the absence of acidity, low amount of metals in the distilled products (metals go to the residue).
4. The process, maximised the salable products to 95–96 per cent, of the feed waste oil (for a typical waste lube oil from Oman, containing 1–2 per cent water and 1–2 per cent process loss). This figure is at least 15–20 per cent higher than the figures in the case of acid-clay technology. The high yield in the process is due to the processing of the waste oil in high vacuum molecular distillation plants.
5. The process is modular and operationally very flexible. It is ideal for waste lube oils containing high amount of additives, synthetic oils etc. which otherwise is very difficult to process by acid-clay technology. Besides the plant can handle with ease, used transformer oils, hydraulic oils etc. The plant is modular in construction and to increase the capacity of the plant, with common storage, infrastructure, testing facilities, molecular distillation modules can be added.

Characteristics of the individual products

Light distillate: this contain mainly, diesel, very light fraction of lube oil, water if any etc. After separation of water by settling, the balance material can be used as diesel fuel for the boiler or sold as such with or without bleaching. The characteristics of various products are given in Tables 21.1, 21.2 and 21.3.

Table 21.1. Product quality of light distillate.

Properties	*Typical range*	*Specification IS: 1448*
Appearance	Clear, bright	–
Colour	Bright yellow	–
Specific gravity	0.84 to 0.86	–
Water content	Nil	–
Flash point	60–80°C	32 minimum
Viscosity cst		
40°C	2.5–10	2 to 5 minimum
100°C	–	
Sulphated ash % w/w	0.01	0.01 maximum
Pentane insoluble % w/w	Nil	–
Total acid no. (mg KOH/gm)	0.05	0.5 maximum

(Contd...)

Properties	Typical range	Specification IS: 1448
Concentration of carbon % wt.	0.15	0.2 maximum
Pour point (°C)	0	6 maximum
Total sulphur % wts	0.05	1 maximum

Table 21.2. Product quality of spindle oil.

Characteristics	*Acid-clay process*	*Oil and vacuum distillation process*	*Suggested/specified*
Appearance	Clear with slight smell	Clear and bright (no smell)	Clear and bright
Viscosity, cst			
At 40°C	10 to 15	15 to 20	To be reported
At 100°C	4 to 5	5 to 6	
Viscosity index	95	95	90 mininum
Flash point (°C)	180	200	200 minimum
Ash sulphated, % wt.	0.008	0.004	0.01 maximum
Carbon residue, % wt (Conradson)	0.35	0.32	0.4 maximum
Total acid no. (mg/KOH/gm)	0.05	0.02	0.05 maximum
Strong acid no. (mg/KOH/gm)	Nil	Nil	Nil
Copper corrosion	>1	>1	Not worse than 1
Moisture content (%w/w)	Nil	Nil	Nil

Table 21.3. Comparative analysis of base oil.

Characteristics	*Acid-clay process*	*Vacuum distillation process*	*Suggested/specified*
Apperance	Clear and bright	Clear and bright	Clear and bright
Viscosity, cst			
At 40°C	60 to 70	80 to 90	To be reported
At 100°C	8.5 to 10	10 to 11.50	
Viscosity index	95	95	90 minimum
Flash point °C (COC)	200	220	Min. 200 (SAE–20) Min. 215 (SAE–30)
Ash sulphated, % wt.	0.008	0.004	0.01 maximum
Carbon residue, % wt. (Conradson)	0.35	0.32 max	0.4 maximum
Total acid no. (mg/KOH/gm)	0.05	0.2	0.05 maximum
Strong acid no. (mg/KOH/gm)	Nil	Nil	Nil
Copper corrosion	>1	>1	Not worse than 1
Moisture content (% w/w)	Nil	Nil	Nil

Installed Capacity: The installed capacity of the project is 1500 kl of waste oil processing per annum.

Land and building: A land area of one acre and a built-up area of 3400 sq.ft., is required for the project.

Plant and machinery: The required plant and machinery can be supplied by Chem Vac Engineers. Mumbai.

Raw materials

The main raw material required is used lube oil, which is available indigenously in huge quantities. The typical characteristics of the used oil is given hereunder:

Constituent	*Content %*	*Viscosity*
Waste lube oil	Lube 70–90	8–12
Light distillate (with water, some diesel etc.)	6–8	1–3
Light oil	10–12	7–9
Medium base oil	52–55	11–15
Heavy gear oil	13–15	18–24
Asphalt	7–8	–

Pollution control aspects

It should be ensured that solid, liquid and gaseous effluents generated and disposed in the project meet the standards laid down by the state pollution control boards, liquid wastes should be kept to the minimum by recycling them back into the process.

Market Aspects

Lube oils are the lifelines for industrial and automotive equipment as also varius consumer appliances. Since virgin lube oils are petroleum based product, any increase in crude oil price automatically increases its cost of production. Also crude oil reserves are fast depleting. In the light of the above scenario, re-refining of used lube oil has gained great momentum.

In India, the consumption of lube oils are ever increasing in view of consistent growth in automobile production. Hence the market demand for re-refined oil would not be a problem. The oil can be sold in bulk to oil companies and major lube oil blenders.

RE-REFINING OF USED LUBRICATING OIL FOR PETROCHEMICAL INDUSTRY

Due to the sharp increase in the oil prices and the resultant energy crisis, conservation of petroleum products has assumed global importance. It is evident that millions of gallons of used oil, which could provide a part of the solution to the energy problems, is being disposed off in ways that pollutes water, air and land. In most of the developed countries it has either become mandatory or it is officially encouraged to re-refine the waste oils and make it fit for re-use either as such or after blending with virgin oils and additives. The importance of oil conservation, in the wake of crude price hikes and heavy drain of foreign exchange on our national exchequer, needs no great elaboration. The age-old reference to crude oil as 'liquid gold' is truer now than ever before. Nearly a quarter of our country's total imports is petroleum and its products, and these consume approximately 35 per cent of our foreign exchange earnings. Increase in the crude oil prices will further enhance the drain on our foreign exchange. For a developing country like ours, it would be well appreciated how important it is to conserve: (i) our available resources; and (ii) every drop of oil imported and put them into best possible use, which should be our national goal.

Lubricating oils are a means of conserving energy by reducing the wear of the metals between two moving surfaces and also reducing the energy needed to obtain the motion. Any theory for the recycling of the lubricating oil is based on the fact that lube oil hardly changes its property on use. On use it gets contaminated but not wasted. We have large quantities of used oil and the problem with which we are faced is to find economical ways of utilising the used oil.

It is reasonable to assume that unless we can find profitable ways of re-using the used oil, from engine pumps, gear boxes, and other systems, this oil will ultimately be dumped into fields (whereby damaging the crops) or to rivers and lakes (to pollute water and environment). A decade back oil was dirt cheap and a barrel of crude oil costed less than \$10.00, the disposal of used oil was more a social problem (prevention of pollution or loss of food crops) than an economical problem.

But, today with the price of crude oil reaching enormous heights, it is worth exploring to get maximum benefits from the used oil. The basis of re-refining/reclaiming used lubricating oil or simply using it in its raw form is due to fact that the lubricating oil does not get wasted but contaminated. Although lubricating oil's constitute only 2 to 4 per cent of total crude oil refined, its contribution to the total cost and profits is much higher. Lubricating oils are used in variety of application and in various grades ranging from heavy duty to light duty and are marketed depending on intended use. Presently, the annual consumption of lubricating oils is 650,000 tonnes, out of which 150,000 tonnes, at an approximate cost of Rs. 50 crores, are imported. Motor oils constitute about 30 per cent of the total lubricating oils consumed and 66 per cent of the motor oils are used for cars and rest for buses, trucks, and railway engines.

After some use the motor oils lose their lubricating properties and have to be discarded. There are various reasons for the deterioration of the oil. The oxidation of the oil causes the formation of the organic acid, asphaltenes and polymer which produce resinous and gummy matter, whereas cracking produces hard deposits of carbonaceous matter and imperfect combustion of fuel causes the oil to become black from fine carbon in the colloidal suspensions. Unburnt ends of fuel mix completely with the oil, simply lowering its viscosity. Solid particles such as dust come from the atmosphere and also from wear and tear of metals add to the contamination.

These used oils can be used either as fuel or re-refined. The spent oil have high ash content, substantial amount of water, gasoline metal (essentially lead) and chlorine. To burn spent oils it will be necessary to remove the light ends and to use a special equipment to take care of low melting point ashes, lead and chlorine. Re-refining is more attractive for countries, which have scarce crude oil resources. The present fuel shortage has rekindled interest in re-refining of used lubricating oils.

Processes for Re-Refining of Used Lubricating Oils

Several processes are available for the re-refining of used lubricating oil as mentioned below:

1. Conventional acid clay.
2. Propane extraction-cum-acid clay treating.
3. Propane extraction-cum-hydrofinishings.
4. IPA extraction-cum-distillation.
5. Distillation/clay.

All the processes have common steps of pre-filtration of solid particles and then pre-flash to remove water and lighter fraction. Conventional process consists of acid treatment to remove polymer, asphalts and additives and is followed by clay treatment to neutralise acid and to improve the colours. Propane extraction-cum-acid clay consists of propane treatment to remove extraction additives, combustion residues and residue coming from the thermal degradation and then acid clay treatment. In distillation cumhydrotreating the used lubricating oil is flashed in a high vacuum distillation column to produce desired lube base stocks. Light ends are removed from the top and contaminants e.g., metal polymerisation products, and asphaltic materials go with the residue. The distilled lube oil is mixed with hydrogen gas under pressure heated in a furnace and is passed through a fixed bed catalytic reactor. Final product is stripped to obtain the desired product. Conventional acid-clay processes have the advantage of low capital investment, simple know-how and are suitable for small scale industries. On the other hand, other two processes need higher investment, indigenous technology is not yet available and are not suitable for small scale industries. In re-refining, a major problem is collection of spent lubricating oils. Due to lack of co-ordination facilities, it is more practical to have large number of small scale units rather than having a few large scale capital intensive plants. For small scale units it is more economical to use conventional acid and clay process.

Conventional acid clay treating process for re-refining of the lubricating oils has been practiced since long time. In this process, the oil is first filtered to remove the solid particle and then flashed to remove the light ends and water. After this, the oil is acid treated and aqueous and organic layer are separated in a settling tank. The aqueous layer is discarded and the oil is clay and/or caustic treated to improve the colour and neutralised the acid present and is then finally filtered to remove any suspended clay particle. Steam distillation has also been used to improve the properties of re-refined oil.

Recycling process consist of vacuum distillation to remove lighter impurities after preliminary mechanical sieve filtration. This is followed by acid treatment and then neutralisation and bleaching by Fuller's earth. After bleaching, filtration through a filter press is done to obtain clean oil, which can be considered equivalent to base oil. Depending on characteristics of the base oil, i.e., amount of additive present, etc. additional blending of additive has to be done to bring the recycle oil to final usable form. However, the problems encountered in using the recycled engine oil can be summarised as follows:

1. Know-how of the process: Many users who have tried recycling have not got the know-how from any tried source. They have just tried centrifuging the used oil. They have used the centrifuged oil to mix with the fresh oil in different proportion for top-up. The ratio of the quantity of the filtered oil used to total fresh oil used is very low.
2. Collection of used oil samples: The situation for the collection of used oil should be improved. And it has also been observed that sometimes used oil sent for recycling contains small parts like circlip, cotter pins, shackle pin, etc. and also cotton waste, foam, etc. Such used oil will have to be filtered through a sieve to remove these things.
3. Transport: The transport cost will depend on the location of the recycling plant, in most of the cases.
4. Location and size of the recycling plant: In order to keep both—capital and transportation–cost within a reasonable limit a portable plant of small size seems to be very worthwhile. For this system of portable recycling plant, the transportation cost will be low now. Here again very small size plant capable of being mounted on a truck will have to be developed.
5. Testing of recycled engine oil: As such some simpler test to enable the evaluation of suitability of recycled oil should be developed.

Some Currently Existing Processes

Superheated steam has been used during the clay treatment and thus decolouring agent requirement was reduced and the quality of oil was better. Different adsorbents, which have been used, are activated alumina, bentonites, tripolis, diatomites, sub-bentonites clay, and Fuller's earth. Many variations of this process have been proposed and patented.

Since acid sludge removal is a major problem, many workers have concentrated their efforts in eliminating the use of acids. Besides sulphuric acid, other chemicals, which have been used as polymerising agent to precipitate and remove decomposition products, including fine dispersed carbon, are aluminium chloride and phosphoric acid. Temperature for all this was 40–70°C. There are other re-refining processes developed, one in which waste oil is treated with zinc chloride followed by alkali treatment and other in which saponification of the alkali soluble contents and passing the oil through an adsorbent is carried out. Acid treatment step can be avoided by heating the used oil with activated Fuller's earth and water followed by vacuum distillation. The use of plant ash (containing 16 to 18 per cent potassium carbonate), sodium phosphate and 1 per cent sodium hydroxide solution followed by treatment with adsorbent has been reported. Other chemicals such as picric acid, sulphuric acid, citronella, charcoal, have also been used to improve the process. The use of various gases to improve the process and properties of the final product has also been investigated. Gases such as carbon dioxide and nitrogen are passed through the heated clay oil mixture to affect the distillation of more readily volatile substances, while ammonia-air mixture followed by ozone-air mixture through gasoline-mixed, filtered use oil and then use of the adsorbent have also been employed.

Various schemes to eliminate or improve the clay treatment have been reported. Calcium sulphate and clay compounded with a mixture of sulphuric acid and powdered silica followed by distillation to remove the lighter fraction have been used to re-refine the used lube oil. Use of acid as well as clay have been eliminated by pre-treating the oil with various chemicals such as aqueous solution of ammonia at 170–210°C, sodium hydroxide at 90°C followed by distillation to remove lighter fractions.

Other processes, which have been studied, use agglomerating agent such as sodium silicate and polymeric coagulant agent. Various solvents have also been tried for re-refining of spent oils such as di/tetrahydrofurfural as solvent to dissolve the impurities. The use of ethanolamine and ethylene diamine to regenerate the oil has also been reported.

Contaminants in Used Lube Oil

For many years, there has been an axiom in the trade that 'good oil does never wear out'. Oil may, however, loose its effectiveness because of usual contaminants. These contaminants are usually: (i) extraneous impurities; and (ii) products of oil deterioration:

1. Extraneous impurities enter from the surrounding air and/or from machine on which the lubricant (or coolant) is being used:-
   (a) Contaminants from the air are dust and dirt—especially in plants where operations and surroundings are dusty such as cement mills, paper mills and sugar refineries, and where crushing, grinding, pulverising etc., are being done.
   (b) Contaminants from machine on which the lubricants are used include metallic particles left from machining operations (viz., cutting/grinding work).
2. Products of oil deterioration occur in oils as a result of physico-chemical changes. A good example of this is motor oil in the crank case of IC engines where the oil is subjected to high temperatures and excessive extraneous contamination. Some oil products are:-
   (a) Sludge-mixtures of oil and carbon particles from incomplete combustion of the fuels or from water, dust, or dirt.
   (b) Lacquers—a hard or gummy substance, also called varnish, which deposits on engine parts.
   (c) Oil soluble and insoluble products.

Selection of the Treatment

Selection of refining agent depends on the availability of equipment, the nature of crude oil and the type of product manufactured. The characteristics of the crude oil establish to a considerable extent the nature of the finished materials. The choice of the reagent is then reduced to a problem in economics.

More than one reagent is often required for re-refining oil. The sequence in which the reagent should be applied may require a special investigation. With only few exceptions the finished oils may have a neutral reaction. A very slight alkalinity, but not acidity might be tolerated. This does not preclude the use of selected organic acids such as additives in the finished oils. Acidic substances may destroy the stability of product and corrode equipment. Alkaline content may increase the ash content of the oil. Acid used in refining petroleum includes sulphuric acid, nitric acid, hydrofluoric acid, hydrochloric acid, boric acid and organic acids and their derivatives. Somè of these acids are of historic interest only. Following are the advantages and disadvantages of this treatment.

Advantages

1. Low capital investment.
2. Technology simple and available.
3. Equipment available indigenously.
4. Material of construction is generally carbon steel.
5. Chemicals—common and available.

Disadvantages

1. Low yields.
2. Possibility of corrosion problems.
3. Acid sludge disposal problem.
4. Oil-clay waste disposal problem.

New Methods of Disposal of By-Product

Disposal of spent acid sludge is always an intricate problem in a lube oil re-refining plant. Acid sludges have usually been disposed off by incineration, acid recovery, land-filling or neutralisation with spent lime or caustic. However, these disposal techniques are not acceptable because of stringent environmental regulation, and economically not feasible because of insufficient quality of available acid sludge. Conventional methods of disposal which are in use now-a-days are enlisted below:

1. Petroleum refineries have disposal of acid sludges by incineration, acid recovery, land-fill and neutralisation with disposal to waste-water.
2. Land-fill of acid sludges, the most common disposal method, appears to be reasonable though not an ideal method of disposal. The sludge often mixes with refuse at the land site. The soluble free acid probably leaches through this oil and alkaline soil are finally converted to sulphate salts entering the ground water or near by streams.
3. Combustion or incineration even after neutralisation is undesirable due to the major emissions of sulphur dioxide and very fine metal-containing particulates which would be emitted.
4. Production of sulphuric acid by sludge combustion is feasible, provided significant amount of sludge is available. Also it is too expensive on a small scale. Disposal to waste-water can only be practised where high volumes of waste-water treatment facilities are available for dilution.

The availability of the crude oil at cheaper prices is now a thing of the past. The trend is now constantly upwards and has reached a point of no return. Its impact on the economy, specially with reference to developing non-oil-producing countries, has jolted them from being wasteful in the use of petroleum products and made them conscious about optimisation in application and re-use wherever possible. Re-refining, therefore, is the 'in thing' and has come to stay, perhaps with a bang.

COMBUSTION EFFICIENCIES OF OIL INDUSTRY WASTE AND PROSPECTS FOR USE AS FEED FOR FLUIDISED BED BOILERS

Oil/gas fired boilers till a few decades ago, generally met the steam requirements of industries for their process, captive power and co-generation needs. However, as about 40 per cent of these reserves have higher ash, conventionally fired steam generators faced operational and maintenance problems. Bubbling fluidised bed combustion boilers, commonly known as atmospheric fluidised bed combustion (AFBC) boilers, were preferred by industries as they could effectively utilise these low and varying quality coals, besides providing higher efficiencies than the waste from sponge iron plants, electrostatic precipitator dust, etc. These fuels have a large percentage of ash, very low volatiles and have a wide size range, which make them practically impossible for utilisation in conventional boilers. AFBC boilers can effectively and efficiently utilise these fuels in an environmentally clean way.

In this study we have taken samples of neem oil cake from the nearby industry and utilised it as feed to AFBC. Fly ash, due to insufficient facilities for disposal, poses a serious pollution threat. This case study tries to analyse the feasibility of using oil industry waste as a source of fuel in fluidised bed boilers. The study envisages a verification of the essential factors such as the proximate analysis of the fuel, particle size distribution of incoming fuel, etc.

AFBC Process

Bubbling fluidised bed belongs to a class of gas-solid contact systems, in which air or gas is passed through an inert bed of solid particles, such as sand or crushed refractory, supported on a grid. Fuel is injected into the bed. The gas will initially seek a path of least resistance and pass upward through the bed, leaving it undisturbed. However, when the gas flow is further increased, it starts bubbling through the bed and the particles attain a state of higher turbulence. Under these conditions, the bed exhibits the properties associated with a fluid and is hence termed as a bubbling fluidised bed.

The name bubbling is derived from the fact that a fluidised bed of this kind has rising gas bubbles. The gas escapes through the incipiently fluidised bed (called the emulsion phase) as bubbles. The bubbling action causes intense stirring of the bed particles, which causes good mixing of fuel and air and uniform temperature, thus creating an ideal environment for excellent combustion.

Advantages of AFBC boilers

The AFBC boilers have many advantages. However, the principal among them include:

1. More than any other type, AFBC boilers have the ability to effectively burn a wide variety of fuels (Table 21.4), which include coal washers.
2. AFBC boilers do not require oil for start-up or for load support.
3. Large amount of ignition energy in the bed, the uniform distribution of air, and the thorough mixing of the fuel and air, enable AFBC boilers to achieve higher efficiency (about 86 per cent). As excess air levels can be maintained at part loads also, the efficiency at part loads is also high.
4. The high heat transfer rates in the bed result in a compact AFBC boiler.

5. AFBC boilers require less maintenance, as compared to conventional boilers, due to the absence of pulverisers and moving grates in the hot furnace zone.
6. As the combustible content of the fluidised bed at any point of time is very small (<1 per cent) and the combustion is flameless, AFBC boilers do not require expensive burner management systems.

Table 21.4. Proximate analysis of various conventional fuels used as a feed to AFBC (%).

Fuel	*Volatile matter*	*Fixed carbon*	*Ash*	*Moisture*	*HHV (kcal/kg)*
Coal	20–25	25–30	40–50	9–20	2500–3105
Lignite	25–30	20–25	2–3.5	50–56	2600–2700
Washery rejects	15–17	13–15	55–73	1–12	1900–2100
Sponge iron kiln wastes					
DRI ash	0.5–1	20–25	50–70	5–12	2400–2600
ESP dust	2–7	22–28	70–80	4–8	2600–2800
Char	2–7	40–45	44–52	3–4	3300–3950

Design Considerations of the Fluidised Bed

The phenomenon of fluidisation and its related application to the bed design needed an in-depth study. A detailed investigation has helped to resolve major issues relative to in-bed and above-bed combustion and heat transfer. The following are the salient features:

1. Bed temperatures.
2. Fluidisation velocity.
3. Bed height.
4. Freeboard combustion.
5. Fuel size.

Table 21.4 gives a proximate analysis of various conventional fuels used as a feed to AFBC.

Thus the case study envisaged the use of fly ash obtained from cement factories and neem oil cake obtained from oil extraction unit to test their efficiency for use as fuel in fluidised bed combustion boilers. The preliminary parameters for determining their use are the fixed carbon, volatile matter content determined by proximate analysis. The tests were carried out on fly ash and the neem oil cake samples and the results have been tabulated (Table 21.5).

Table 21.5. Proximate analysis of neem oil cake taken for case study.

Feed	*Moisture*	*Volatile matter*	*Ash content*	*Fixed carbon*
Fly ash	–	0.61	79.36	0.03
Neem oil cake (oil industry waste)	14.06	77.69	6.55	1.7

The results obtained indicate the presence of a favourable range of volatile matter and fixed carbon content in the neem oil cake sample and in the case of fly ash, the carbon content does not satisfy the preliminary requirement (fixed carbon should at least be greater than 1 per cent), but stands a good chance for use as a mixed fuel as the high ash content is sufficient to supply the bed material for the fluidised bed boiler. The neem oil cake, satisfying the preliminary requirements, is obtained in a pulverised state and hence has a favourable particle size determination range. The ash analysis of the sample and the ash initial deformation temperature are to be determined before the sample can be put to use as fuel. The major orientation of this case study is to develop a complete procedure and eventually design a fluidised bed combustion boiler for use by oil extraction industries to meet their power needs, which is growing higher and higher each day. However, it requires a disquisition into various factors as bed height, feed inlet point, the fluidisation velocity etc., which have to be calculated for the given type of fuel and the cost analysis made. This, in turn, provides good scope for pilot plant studies in using the oil cake as fuel and the extent to which pollution, mainly SO_x and NO_x emissions can be reduced. The present utility of the neem oil cake is mainly in agriculture, but it has a much greater end usage as an energy source at a time when cost effectiveness has become important.

22

Reconditioning and Reclamation of Transformer (Insulating) Oil

INTRODUCTION

Mineral oil used as a dielectric and coolant media in transformers and switchgear are essentially derivatives of crude petroleum from which they are obtained by various processes of refining. Before being accepted for service these oils must conform to the Indian Standard no. 335 (1963) 'Specification for Insulating Oils for Transformers and Switchgear'. The basic requirements for an oil to ensure a long and trouble free service are that it must be free from all extraneous matter and moisture, and must also possess sufficient stability against chemical deterioration. Transformer oil is a liquid having the property of insulating the coils of transformers, both electrically and thermally. The natural type includes refined mineral oils (petroleum fractions), which have low viscosity and high chemical oxidative stability.

When put into service insulating oils deteriorate due to contamination and oxidation of their hydrocarbon components giving rise to undesirable products like sludge and acids. These products adversely affect proper functioning of the equipment and may finally lead to an undesirable breakdown. It is therefore of prime importance that the oil is properly maintained during service and suitable action taken if the concentration of the deterioration products exceeds the specified limits. It is observed that the state electricity boards and other electricity undertakings often find it difficult to follow a strict schedule of oil maintenance.

Also after the oil has rendered its useful life it is discarded and replaced by new oil. Even mixing of varius types of oils is rerely done—and instead of topping up equipment with a different brand, the entire quantity is replaced resulting in extra expenditure.

An insulating oil which has been found unfit for service on the basis of the tests conducted to assess its quality, is liable to contain one or more of the following types of contaminants:

1. Solid matter like dust, fibres, metallic particles and sludge.
2. Free and dissolved water.
3. Dissolved gases.
4. Organic acids, soaps, and other products of chemical deterioration.

When the concentration of such impurities in the oil is not very large (particularly the acids), the oil can be made fit for service again by filtration preferably accompanied by heat and vacuum treatment. This process called re-conditioning is often carried out while the oil is still in the equipment, in order to effect savings in expenditure and also protect the equipment from the ingress of moisture during draining and refilling operations.

When the concentration of the impurities is large and the oil contains appreciable quantities of acids, soluble soaps etc., which cannot be removed by filtration alone, there is no option but to remove it from the equipment. At this stage there are two alternatives available to the operational staff:

1. The oil be discarded and replaced by new oil.
2. The oil be reclaimed and used again.

The first solution is what most of the power supply undertakings in India follow at present. The fact that the oil forms an expensive item of the electrical industry, clearly stresses the need to reclaim the oil so that it gives a fairly long if not the original amount of service. However, apart from the considerations of large savings that may result, it is of equal importance that the reclamation process should be economical when compared to the cost of the new oil and the quality of the reclaimed oil should also compare favourably with that of new oil. Several methods have been developed during the last few decades to efficiently re-condition or reclaim the oil unfit for service. These are:

1. Re-conditioning:-
   (a) Simple filtration.
   (b) Filtration-cum-vacuum dehydration.
2. Reclamation (or regeneration):-
   (a) Chemical processes.
   (b) Adsorption processes.

A brief description of these procedures, and the equipment available indigenously for the same is given below.

RE-CONDITIONING

Simple Filtration

If the impurities present in the oil are only of the nature of well defined solid matter e.g., dirt, fibres, metal particles etc., these can be removed by filtering the oil. The process includes gravitational or forced filtration through paper, cloth, fine metal gauge, porous or sintered ceramic filter plate or through a bed of inert mineral substance such as a filter cell. This method does not remove colloidal particles or soluble impurities such as acids, asphaltenes and water. Several filter presses are available commercially and some of them have provisions for circulating the oil through the filter bed while the oil is still in the equipment. As such these are more useful for the routine maintenance rather than re-conditioning of contaminated oils having more than one type of impurities listed earlier.

Filtration-cum-Vacuum Dehydration

This process involves two distinct stages. Firstly the heated oil is either filtered through a specially designed filter pack or the solid impurities are removed by centrifuging and secondly the filtered oil is subjected to vacuum treatment which dehydrates and degasses the oil completely. The treated oil is thus free from all solid and colloidal matter, water, dissolved gases and volatile acids. Many commercial plants based on this principle are available and a few of these are:

Streamline filter

The streamline filter has, for many years been recognised as the most perfect practical method of removing all traces of solid matter from liquids. For insulating oils where removal of moisture is of utmost importance the filter has another advantage in that oil passes through it in layers so thin that finely dispersed water and even dissolved water are evaporated by the application of moderate heat and vacuum. If discarded and clean oil storage tanks are provided the filter can be coupled up so that with its own pumps it takes oil from the discarded oil tank and delivers re-conditioned oil to the clean oil tank, a single passage being sufficient for complete purification. The plant can also be connected to transformers on line and the oil purified by circulation. In such cases the connections should be made so that the deteriorated oil comes to the filter from the lowest connection on the transformer and is returned to the highest connection on the transformer. The filter is kept full of oil when not in use, so that it can be put to work to circulate without inducing a change of level in the transformer.

The discarded oil entering the plant passes into a chamber where it is heated by means of immersion heaters. The heaters are of ample surface area and the circulation of the oil is so arranged, that it is exposed to the effect of heating for a very short time and no harmful effects due of local overheating are possible. From the heater chamber the contaminated oil is drawn by means of metering pump and passed through the streamline filter at a constant pre-determined rate. In the heater the oil is raised to a suitable temperature for effecting dehydration when later exposed to vacuum. The filter consists of a number of streamline packs built up of specially prepared paper and maintained under end compression by spring loading, so that only interstices of infinitesimal size remain between the discs. The oil passes through these interstices in layers of almost molecular thickness, leaving the solid impurities in their entirety deposited at the paper edges.

Unique fineness of filtration is thus obtained, it being possible to eliminate even sub-micronic particles. All solid impurities in the oil are thus easily removed, even including the colloidal carbon formed by arcing in a circuit-breaker. At the same time the minute sub-division of the liquid permits every drop of moisture to take up the latent heat necessary to convert it into steam as it passes through the pack and the pressure falls to the vacuum maintained inside the packs. This ‘thin layer’ effect is most important and peculiar to the streamline filter.

The oil passes on from the filter into a vacuum receiver and being under vacuum from the time of passing the paper edges, it is subjected to an intensive drying and deaerating action. At the degree of vacuum prevailing the water is completely evaporated, and in the vacuum receiver the water vapour, air and other gases are mechanically separated from the oil in passing over a series of shallow trays.

Two pumps are coupled upto the vacuum receiver. A dry vacuum pump withdraws the water vapour and gases from the top of the receiver and maintains a vacuum of the order of 71 cm of mercury. A special form of rotary pump extracts the clean dry oil from the bottom of the receiver and delivers it to the clean oil tank, or back into the transformer when circulating. A float controlled valve mechanism maintains the oil level in the vacuum receiver so as to provide a liquid seal to prevent water vapour and gases from passing out with the oil. The same float mechanism also breaks the vacuum, should for any reason the liquid level rise in the vacuum receiver beyond a certain point. This is to prevent damage to the dry vacuum pump. The filter can be used for continuous dehydration without any necessity for stoppages or renewals.

After use the solids deposited on the filter packs can be cleaned easily by the simple process of reverse flow with compressed air. There is no need to open up the filter; the filter body is first drained and in the mean time a simple valve manipulation cause the pump normally creating vacuum to charge a suitable air bottle to 28 tonnes/sq.m. The cleaning operation is then effected by turning a cock which causes the compressed air to discharge through the packs. The accumulation of solids is thereby dislodged from the packs and discharged from the sludge outlet. The whole cleaning process is carried out in a tew minutes merely by movement of the control levers on the exterior of the apparatus which actuate the valves as described.

A valve is also provided for the purpose of circulating the oil when the plant is started up. The oil is circulated internally until the correct temperature and conditions are obtained, after which by moving the valves to the running position the discharge of the clean oil commences.

The filter body and vacuum dehydrator are heavily lagged to conserve heat and the whole set up is mounted on a suitable frame fitted with small wheels or castors for easy transportation. The filter packs, which are specially prepared by patented processes last a very long time as leaning by reversal processes produces no abrasive action. Even if replacement becomes necessary the packs do not constitute an expensive item. The cost of operating a stream-line filter is neligibly small. The average consumption of a medium size unit capable of filtering 550 litres per hour is about 16.3 kW for the motor and heaters. The filter requires only about 15 minutes of unskilled attention during an 8 hour run. The streamline insulating oil filters are the most popular filtration units.

Sharples vacuum ultrafilter

The sharples vacuum ultrafilters having a rated filtration capacity ranging from 300 to 600 litres per hour. The equipment consists of a set of filter packs comprising a number of filter elements duly compressed through which the oil pre-heated to a determined temperature is filtered. The filtered oil is then fed to a vacuum chamber where it flow down in the form of a thin layer over a very large surface area. During the passage of the oil down the heated walls the moisture in the dissolved as well as in the undissolved form is removed together with any gases incorporated in the oil due to products of decomposition.

The vacuum chamber is equipped with a level control device by means of which it is possible to maintain the level of the oil at a pre-determined flow rate once the gate valve has been adjusted. The oil is sucked from the bottom of the vacuum chamber by means of a suitable pump and is then discharged. A single motor drives this pump as well as the one maintaining vacuum inside the vacuum chamber.

The sharples vacuum ultrafilter utilises an indirect type of heater for heating the oil. This has got an advantage since it avoids localise heating and eliminates charring. The heat is applied to the oil exactly in the zone in which vapourisation of the moisture is taking place. The equipment also includes a vacuum gauge, a thermometer for noting the temperature of the oil and a thermometer for controlling the temperature of the oil at a pre-determined level and a flow indicator.

The equipment can also be used to filter the oil in a transformer on line where the oil is continuously circulated through the machine and back into the transformer (similar to the streamline filter).

The sludge accumulated on the edges of the filter stack can be very easily removed by manipulation of the valves 3 and 2 (Fig. 22.1). By opening valve 8 and closing valves 2 and 1 the flow of the oil is changed in the reverse direction and in this way the sludge deposited on the edges of the filter pack is drawn out and is discharged via valve 6 which is kept in open position to the sludge tank from where it can be removed. The filter is successful in removing all types of solid matter, moisture and gases from the contaminated oil. In case the oil contains a large amount of free water, a centrifuge accessory is also available which when coupled with the filter removes the excess water before the oil enters the filter, thus preventing damage to the filtration (paper) stacks.

De-laval centrifugal purifier

The De-laval oil purifiers are based on the principle that impurities having different densities from that of the pure oil can be separated from the latter by the application of centrifugal force.

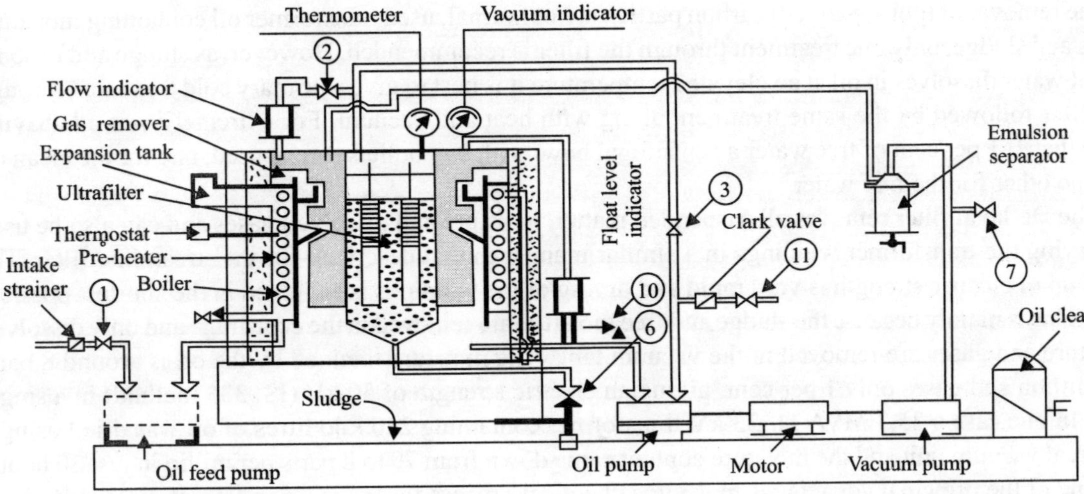

Fig. 22.1. Operating diagram of the sharples vacuum ultrafilter.

Oil normally taken from the bottom of a transformer or a separate tank is transported by the inlet pump via a pre-strainer into the electric oil heater where the oil is heated to a temperature of about 60° to 65°C. From the heater the oil is pressed into the centre of the centrifugal bowl assembled as clarifier. The discarded oil containing solid impurities and moisture is fed in at the top bowl inlet. The feed material is forced upwards through the spaces between the intermediate discs, where the centrifugal action separates it into a light and a heavy phase. The latter comprising solids and free water is thrown indirectly to the bowl wall. The light phase consisting of clean oil with dissolved moisture and gases is moved inwards between the intermediate discs and upwards through the space around the inlet tube to the light phase discharge.

This unique combination of an optimum centrifugal force and the stack of conical discs, the intermediate spaces of which serve as an individual centrifuge is successful in attaining continuous and effective separation of extremely small particles of insignificant weight difference from the oil. The centrifuging action and the spraying of the clarified oil leaving the bowl take place at a vacuum of about 76 to 38 mm of mercury absolute pressure corresponding to a boiling point of water of 45° to 32°C. Tests have shown that the oil droplet size leaving the centrifuge bowl is in the ratio of 20–100 microns giving surface of 250–265 sq. metres to 1 kilogram of oil. The warm oil droplets containing dissolved water and gases, under the influence of heat and vacuum expand and burst resulting in free steam and gas which are instantaneously evaporated by the vacuum pump. Unburst droplets explode on reaching the collecting cover wall. The drying and degassing of the oil is completed in the vacuum tank where the oil remains for several minutes. The vacuum tank is equipped with a constant level device which secures a sufficient holding time during the degassifying process and completes the automatisation of the unit in service. Thus after the starting-up period no stand-by operator is required.

From the vacuum tank the oil is transported back to the transformer or a clean oil tank via the discharge pump. In case of heavily carbonised swithchgear oil, the treated oil is finally passed through a filter press for the removal of (not separated) carbon particles. For normal, used transformer oil containing moisture, gases and sludge, only one treatment through the filter is recommended. However, as sludge and to some extent water dissolves in oil at an elevated temperature it is best to give a primary cold treatment through the filter followed by the same treatment along with heat and vacuum. For extremely wet oils having more than 0.1 per cent of free water a centrifugal bowl with two outlets can be used, one for the clean oil and the other for the free water.

The De-laval filter removes all suspended matter, moisture and dissolved gases and can also be used for drying the transformer windings in a similar manner to that described for the streamline filter. The build up of electric strength is very rapid and drying of the windings is achieved in the shortest possible time. This is mainly because the sludge and free moisture are removed in the centrifuge and only dissolved moisture and gases are removed in the vacuum tank. The moisture retained by the oil is around 8 parts per million and gases only 1 per cent giving an electric strength of 50 kV (IS: 335 method) in a single pass. In one case a 35.3 MVA 11–33 kV transformer containing 250 kilo litres of oil was dried using a De-laval vacuum unit and the moisture content came down from 70 to 8 parts per millio in just 30 hours.

One of the principal advantages in using a chemical process for the regeneration of discarded oils is that it is applicable to all types of oils including the badly burnt or heavily oxidised oils, even circuit breaker oils having a lot of colloidal carbon etc., can be brought back to a renewed state of usefulness and stability. However, the chemical processes have not found any popularity in our country due to the heavy initial investment involved in setting up of the pilot plants.

Adsorption Processes

The adsorption processes which actually consist of an adsorption and a filtration stage utilise certain finely divided, microporous substances which by virtue of their large surface area are capable of adsorbing the impurities from the oil. Some of the commonly employed adsorbents are, Fuller's earth, activated alumina, activated silica, keisulghur, bentonite, montmorillonite and other bleaching earths. The process consists of mixing a pre-calculated amount of the adsorbent to the oil and agitating the mixture. After a specified interval of time the oil is decanted and filtered hot preferably with vacuum, to remove the suspended impurities and moisture. These adsorbents, although varying in their composition are essentially chemically inert. When activated by processes which involve heating, crushing and in some cases acid treatment they possess an active surface. The active surface has certain attractive forces which hold the polar contaminants in the oil like acids and other products of chemical deterioration.

Many commercial plants based on this principle are in vogue. Two of the popular ones are:

Nigo's re-conditioner

This plant consists of a main tank in which the discarded oil is heated upto a temperature of 150°C. The heating can be done externally by fuel or internally by immersion heaters. After the oil has attained the requisite temperature it is treated with a patented powder in a ratio of one specified measure for every 3 litres of the oil. The mixture is agitated by means of a mechanical stirrer and the heating is stopped after homogeneity is attained. Air tight covers are now put on the tank and air is forced in through an air inlet tube to a maximum pressure of 20 tonnes/sq. m. The bottom of the tank contains specially designed membranes through which the oil is filtered and it issues through the clean oil outlet. The plant removes all suspended matter, moisture and other dissolved impurities including acids. The membranes need continuous changing and the complete plant is portable.

Sparkler filter unit

The sparkler units are extremely compact in size and light in weight. They make use of activated alumina and silica as adsorbents. The main features of the unit include a horizontal plate totally enclosed air sealed purification unit which forms the main body of the plant. The plates in the tank are first covered with a special type of filter paper and then pre-coated by passing a slurry of silica or alumina powder in dry new oil. The discarded oil is treated with activated alumina in a ratio depending upon the acidity of the oil. After thorough mixing, the oil is filtered through the pre-coated papers, by applying vacuum on the delivery side. The free water present in the oil cannot pass through the filter packs because they are preferentially wetted with dry oil and the dissolved water is absorbed by the silica and alumina. The sludges and other solid impurities are also removed by filtration. The acids, and other polar contaminants are adsorbed by the activated alumina. The dissolved air and gases are removed by the air-impermeable barrier formed by the silica pre-coat.

In short all the impurities are got rid off by adsorption-cum-filtration without having to resort to heating. The unit can also be connected to transformers on line for re-conditioning the oil including reduction of acidity.

A process based on selective adsorption has been developed at the Central Power Research Institute, (CPRI) Bangalore, which makes use of indigenously available Fuller's earth.

The activated earth (UL grade) is heated at 90° to 100°C for 2 to 3 hours to drive out the moisture it may have absorbed during transit and handling. The drying should be done just before use. The dried earth is mixed with the discarded oil in a drum in a ratio depending upon the acidity of the oil. Usually oils having a neutralisation number of 1.0 mg/KOH/g or more require 250 g of the earth per litre of the oil. The drum is then to be agitated or the oil mechanically stirred to ensure complete mixing of the earth. It is then allowed to stand for about 16 hours (say overnight). The earth along with the sludge and adsorbed impurities settles down. The oil is then decanted and filtered. Any type of filter capable of removing solid impurities, and moisture can be used.

This process when used in conjunction with a streamline filter yields reclaimed oil at 1/10 the cost of the new oil. The process is simple and is suitable for use in industrial establishments and state electricity boards. The used earth can be re-activated by igniting in a furnace at 200°C to burn off the oil. The results of tests carried out on an oil before and after reclamation are given in Table 22.1. A flow sheet of the process is given in Fig. 22.2.

Table 22.1. Properties of discarded and reclaimed oil using the process developed at CPRI (Bangalore).

Property.	*Discarded*	*Reclaimed*
Colour	Brown opaque with sludge	Clear bright yellow
Density g/cc 25°C	0.8588	0.8568
Viscosity 35°C Cst	16.67	16.30
Neutralisation no. mg/KOH/g	1.12	0.05
Saponification no. mg/KOH/g	2.55	0.71
Dielectric strength kV	45	Greater than 50
Dissipation factor	0.003	0.002
Refractive Index	1.4708	1.4691
Crackle test	Passes test	Passes test

Some novel processes developed in the USSR make use of silica-gel for adsorbing the impurities. The gel either plain or activated with ammonia gas is placed in tanks through which the oil is percolated. The dessicant action of the gel removes moisture and acids to a large extent. The oil can then be filtered in the normal manner to remove the suspended impurities.

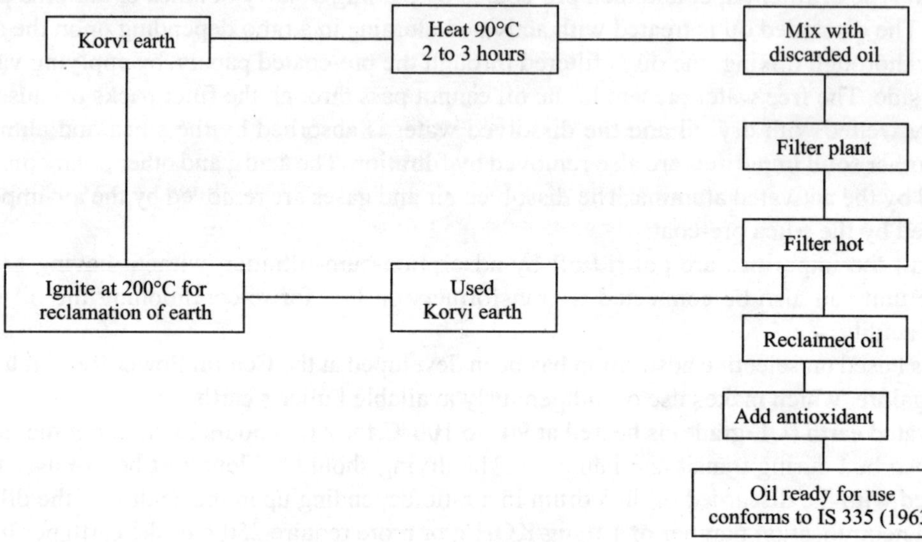

Fig. 22.2. Flow diagram for reclamation of discarded transformer oil.

Quality of Reclaimed Oil

It is observed that oil regenerated by any of these processes usually has all the characteristics required for a new oil except the right amount of stability against chemical deterioration. In fact it has been proved that the reclaimed oils often deteriorate at a faster rate when put into service. This is due to the fact that many hydrocarbons and natural antioxidants present in the new oil are either removed by these processes or are lost during the service life of the oil.

To overcome this, a fresh quantity of synthetic antioxidants should be added to the reclaimed oil. Some of the commercial plants have inhibiting tanks as the last stage where a calculated amount of the antioxidant is dissolved in the oil. Reclaimed oils with antioxidants give service almost equal to that of new oils. The most common antioxidants used for this purpose is 2:6 ditertiary butyl para cresol (DBPC) which is added in a ratio of 0.3 per cent by weight to give the oil an exceptional stability against oxidation without harming its properties.

23

Control of Gaseous Wastes and Odours

INTRODUCTION

The control of gaseous wastes, organic solvent vapours and fumes, and objectionable odours is one of the most complex of waste disposal problems, causing much concern to environmental pollution control authorities.

Processes which involve the evaporation of organic solvents are largely responsible for the emissions of light volatile hydrocarbons and solvent fumes. These may be produced from operations such as:

1. Petroleum refining and the petrochemical industries.
2. Paint, varnish, enamel and lacquer manufacturing, drying and storing.
3. Printing and lithographing.
4. Curing and polymerising.
5. Certain chemical manufacturing industries.
6. The disposal of sewage and refuse.

Objectionable odours arise from certain industries and 'the offensive trades' such as the manufacture of soap and detergents, paper making, coffee roasting, fish manure and fertiliser plants, food processing, tanneries, fish markets, abattoirs, fat rendering processes, pigsties, the animal by-products industries, and the like. Such odours are often associated with the presence of sulphides, mercaptans, phenolic compounds, aldehydes, amines, amides, esters and various other organic compounds and acids containing nitrogen or sulphur. That many of these are objectionable, most people will agree. But often, it is difficult to determine the extent to which an odour is objectionable to the community: response by people is very subjective, there are tremendous variations among individuals, people get accustomed to familiar odours and complain about new ones.

Because it is difficult to measure and characterise odours quantitatively (ASTM D–1391–57) and because odour thresholds in air for certain objectionable compounds are as low as one part in a million parts to one part in 1000 million by volume (Table 23.1), detection by the human nose is still more sensitive and more reliable than that by most instruments.

As the actual amount of contaminant present in the effluent gas is usually very small, much lower than the threshold for toxicity, odours rarely have an adverse effect on health; they are important as pollutants mainly because of their nuisance value. Reactive organic vapours (solvent emissions) and fumes, on the other hand, may be serious health hazards and may play also a significant part in the formation of secondary atmospheric pollutants and in photochemical smog reactions, with all their associated and familiar consequences. This group of emission consists principally of compounds of carbon and hydrogen, and their derivatives. These include all classes of hydrocarbons (olefins, paraffins and aromatics) and the compounds formed when some of the hydrogen in the original compounds is replaced by oxygen, halogens, nitro- or other substituent group.

Table 23.1. Odour thresholds in air (ppm)*.

Compound	*ppm*	*Compound*	*ppm*
Acetaldehyde	0.21	Ethyl acrylate	0.00047
Acetic acid	1.0	Ethyl mercaptan	0.001
Acetone	100.0	Formaldehyde	1.0
Acrolein	0.21	Hydrochloric acid gas	10.0
Acrylonitrile	21.4	Methanol	100.0
Allyl chloride	0.47	Methyl chloride	(above 10 ppm)
Amine, dimethyl	0.047	Methylene chloride	214.0
Amine, monomethyl	0.021	Methyl ethyl ketone	10.0
Amine, trimethyl	0.00021	Methyl isobutyl ketone	0.47
Ammonia	46.8	Methyl mercaptan	0.0021
Aniline	1.0	Methyl methacrylate	0.21
Benzene	4.68	Monochlorobenzene	0.21
Benzyl chloride	0.047	Nitrobenzene	0.0047
Benzyl sulphide	0.0021	Paracresol	0.001
Bromine	0.047	Paraxylene	0.47
Butyric acid	0.001	Perchloroethylene	4.68
Carbon disulphide	0.21	Phenol	0.047
Carbon tetrachloride (chlroination of CS_2)	21.4	Phosgene	1.0
Carbon tetrachloride (chlorination of CH_4)	100.0	Phosphine	0.021
Chloral	0.047	Pyridine	0.021
Chlorine	0.314	Styrene (inhibited)	0.1
Dimethylacetamide	46.8	Styrene (uninhibited)	0.0047
Dimethylformamide	100.0	Sulphur dichloride	0.001
Dimethyl sulphide	0.001	Sulphur dioxide	0.47
Diphenyl ether (perfume grade)	0.1	Toluene (from coke)	4.68
Diphenyl sulphide	0.0047	Toluene (from petroleum)	2.14
Ethanol (synthetic)	10.0	Tolylene diisocyanate	2.14
		Trichloroethylene	21.4

* These recognition odour thresholds for 53 commercially-important chemicals have been determined in the laboratory by an expert panel under standardised conditions.

CONTROL METHODS

The disposal of gaseous wastes usually involves the handling of very large gas volumes. In turn, the gas volume determines the physical size and cost of the treatment equipment, even though the quantity of actual contaminant may be extremely small.

Hence, the first step in any control operation is the containment and capture of the offensive pollutants in order to prevent their further dilution. This is achieved by effective design of hoods, ducts and exhaust systems. The separation, conversion or destruction of the gaseous wastes may then be carried out by one or more of the following conventional techniques.

Reduction at Source

By better housekeeping in plant operation and by process variation and modifications. This depends on the individual processes, but often by improved control or modification of the production process it is possible to eliminate or reduce solvent fumes or odours to an acceptable level; sometimes this may be achieved by investigation and modifying the properties of the raw materials of production, e.g., by elimination of reactive solvents or elimination of combustible solvents (use of water emulsion).

Adsorption of odorous chemicals and organics

Absorption of odorous chemicals and organics on a solid material of suitable active properties, i.e., a porous solid which has a very high surface area per unit mass. The adsorbent may be subsequently re-activated by heating or by a solvent, and the odourant recovered. The most useful and most common adsorbent for odour control is active carbon or charcoal in granular form, supported in porous beds. It has the advantage that its effectiveness is not weakened by the presence of moisture. This can be an effective and economical control method for emissions with low concentration of odorous compounds, or for the recovery of hydrocarbon solvents.

Absorption

Absorption in spray towers or by wet scrubbing in a suitable liquid. The design of absorption equipment is a well-developed chemical engineering technique. Water is a cheap and commonly-used solvent. It may be used, for example, for the removal of hydrochloric or hydrofluoric acid gases, nitrogen oxides, chlorides, and silicon in water. Some contaminants and odorants, however, are insoluble in water or need to be reacted chemically in solution, as with sulphur dioxide and mercaptans removal in an alkaline solution; hydrogen sulphide (H_2S) or carbon disulphide (CS_2) in an ethanol amine aqueous solution in a bubble column; aldehydes by sodium (or calcium) bisulphite; and amines by hydrochloric (or sulphuric) acid. However, it should be noted that disposal of the contaminated waste-water from such a process may be a serious problem.

Surface and contact condensation

Surface and contact condensation techniques operate on the principle of removing heat from the vapour, thus allowing condensation of the offensive vapours, or condensation of the water vapour with the odour-producing substance.

Odour masking or odour counteraction

Odour masking or odour counteraction is achieved by mixing or superimposing a stronger and more pleasant odour with the objectionable odour. For example, diesel exhaust odours are reported to be controlled by one transport company by vapourising a compounds and reodourising the gases to given them a 'fresh woody scent', and at least one local council in Sydney has been experimenting with the spraying of streets with pleasant-smelling odorous compounds.

However, pungent odours are not usually amenable to such treatment and the disadvantage of this technique is that the odorous compounds are not removed and that the additives themselves may become air pollutants.

Chemical oxidation

Chemical oxidation by reacting the undesirable compounds with chemical agents such as:

1. Ozone—suitable for waste-water treatment processes but not yet widely used.
2. Potassium permanganate—effective, but costly; sometimes used for feed lot odour control.
3. Chlorine, as:-
   (a) A bleach liquor from chloride of lime.
   (b) Sodium hypochlorite solution.
   (c) Chlorine gas.

For example, the use of chlorine in the sewage industry to control odours, is illustrated by the reaction:

$$H_2S + Cl_2 \longrightarrow 2HCl + S$$

Chlorination, as an odour control method, is simple, efficient and comparatively low cost.

Dispersal and dilution

Dispersal and dilution from high chimneys to reduce the odour intensity and its local nuisance. This is relatively cheap as there are no operating costs but must allow for a large factor of safety, taking into account all likely meteorological conditions, i.e., for acceptable odour control, peak rather than average concentrations are important.

Incineration

High temperature oxidation of the organic and toxic compounds present in gaseous wastes is employed to destroy the offensive effluents and odours by converting them into harmless gases, i.e., carbon dioxide and water vapour. This is by far the most effective control method at present available for the disposal of gaseous organic wastes and odours. Its successful applications needed for satisfying the basic principles of combustion: good contact of the contaminated gas with the burner flame, adequate gas velocities and turbulence to ensure good mixing and adequate time and sufficiently high temperature for combustion to be completed. Incineration should be considered wherever there is a problem of pollution by combustible matter. It should be noted, however, that with certain compounds, e.g., degreasing solvents which contain trichlorethylene, perchlorethylene or other halogenated compounds, special care is needed as these may, on reaction with a direct flame, produce a poisonous gas, e.g., phosgene. Also, in some cases, if combustion is not complete, the intermediate products produced may be even more undesirable than those which we are trying to eliminate.

Basic Types of Incineration Systems

For convenience the basic types of incineration systems suitable for the oxidation of gases and organic vapours may be considered under the following three main groups (in order of decreasing concentrations of combustibles):

1. Direct flame or direct-fired incineration systems, at temperatures above 1200°C.
2. Thermal incineration, in a temperature range of about 600 to 900°C.
3. Catalytic oxidation, at temperature of 300 to 600°C.

Direct-flame or direct-fired incinerator systems

This method is generally used for burning gases of low calorific value when the gases are at or near the lower combustible limit when mixed with air, or when the waste gas is a combustible mixture without air addition. Direct-flame incineration usually takes place at temperatures above 1200°C.

Gases of about 950 kcal/nm^3 calorific value or above may be burnt without auxiliary fuel; if the calorific value is below, as in the case of blast furnace gas, autothermic combustion is still possible provided the gas is pre-heated. Such wastes may be treated almost as a fuel, and often, standard fuel burning equipment can be used for incineration.

If the combustible component in the mixture is below the lower flammability limit (about 500 kcal/nm^3) then natural gas or other auxiliary fuel must be added to sustain combustion. Hydrogen sulphide, hydrogen cyanide and carbon monoxide produced as by-products from reactions, or solvent organic vapour mixed with air, may be burned in this way. Generally, direct flame incineration is employed when little auxiliary fuel is required and the combustible waste gas supplies at least 50 per cent of the total fuel value. The key criterion for success in such a system is the satisfactory provision in the design and operation of the basic requirements for combustion, to which a reference has already been made.

This type of combustion system is exemplified by:

1. Conventional industrial burners fired in the combustion chamber of industrial boilers and furnaces operating on forced and/or induced draught; the mixture is injected into the burner or combustor, mixed with sufficient air to achieve complete combustion, and burned as a fuel.
2. The flare-type burner, as used in petroleum refineries, petrochemical plants and chemical manufacturing industries: in this case the fuel-air mixture is ignited at the tip of the flare and burns in the surrounding air.

The latter is used for incineration of economically unrecoverable mixtures of refinery waste gases or for disposal of excess gases and vapours, but care is needed when burning highly odourous gaseous impurities (e.g., H_2S), for which an upper concentration limit may have to be imposed, since complete combustion in flare is difficult to achieve. Refinery flares may have a capacity of more than 50 tonnes of waste gas per hour. A major control problem in such cases is often the wide variation of the gas flow rates. In this system, steam or air jets are usually incorporated to promote rapid mixing and to control smoke emissions, particularly when hydrocarbons of high C/H ratio are burnt. As a result, the noise sometimes associated with such flares and the light from the flame may give rise to a nuisance problem. In a typical flare design the steam jets surround the flame and introduce air together with the steam to assist rapid combustion.

Velocities in the pipe carrying the gases should be kept above the flame propagation of the mixture, to prevent flashbacks; flame arrestors may be placed in the line but they are not a complete protection. A pilot flame is also provided. Some of the demands which must be met in order to ensure that the operation of a flare does not constitute an air pollution.

An interesting new development is the Coanda flare tip. It makes use of the wall adhesion effect to achieve air aspiration and hence good combustion characteristics. In the B.P. Coanda flare tip, gas at high pressure (around 2 atm) is ejected from an annular slot over a curved surface to which the gas stream clings. The resulting change in momentum causes air to be induced into the gas stream and the gas/air mixture burns at the end of the Coanda surface. Low pressure, smoke-producing gas (propane, butane, etc.) is injected into the aerated flame produced by the Coanda injector, and is consumed smokelessly (limit ratio for low pressure to high pressure gas should not exceed 1:3). This burner is small in size, simple and cheap to construct and easy to install.

Thermal incineration

Thermal afterburners are suitable for the treatment of non-flammable weak mixtures of organic vapours, solvent fumes or other odourous emissions. The calorific value of such gases may be in the range of 10 to 200 kcal/nm^3 (say, concentrations upto about 25 per cent of the lower explosive limit) and hence if incinerated directly they would require too much additional thermal energy.

The gases are destroyed by combustion in an independently-heated enclosed refractory-lined combustion chamber or afterburner, at temperatures usually in the range of 700°C to 900°C, but sometimes as low as 550°C, in the presence of flame. Natural gas or other fuel may be used in an industrial burner to pre-heat the chamber, preferably to a temperature of over 760°C. The contaminated gas is injected into this chamber (at velocities of 4.5 to 9.0 m/sec.) and allowed a retention time of from 0.3 to 1.0 sec at this maximum temperature; for effective control a minimum of 0.5 sec is often specified. Such systems are capable of operating continuously at efficiencies of 90 to 99 + per cent.

An effluent gas of high oxygen content (say, over 16 per cent), can be used as either primary or secondary air in the combustion process. The hot combustion gases are usually passed through a heat exchanger, where much of their sensible heat (50 to 80 per cent) may be recovered by direct recuperation, with substantial savings in fuel costs. The amount of heat recovered is largely a function of the heat exchanger costs and its design. A major advantage of thermal afterburners is their flexibility and the ease with which they can be adapted to meet changing emission standards, or emission composition, by simply adjusting their operating temperature. However, sufficient care must be exercised to ensure that such adjustments are made within the design temperature range and capabilities of the furnace refractories. They can also be readily adapted for automatic temperature control.

If a boiler is available as part of the plant, it may be more expedient to discharge the waste gases directly into the boiler furnace as part of the combustion air supply, provided sufficient dwell-time and adequate temperatures for the destruction of the unwanted gaseous pollutants could be ensured at all times. The desired afterburner temperature varies with the design conditions, the nature of the contaminants and the efficiency of cleaning required. High afterburner efficiencies generally demand higher discharge temperatures and longer residence times. Recommended minimum afterburner operating temperatures for various types of industrial operations are indicated in Table 23.2.

Table 23.2. Recommended minimum afterburner operating temperatures for various type of industrial operations.

Operation	*Recommended temperature °C*	*Operation*	*Recommended temperature °C*
Carpet laminating	650–760	Core oven	760
Cloth carbonisation	980	Deep fat fryers	650
General opacity problems	650–760	Odour control	700–820
Oil and grease smoke	650–760	Paint bake ovens	650–820
Pipe wrapping	760	Rendering operations	650
Smokehouse	650	Solvent control	700–820
Varnish cookers	650	Vinyl plastisol curing	650–760

For the destruction of odours from sewage sludge, slightly higher temperatures may be required. On testing 29 sludge samples for 174 odour evaluations it has been found that only a few sludges were adequately deodourised at 680°C. Most sludges required temperature of 700 to 735°C, while a few required 760°C, and over. The intensity of odour was appreciable at 30°C below the threshold level and severe—at 60°C below. Temperature required for the destruction of odours from volatile, low-molecular weight fatty acids commonly present in sewage sludge are as given in Table 23.3.

Table 23.3. Low molecular weight fatty acids commonly present in sewage sludge.

Organic acid	*Temperature for destruction of odour °C*
Acetic	745
Propionic	745
Butyric	775
Caproic	775

Catalytic oxidation

Catalytic afterburners, or catalytic vapour incinerators as they are sometimes called, are particularly useful for the destruction of gaseous wastes and odours containing low combustible concentrations of organic effluents (say, below 10 kcal/nm^3) which would not directly support combustion and would otherwise require large quantities of additional fuel to complete their combustion. The principle of catalytic burning may also be applied to higher calorific value gases (e.g., 280 kcal/nm^3) with energy production. While this is feasible, the adiabatic temperature rise of 880°C as complete combustion gives rise to a number of engineering problems.

Here we are concerned with a heterogeneous chemical reaction, which is surface limited, and the function of the catalytic agent is to provide a large surface area so as to help accelerate the surface reactions. As a result the combustion process is completed at substantially lower temperatures (say 300 to 600°C) and with less excess air; hence auxiliary fuel requirements are much lower.

In this system, the odourous gases or organic fumes are passed through a heat exchanger where they are pre-heated by the outgoing hot deodourised gases, before entering the combustion chamber. In the combustion chamber, they are further heated to the temperature required to achieve catalytic oxidation (gas, oil or even electric heating may be used) and then are passed through the catalyst bed. The 'clean' gases are discharged through the heat exchanger. Generally, catalytic oxidation units are smaller and weigh less than comparable thermal incinerators, since the process is carried out at lower temperature and at much shorter residence time (say 1/30 sec.) the combustion chamber is usually smaller and constructed of light weight material. This gives a low heat capacity with a fast warm up to full operating conditions after shut-down. The catalysts usually employed are noble metals such as Pt, Pd or Pt Rh, or the oxides of various base metals (Cu, Cr, Mn, Ni or Co), finely divided on various supports, such as a heat resistant steel or a ceramic body, e.g., gamma alumina.

The various catalysts required different operating conditions with regard to temperature and residence time. A good catalyst support needs to be a material of high intrinsic surface area that is shaped to give a high geometric surface area and a configuration that gives a low pressure drop.

Commercial catalyst may be supplied in many different forms in a manner designed to ensure a low pressure drop and a higher surface to volume ratio. The following are four basic configurations commonly used are given below:

1. Nichrome wire screen support, with a coating of the precious metal.
2. A number of air foil-shaped alumina rods on which a noble metal catalyst has been deposited.

3. A bed of spheres of pellets coated with the catalytic agent.
4. An alumina honeycomb on which the catalyst has been deposited.

These latter have high resistance to thermal shock, high mechanical strength and freedom from wear by attrition. Because of the cost of the catalysts and of the additional equipment required, catalytic afterburners require a higher initial capital outlay than thermal afterburners but the savings in operational cost in terms of fuel requirements may be substantial, particularly if a long catalyst-life period can be achieved. It is important to ensure that temperatures are kept below the safe limit of the catalyst, say below 800°C maximum, and that the effluent gases do not contain impurities (e.g., arsenic, lead, zinc, mercury or phosphates, sulphur and chlorine) which will 'poison' and deactivate the catalyst.

This system appears to have been most successful for application to gases containing low concentrations of organic contaminants or particulates such as those from—food preparation processes, oxidation of paint solvents, wire enamelling ovens, lithographing ovens, chemical manufacturing processes and the like. However, wider use of this method must await the development of more economical, robust, and poison, resistant catalysts.

Case Histories of Practical Control

The following successful applications of afterburners to destroy industrial odours has been investigated:

1. Exhaust from coffee and chicory roasters, containing complex odourous organic compounds (mainly aldehydes and organic acids), was effectively deodourised in a propane-heated catalytic afterburner at temperatures of 380°C (for the coffee) and 480°C (for the chicory).
2. Exhaust from an oven, drying phenol formaldehyde-bonded plastic sheeting, gave completely satisfactory performance (over 98 per cent reduction in phenol concentration) after passing through a catalytic afterburner at an outlet temperature of 590°C.
3. Exhaust gases containing aldehydes (particularly acrolein) and aerosols from a foundry core oven was passed through a thermal afterburner in which at temperatures above 800°C the aerosols were effectively removed, but the removal of the aldehydes was less efficient.

The relative economics of three different types of direct flame afterburners, treating gases has been reported as under:

1. From tinplate lacquer baking ovens, in an oil-fired (operating temperature 760°C) unit with an elaborate heat exchanger.
2. From a PVC coating plant, using an oil-fired unit with a simple heat exchanger.
3. From another PVC coating plant, but with a gas-fired unit, and also with a simple heat exchanger.

It was shown that the major operating cost factor was the fuel consumption.

Catalytic combustion was not practicable for the latter two cases because of the presence of an organic plasticiser in the gases. However, data is given for two catalytic combustion afterburners which are used for:

1. Incineration of formaldehyde and associated fumes from the manufacture of laminated resin-impregnated-paper surfacing material at about 360°C.
2. Control of odours and heat recovery from wire-enamelling ovens.

Thus thermal afterburners are quite high in operating costs, and that catalytic combustion, is cheaper, particularly when effective heat exchangers are used.

Thermal incineration is said to be the most effective method of controlling varnish kettle emissions, which may contain several thousand compounds. The afterburners are operated at 680°C with 0.67 second residence time. The development of certain improvements involving the use of an integral-blower burner or a non-powered raw gas burner are claimed to reduce installation, operating and maintenance costs. The following data are given for the performance of an improved type-burner in the fume stream:

1. Wire enamelling oven (0.3 sec. residence time).

Outlet temperature °C (combustion chamber)	*Hydrocarbon, ppm In*	*Out*	*Conversions per cent*
490	1400	480	66
590	1410	300	78
680	1420	130	91
700	1400	110	92

2. Metal coating oven (0.3 sec. residence time).

Outlet temperature °C (combustion chamber)	*Hydrocarbon, ppm In*	*Out*	*Conversions per cent*
425	3300	1400	58
540	3200	480	85
650	3300	250	92
730	3300	90	97

3. Plastic coating machine.

Outlet temperature °C (combustion chamber)	*Efficiency, % conversion hydrocarbon only*
760	99.9
720	99.3
695	97.7

Thus complete answers to many objectionable industrial odour, solvent and fume problems do not exist. Design criteria are not easy to detail, as knowledge is limited and much of the data is proprietary.

With regard to odour control, there is a need for standardisation of measurement techniques, instruments and odour panel procedures in order to be able to evaluate reliably the effectiveness of odour removal by various control devices. Catalytic oxidation is not universally accepted at present.

The simplest equipment for direct incineration is a combustion chamber in which the process gases mix turbulently with a flame produced by an oil- or a gas-fired burner.

Thermal incineration is capable of giving a sustained performance level if the flows, temperatures and operating conditions are held constant, and appears to be becoming the main control technique for solvent and other combustible gaseous wastes. However, fuel costs are critical in many applications, and it is important to determine accurately the conditions needed for optimum combustion.

REMOVAL OF METHANOL BY PERVAPOURATION

The concentration of final products after synthesis is one of the most important steps in chemical engineering, because the value of the products is normally directly related to their purity. Distillation is the process most commonly used for this separation step. It is based on the boiling point differences of the components to be separated. However, some mixtures reach a certain composition where no further separation occurs (so-called azeotropes), and these cannot be separated through normal distillation without adding a third component. Ethanol-water for example forms an azeotrope at 96 per cent ethanol. In pervapouration, the separation mechanism is the difference in the diffusion rate of the components through a semi-permeable membrane. This technique works well to separate azeotropic mixtures. The separation of water from organic solvents by pervapouration is by no means new. Sulzer Chemtech is now also applying this process successfully for the separation of methanol from other organic solvents (Fig. 23.1).

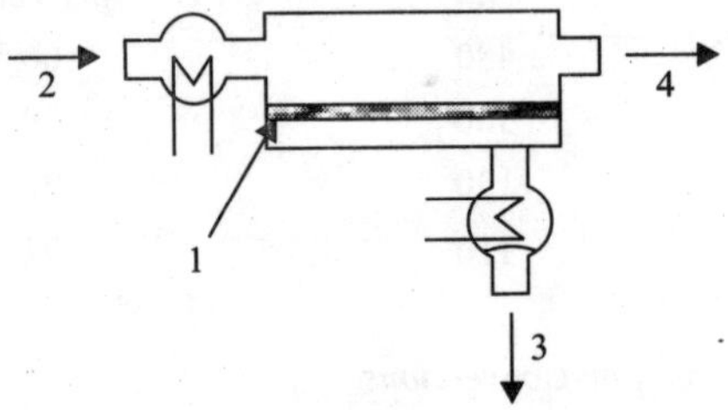

Fig. 23.1. Principle of pervapouration.

DISADVANTAGES OF CONVENTIONAL PROCESSES

Methanol is found as a reactant or reaction product in many different processes (esterification, trans-esterification) where it is more difficult to separate it from other organic compounds than from water. Methanol forms azeotropes with numerous organic compounds. There are various conventional processes for the removal of methanol. Selection of the process is dependent on the specific characteristics of the mixture to be separated. They all have greater or lesser disadvantages:

1. In the case of water-washing followed by distillation, the water is introduced into a previously water-free mixture and cannot be removed completely afterwards. In addition, this process requires three distillation columns and thus a relatively large amount of space. Furthermore, it consumes a large amount of energy and generates waste-water that has to be treated.
2. Although it is simple to extract methanol with alkali salts, it entails a risk with regard to the product quality, because the presence of a certain amount of salt in the final product cannot be avoided. The treatment of the separated salty phase also necessitates additional equipment and energy.

Moreover, it is labour-intensive due to the handling of solids, causes corrosion and creates environmental problems.

Development of a New Membrane

Pervapouration avoids the aforementioned disadvantages. It is now regarded as the latest state-of-the-art in the chemical industry, where separation is concerned. The most frequent application is the removal of water from organic compounds with hydrophilic membranes. Typical applications are the splitting of azeotropes and the final dehydration of the product. The separation principle of pervapouration is based on the difference in the polarity of the compounds to be separated, their molecular size, and the affinity of the most polar substances for the interface of the membrane. The more polar and smaller a molecule is, the easier it can be absorbed and passed through the membrane. Since water is the smallest and most polar molecule, the separation of water from organic compounds functions the best. The driving force for the separation is the difference in the partial pressure of the water vapour between the feed and rear side of the membrane. The necessary partial pressure difference is attained by feeding the mixture to be separated to the membrane preferably with a high temperature, while a vacuum at low temperature is created on the rear side of the membrane. The substance passing through the membrane—the so called permeate—is then condensed there.

As far as the polarity is concerned, methanol is water's closest neighbour. Based on the experience gained with the separation of water, Sulzer chemtech has developed a new generation of membranes which can remove methanol from organic substances. This constitutes a further field of application for pervaporation—the first plants are already in operation. The use of pervaporation dramatically simplifies the separation of mixtures containing methanol. The azeotrope is fed to the membrane unit, and the methanol selectivity permeates the membrane.

The retentate—depleted in methanol is either condensed (providing the methanol concentration is low enough) or passed as a vapour to a downstream column, where even higher purities are attained. The methanol-rich permeate is recycled back to the process which produces the azeotrope to remove the methanol. This process can be employed for new plants and also for the revamping of existing installations.

High Purity—Lower Energy Consumption

The advantages gained through the separation of methanol by pervapouration are obvious: No additional water or salts that have a negative effect on the product quality are required. The membranes are inert and do not release any substances to the solvent mixture. There is no impact on the environment, since neither waste-water nor salt solutions containing methanol have to be treated. Furthermore, the membrane process is characterised by its low energy consumption. If the mixture can be taken directly off a column as a saturated vapour with a temperature of approximately 90°C, no additional energy is required to pass the methanol through the membrane. And should an evaporator have to be installed, the retentate can be fed directly to the column as a vapour to save energy.

Successful Operation

Pervapouration plants are delivered as skid-mounted units. They contains all the necessary components (heat exchangers, pumps, valves, etc.), as well as the internal pipework and the instrumentation. This minimises the installation costs and the time needed for the installation on the site. Since the process is simple, the skids are easy to install and require relatively little space. Sulzer Chemtech has already installed two of these plants, which are both being operated successfully by the customer. The first plant was installed in a fine-chemicals plants in the Netherlands in 1997 to accommodate a capacity expansion as well as replace the existing salt extraction.

Since then, the customer has been using the fully automatic process and achieves a higher product quality. The second plant, which separates methanol from methyl acetate, was installed in Spain at the end of 2000. The customer insisted on a water-free end product with a specific methanol content. The membrane process was able to fulfill both these requirements.

24

Recycling and Purification of Solvents for the Laboratory

INTRODUCTION

Solvent recycling, at first, was adopted by companies which wanted to 'do the right thing' and protect the environment from harmful waste solvents. But with time it became more and more obvious that these companies were saving large sums of money. They did not have to buy new solvents but they were reducing other categories of cost such as purchasing, administration, storage, shipping and disposal.

The impressive cost savings from recycling solvents gave this idea widespread appeal. Today, various companies and organisations recycle waste laboratory solvents on a daily basis all over the world.

This section describes the implementation of a solvent recycling programme for a laboratory and give some common examples of solvent recycling scenarios.

Solvents can be purified or recycled through the process of distillation. By taking advantage of the difference in boiling points between a solvent and its contaminants, it is possible to purify a solvent by distillation. Three main strategies are used to take advantage of this:

1. A waste solvent can often be purified back to its original purity so that the solvent can be re-used in the original process.
2. The solvent may be purified to a quality which allows it to be used for purposes other than the original use.
3. It is possible to purchase lower grade solvents and purify them to a higher grade.

All of these strategies can result in significant saving in solvent costs.

DEVELOPMENT OF A SOLVENT RECYCLING PROGRAMME

The development of a solvent recycling programme requires several components, such as:

1. An analysis of waste streams and their composition.
2. An analysis of solvent purity requirements.
3. An investigation into the type of distillation equipment required for the application.
4. The development of solvent recycling protocols for the waste stream.

Waste Solvent

Before looking into equipment to recycle waste solvents, it is important to study the waste mixture. The more complicated the waste mixture, the more difficult it is to recycle. Therefore, the most important component in any waste solvent recycling programme is segregation of wastes. Waste solvents produced from different activities in the laboratory need to be carefully segregated to keep the waste mixture to a manageable level. Next, it is important to understand the composition of the waste to be recycled. This understanding usually requires research into the processes which generate the waste and may requires some analysis of the waste itself.

Lastly, the volume of waste to be recycled per week or per month must be determined so that equipment with the appropriate throughput (volume distilled per unit time) can be determined.

Solvent Purity Requirements

Likewise, thought must also be given to desired quality of the recycled solvent. One approach is to require the same specifications of the recycled solvent as the purchased solvent. This approach allows good confidence in the recycled materials performance. Another approach is to develop performance based criteria which examine those solvent characteristics which effect the process using the solvent. The distillation procedure can be focused on removal of the relevant contaminants, allowing higher throughputs and less stringent conditions.

Equipment

Distillation is an ancient process. The wide variety of distillation equipment available reflects this long history of use. Choosing the correct distillation equipment without expert guidance can often be a difficult experience, involving a great deal of trial and error. Consideration must be given to the difficulty of separation, throughput materials compatibility, degree of automation and cost of equipment.

Types of distillation

The three types of distillation columns which are available for laboratory and light industrial use are:

1. Simple distillation columns.
2. Packed columns.
3. Spinning band distillation columns.

Simple distillation units consist of a boiler connected directly to a condenser. These units are generally used for quick, gross separation of solvent from solids or non-volatile contaminants. These units are generally not suitable for separating a mixture of solvents.

Packed columns consist of an open column randomly packed with material (packing) which maximises contact between the rising vapours and the descending condensate. The intimate contact between the descending condensate and the rising vapours causes an enrichment of the more volatile components in the vapour phase, which facilitates fractionation. These columns are generally well suited for separating solvents which do not boil too closely together. Packed columns give good comparative throughputs but are generally dedicated to recycling a single waste stream. A significant amount of solvent is held up in the column packing and cannot be recovered. Re-distilling successive batches of the same waste solvent in the column can be done without any special preparation. However, to distill different waste solvents in the same column generally requires a thorough cleaning of the column and packing between batches.

Spinning band columns consist of an open column containing a helix made of Teflon or metal, spinning at several thousand rpm. The action of the helix on the rising vapours and the descending condensate gives a great deal of intimate contact between the phases which leads to very good fractionation. These columns are generally well suited for separating solvents which boil as close as 5°C. However, spinning band columns do not give as high throughput as packed columns in some cases. Spinning band columns also have the unique characteristic of having very low hold-up. Once a distillation is complete, less than 5 ml of solvent is left in the distillation column. This makes the spinning band column ideal for distilling different waste streams with the same unit.

ROUTINE LABORATORY WASTES

Acetone and toluene are routinely used in laboratories for cleaning glassware and other general uses. These solvents are easily recycled using a packed column or a spinning band column. In some laboratories both these materials are purified and recycled for the production facility. These solvent are used to clean glassware in the glassware fabrication shop before certain procedures which require extreme cleanliness. Industrial grade acetone and toluene are purchased and then distilled using spinning band columns. Water, particulates, metals and organic contaminants are removed, giving a very high quality solvent. After the solvent is used to clean the glassware, it is collected and then distilled using a spinning band column, after which it is suitable for re-use in the process.

WASTE FROM HPLC

High performance liquid chromatography (HPLC) uses solvents such as acetonitrile in the analysis.

While new HPLC column technology has allowed for the reduction of solvent wastes, the high cost of the purchased solvent makes recycling a very attractive alternative to disposal of the toxic material. Many acetonitrile waste streams can be recycled for re-use in the chromatographic procedures. Waste steams of acetonitrile—4 per cent aqueous ammonia or acetonitrile—aqueous acetate buffers from typical drug analyses (theophylin or opium derivatives) have been routinely recovered using the following spinning band conditions: Equilibration for 2 hours and distillation at 800 ml/hr throughput.

A 10 per cent forecut containing the acetonitrile-water azeotrope is removed before collecting a heart cut, boiling point 81–82°C of excellent UV quality (UV cutoff < 190 mm) and GC purity. The dynamic HPLC chromatograms and the pre-concentrated UV spectra for HPLC grade and recovered solvent are seen in Figs. 24.1 and 24.2. In many procedures, a lower acetonitrile content may be present in the waste stream. Distillation will afford the acetonitrile-water azeotrope (84.3 : 15.7) collected at the boiling point = 77–78°C. While this distillate is not the pure solvent, it can be re-used with the proper chromatographic solvent mixing system. An added advantage in these situations is that after the azeotrope has been collected, the remainder is water containing the analyte. Usually, this can then be conveniently disposed resulting in a very large waste reduction.

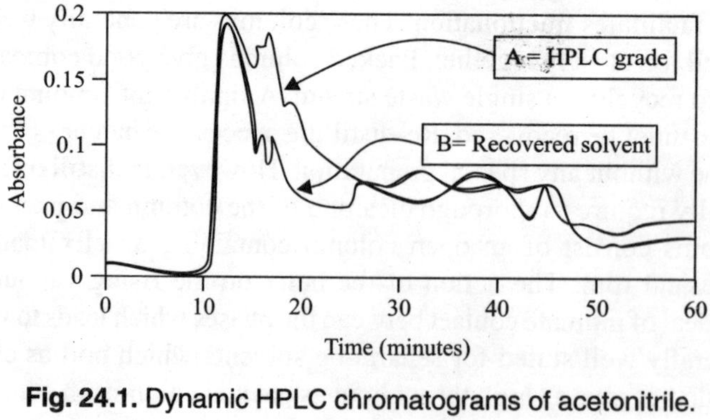

Fig. 24.1. Dynamic HPLC chromatograms of acetonitrile.

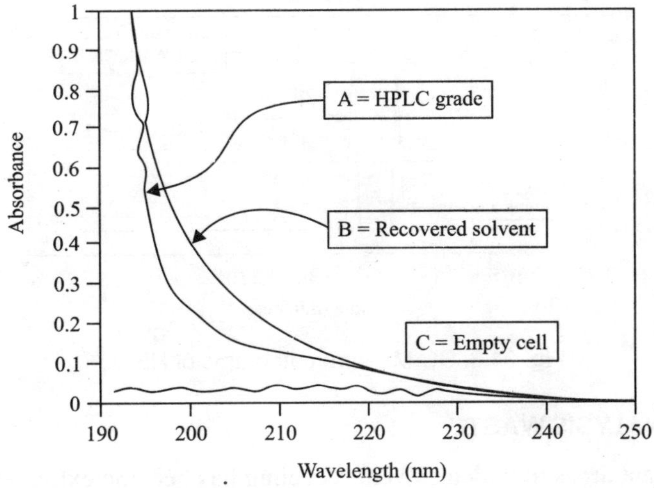

Fig. 24.2. Pre-concentrated UV spectra of acetonitrile.

Acetonitrile wastes containing aqueous methanol lead to a more complex situation. The presence of a constant boiling mixture containing all three components mitigates against complete separation.

It turns out, however, that regardless of the starting composition, material collected upto 70°C contains a 44:52:4 acetonitrile/methanol/water mixture with a purity > 99.9 per cent. While this recovery is certainly more involved, it might be justified under certain circumstances.

Aqueous methanol and isopropanol mobile phases commonly used in reverse-phase HPLC analyses can easily be recovered at high throughputs with excellent analytical purity. Methanol wastes are recovered as the pure solvent, while isopropanol is collected as the 87:13 azeotrope. While the cost of these solvents is less than acetonitrile, the ease of recovery and its reuse mitigates against its casual disposal.

Other High Purity Solvent Waste Recoveries

There are a number of other waste streams that can be conveniently recycled to give an excellent recovery of high purity solvent. Tetrahydrofuran (THF) and hexafluoroiopropanol (HFIP) are used extensively in large volume applications of size exclusion chromatography. Both of these can be easily recycled.

Data from a Distillation of Pot Charge of the Waste

The following data was obtained from a distillation of an 8 litre pot charge of THF waste from analysis of polystyrene polymers. Distillation was conducted on a spinning band system at a 900 ml/hr throughput after equilibrating the mixture for an hour. A 90 per cent recovery of a very high purity solvent (GC purity 99.8 per cent and UV comparable to unstabilised HPLC grade) was obtained. The recovered material must be used immediately after recycling or must be stabilised since air oxidation produces a peroxide leading to a deterioration of the UV quality of the solvent. HFIP has a purchase price that is extremely high and its disposal is both environmentally difficult and costly. HFIP recovered from spinning band distillation produced a heart cut (about 89 per cent recovery) of material boiling 57.6–58.2°C. This material was analysed using GC-FID and GC-MS, both of which showed a purity of > 99.5 per cent. Fig 24.3 presents the GC-MS chromatograms of both the initial waste and the recovered solvent.

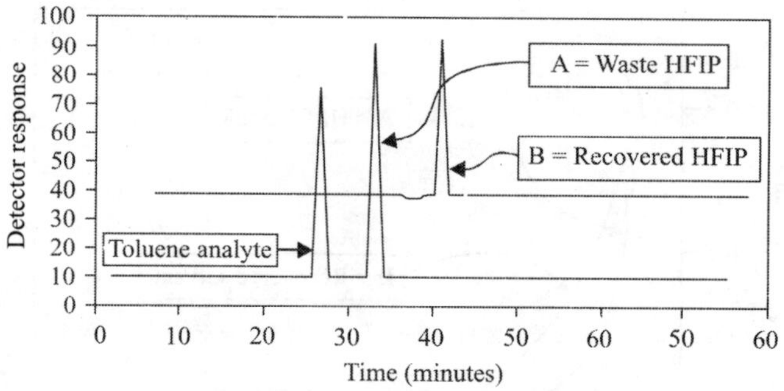

Fig. 24.3. GC-MS chromatograms of HFIP.

ENVIRONMENTAL ANALYSIS WASTE

One of the most important areas in which solvent recycling has become extremely significant is in the environmental analysis laboratory. There are numerous facilities in USA that are currently analysing and evaluating large numbers of samples. Most of the standard methods require large volumes of DCM, freon and hexane. All of these can be successfully recycled. DCM from waste containing methanol gives a 93:7 azeotrope upon direct distillation. If on the other hand, the waste is first washed with water to remove the methanol and the DCM dried and distilled, a high purity DCM (>99.9 per cent) is obtained. DCM from pesticide analysis has been extensively studied. 'Original experiment with DCM spiked with Lindane, 4,4′-DDD, and 4,4′-DDE at levels of 62,000 ng/ml each gave a distilled solvent with a pesticide carryover of 9–48 ng/ml. Recent work with actual wastes with much lower pesticide contamination has shown a considerable reduction of the carryover level.

Another important and environmentally sensitive solvent is freon. Initial studies have resulted in the use of 'spinning band recovery of used freon from tests for oil and grease and hydrocarbons in water, showed a hydrocarbon content of < 0.001 mg/litre. For wastes containing higher initial hydrocarbon content, double distillation (distillation of the heart cut) is recommended.

Thus, solvent recycling of a large array of different waste mixtures can be accomplished and will produce a recovered material of high purity. Because of the inflammability and toxicity of organic solvents, one of the overriding concerns with regard to solvent recycling is the safety of the process. To begin with, the involved laboratory must keep accurate records of their solvent waste and the recycled material in order to insure against the admixing of different waste streams. Laboratory personnel must be made aware of the characteristics of the materials and must ascertain that no chemical reactivity may result from the distillation process. While this is not the case with most solvents, there is the possibility that some unusual mixtures may be encountered. The other concern is the safety of the equipment. spinning band units used for recovery are now the state-of the-art in safe, efficient, distillation systems at an affordable price. The overall benefit of a solvent recovery programme is tremendous, both in the savings involved in purchased material and in its disposal and in the positive environmental impact of such a programme. It is expected that as the cost of these solvents spiral upwards, more and more laboratories will routinely incorporate recycling into their procedures.

25

Liquid Chemical and Chlorinated Waste Treatment

INTRODUCTION

Industrial liquid trade wastes have become a problem. Municipal authorities are not eager to accept these pollutants into the sewage system; land disposal is becoming difficult. Incineration is now suggested as a means of disposal. Two designs of incinerator are described, using the same principle of tangential feed admixed with the necessary combustion air.

The horizontal design unit is for large quantities of waste, whereas the vertical unit can cater for factory or ship effluent. In some cases the recovery of valuable ingredients of the waste (oil, solvents, etc.) before destruction should be considered. Inorganic waste, especially toxic or containing heavy metals, should be chemical treated before incineration. Liquid wastes containing chlorides have been successfully treated in these incinerators.

LIQUID CHEMICAL WASTE DESTRUCTION

Liquid chemical residues for which the producer has no further use can be termed waste products of that process: they must be removed from the site and disposed off by some means. This does not imply that there is no ultimate use for this waste, after treatment by a recovery process. However, to put these waste into the sewer is illegal and heavy penalties are being enforced; to dump these wastes is becoming difficult, as sites are scarce and charges high.

What is to be done? Incineration is suggested; the incinerator is placed at the hub of the waste disposal plant and recommended ancillary equipment built up around it. The choice of the ancillary equipment depends on the purpose of the incinerator, viz., to be used for recovery of solvents, etc., prior to destruction or only for destruction. The type and size of incinerator to be installed will be determined by the quality and quantity of waste to be dealt with.

Basic Principles of Incinerator

The principle of the incinerator is as follows: Waste enters the combustion chamber tangentially together with air via an inlet manifold. The waste is now atomised by turbulence before being ignited by a fuel-oil fired burner. Choking by solid impurities is prevented due to the fact that there is no reduction of flow area between the fed pump and the furnace. The furnace is equipped with a pilot flame nozzle and a secondary nozzle to use additional fuel, if the calorific value of the waste is not sufficient to encourage self-burning. The waste gases are held for a short time in the combustion chamber where the majority of the combustible materials are incinerated; the rotating gases flow into an afterburner chamber where any remaining combustible matter is completely burnt off. Before being exhausted to atmosphere, the gases can be scrubbed and cooled.

The small incinerator unit is used for capacities from 0.04 m^3 to 4.0 m^3 per hour liquid waste having a calorific value of about 5000 kcal/kg. Wide variations within the calorific content of the fuel constitute no objection whatsoever to incineration in the Celleco incinerator. Should the combustion heat value be low, secondary fuel is necessary to maintain complete combustion. The incinerator shown in Fig. 25.1 comprises two modules—the burner body and the control panel with blower. These can be installed either separately or adjacent as one unit. The body consists of two vertically arranged cylindrical sections. The lower combustion section supports the higher afterburner chamber.

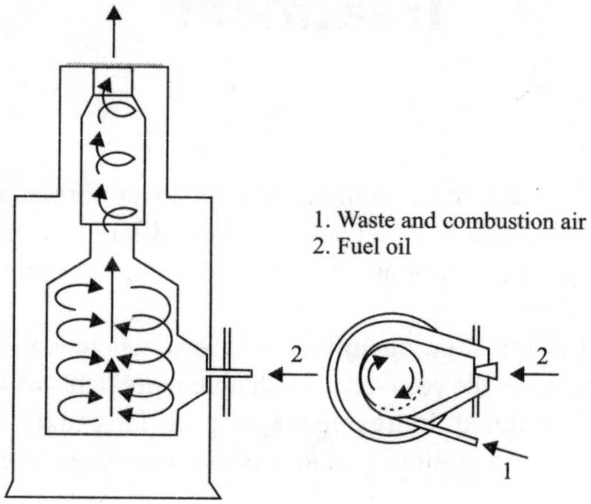

Fig. 25.1. Vertical design incinerator (cut-away view of combustion chamber showing flame path).

This smaller design of unit is capable of a capacity of 60 to 100 kg/hr waste with a calorific value from 0 to 7000 kcal/kg, and is used in factories and ships.

The burner body is air cooled, thus saving expensive lagging. The firebrick insulation is of standard specification to meet the thermal conditions. Combustion temperatures range from 600°C to 1100°C; exhaust gas temperature varies with the pre-treatment but can always be controlled to acceptable limits.

The choice of ancillary equipment to be used with the burner will depend on: (i) nature of liquid waste; (ii) capacity of incinerator; and (iii) object of incineration.

Should the liquid waste contain valuable and easily recoverable liquids, i.e., oils, solvents, etc., recovery of these liquids should be considered for use either as auxiliary fuel in the incinerator or as a by-product of the process. There are plants operating in Europe where the original purpose of the incinerator was destruction, but now there is a small chemical plant complex producing valuable solvents for market. Central depots, based near large ports, can collect waste oil from factories, garages, workshops, and ships. This oil can be recovered to produce a light oil fraction and a fuel oil fraction. A disposal system (Fig. 25.2) which has been developed for this purpose is discussed below.

Celleco ID Disposal Systems—Waste Oil

Receiving department

The waste oil is delivered by trucks into a receiving tank (1) through a strainer (2) for catching of solid impurities. From the receiving tank the oil will be pumped to storage tanks (3) which are provided with steam coils for heating the oil. This makes it easy to handle. From receiving tank (1) and storage tank (3) the water will be let out from the bottom into a collecting tank for water (4). The oil separated on top of the collecting tank (4) will be pumped back into receiving tank (1). Condensate from steam coils and other equipments will be collected in a tank (21).

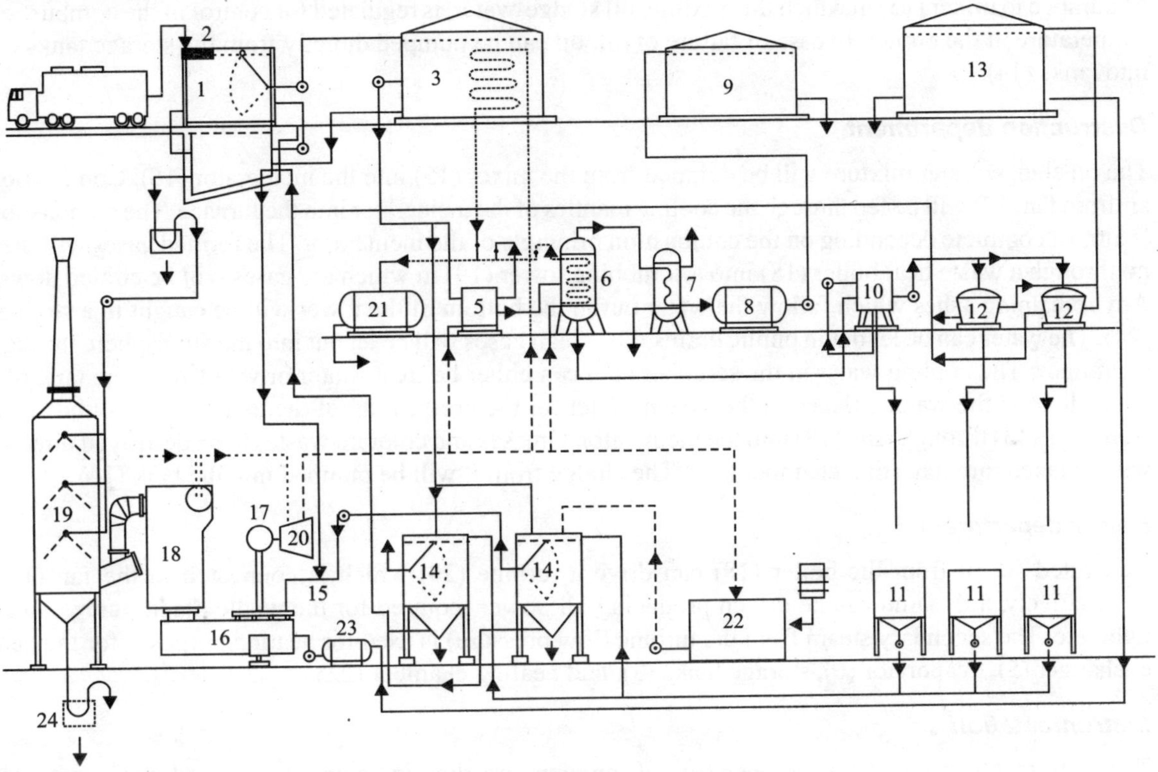

Fig. 25.2. Celleco ID disposal system for waste oil.

Distilling department

From storage tank (3) the oil will be pumped through a heat exchanger (5) into an evaporator (6) provided with steam coils. The lightest gasoline vapour will be let off and cooled down in a water cooled condenser (7). Formed distillate goes into a collecting tank (8) and is then pumped further into the storage tank (delivery tank) for topped gasoline (9).

Separating department

The oil not evaporated in the evaporator (6) will be transported to a self-cleaning filter (10) equipped with a circulating pump. The oil sludge reject will go to the collecting tank (11) and the oil accepted will be pumped to the separators (12). From the separators clean fuel will be pumped to storage tank (delivery tank) (13). The reject products, water and sludge, will be conveyed to respectively water collecting tank (4) and sludge collecting tank (11).

Mixing department

The sludge from collecting tanks (11) will be pumped to storage tanks (14) before destruction. These tanks work as clarifiers from which oil will be pumped from the top to a mixer and water from the bottom to collecting tank (4). Solids can be emptied at the bottom. Water from the collecting tank (4) will be pumped to mixer (15) in which the mixture, oil sludge-water, is regulated for control of the combustion temperature in the burner. In case of failure of oil, oil can be pumped directly from the storage tanks (3) into tanks (14).

Destruction department

The oil sludge-water mixture will be dumped from the mixer (15) into the incinerator (16). Combustion air from fan (17) will be fed through the cooling mantles of the incinerator into the furnace. The combustion is almost complete depending on the combustion principle of the incinerator. The formed gases will then go through a waste heat boiler (18) into a scrubbing tower (19) in which the gases will be cooled down. Any remaining ashes which follow the water out in the bottom of the tower will be caught in a strainer (24). The water can be let out in public drains. The clean gases will be let out into the atmosphere through a chimney. The cooling water in the scrubber (19) can either be fresh water or water from the tank (4), depending on the water balance in the system. Fuel for the pre-heating of the incinerator will be taken from tank (13) through tank (23) into the incinerator. Greases and colorant wastes to be destroyed directly will be taken into a heating chamber (22). The sludge from it will be pumped into the tank (14).

Power department

Generated steam from the boiler (18) can drive a turbine (20) which is connected to the fan (17). Alternatively, a dynamo can be driven producing all power required for the whole plant, pumps, fans, light, etc. The secondary steam from the turbine (low pressure) is used for heating purposes for the heat exchanger (5), evaporator (6), storage tanks (3), and heating chamber (22).

Instrumentation

The instruments required for the operation of the plant are flow indicators for oil and water, pressure indicators for steam and oil, temperature indicators for oil and combustion and flow meter for combustion air as well as a CO_2-meter for the incinerator. Furthermore, the usual instruments for the boiler are required.

Special Types of Liquid Wastes

Should the liquid waste contain heavy metals, it is advisable to treat chemically and remove these metals at the source of waste before incineration. If not, the metals will be deposited with the slag and will be dumped on disposal land where the concentration will build up and could contaminate some water supply. Similarly, liquid wastes containing hazardous or toxic materials should be made less harmful or recovered before incineration. All organic liquid wastes can be treated with confidence. This includes compounds containing chlorides. The resultant hydrogen chloride gas can be scrubbed with lime water and the calcium chloride recovered as a sludge.

Incinerator with Scrubbing Tower and Stack

Fig. 25.3 shows the slightly inclined furnace chamber in which the high temperature oxidation of the waste takes place.

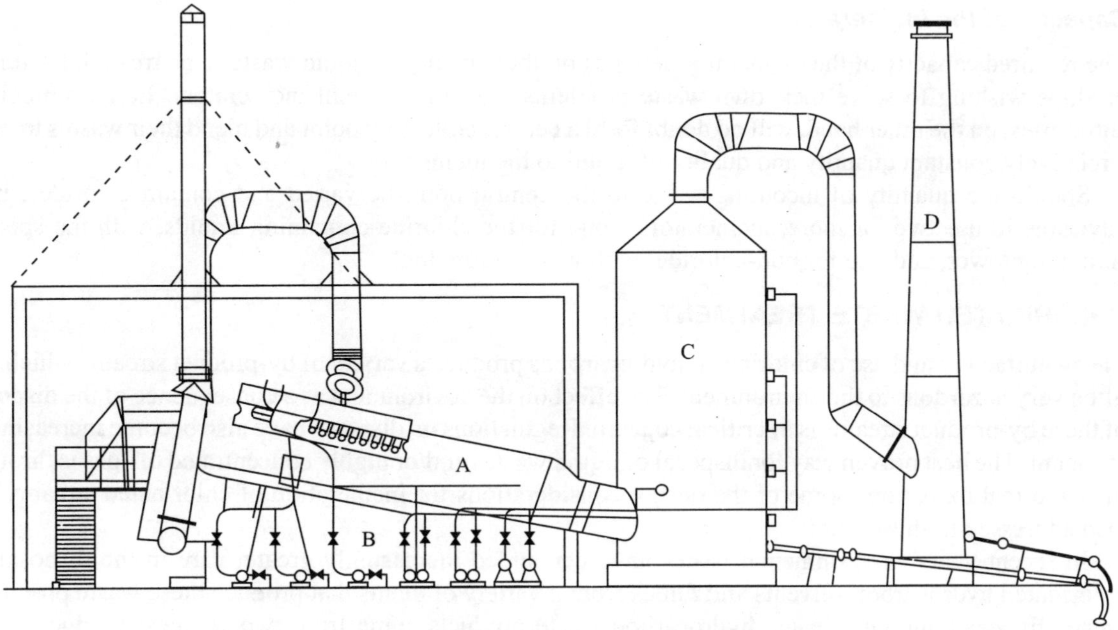

Fig. 25.3. Incinerator with scrubbing tower and stack.

Below the furnace chamber are located the various feed pumps for injecting the combustible wastes.

The scrubbing tower is provided with large volumes of cooling water to remove the water-soluble impurities produced in the combustion process. The tower is constructed in two sections, the lower part being provided with four water sprays to facilitate the primary cleaning of the combustion gases, the upper part comprising a packed section of intolox saddles.

The clean gases then pass to atmosphere through the stack. The liquid discharge from the scrubbing tower and the stack then flows to the acid neutralisation plant. The following is the material specification of the liquid wastes which can be handled by this incinerator.

Chemical Classifications

Hydrocarbons — Aliphatic and aromatic
Alcohols — Esters
Ethers — Ketones
Organic-acids — Amines

Aqueous solutions or suspensions of the above can be accepted.

1. Calorific value — No limitation.
2. Viscosity — Heated reception and blending facilities enable the most viscous wastes to be accepted (upto 3200 sec Redwood approximately).
3. Flash point — Intrinsically safe handling facilities permit the safe reception of the most volatile combustible materials (e.g., diethyl ether).
4. Chlorine content — Blending facilities permit acceptance of all chlorinated wastes, with no restrictions on the chlorine content of individual wastes.
5. Ash content — Limitation on ash content exists; the ash can be removed periodically on shut-down.

Capacity of the incinerator

The required capacity of the incinerator depends on the quantity of liquid waste to be treated. Factories or ships wishing to solve their own waste problems will utilise small incinerators. Large municipal authorities, on the other hand, will no doubt form a central collecting point and blend their wastes to give a relatively constant quantity and quality of liquid to the incinerator.

Should the quantity of incoming waste to the central point be varied and contain chlorides, it is advisable to use two, or more, incinerators—one for the chloride-containing liquids, with the special scrubbing tower, and one for non-chloride, with water scrubbing.

CHLORINATED WASTE TREATMENT

The manufacture and use of chlorinated hydrocarbons produce a variety of by-product streams which are often very hazardous to the environment. The effect on the environment as a consequence of the disposal of these by-product streams is of critical concern. Regulations on disposal have also become increasingly stringent. The best proven way for disposal of liquid wastes and/or highly concentrated off-gas is through their thermal oxidation. Some of the design considerations for incineration of chlorinated streams are also addressed in this section.

In recent years, environmental issues have demanded increasingly greater care in the disposal of chlorinated hydrocarbon solvents and fumes from a variety of plants that produce these waste products. Generally speaking, chlorinated hydrocarbon waste products come from two sources: producers and customers. Examples of some of the plants that produce these waste products are:

Consumers: Metal cleaning, dry cleaning, vapour degreasing (electronics), solvent extraction and textile processing are all operations which consume large quantities of chlorinated hydrocarbons.

Producers: Major sources of chlorinated by-product residues result from the production of chlorinated methanes (methyl chloride, methylene chloride, chloroform, and carbon tetrachloride), chlorinated ethanes and ethylenes (EDC, VCM, 1,1,1–trichlorethane, trichlorethylene, perchloroethylene, etc.), and others (chlorinated aromatics, hexachlorocyclo-pentadiene, etc.).

VCM/EDC facilities are among the largest generators of by-product chlorinated residues.

Overall, it has been estimated that there were approximately 1.3 billion kg of chlorinated chemicals for disposal in 1995, which is expected to increase to 1.5 billion kg in 2010.

These estimations are summarised in Table 25.1.

Table 25.1. Projected world disposal of chlorinated chemicals by region (million kg).

Year	*USA*	*W Europe*	*Japan*	*Other*	*Total*
1995	343	447	196	284	1270
2000	338	413	195	303	1249
2005	360	418	216	337	1331
2010	417	443	254	383	1497

Methods of Disposal

The capacity for the production and use of such chlorinated products is growing and, consequently, the quantity of chlorinated by-products to be disposed off is also increasing. In addition, the requirements for air and water pollution control are becoming more stringent. Consequently, there is a rising demand for efficient disposal methods.

At present there is actually only one such method available: incineration in a specially designed system. It is also important to consider methods for recovering the large quantities of hydrogen chloride that form during burning because their removal by neutralisation could create serious pollution problems by increasing the salt content in the receiving river. A complicating factor in designing a disposal system for such wastes is that not all of the feed components are known in exact detail and, apart from this the composition may vary. In many cases, there are several waste streams coming from different process points at variable rates. In a combustor several off-gases are injected and different streams are collected in storage tanks, sufficient mixing by agitation or recycling is often not achieved. This causes further fluctuations in the composition of the feed. The chlorine content in the waste can reach as much as or more than 80 per cent. Since the calorific value of the waste is inversely proportional to the chlorine content, the composition plays a decisive role in the disposal of such wastes. With maximum chlorine content, the calorific value drops to less than 2000 kcal/kg. (For reference, pure hydrocarbons such as ethylene or benzene have a calorific value of roughly 12,000 kcal/kg). Fig. 25.4 shows the trend of calorific value versus chlorine content.

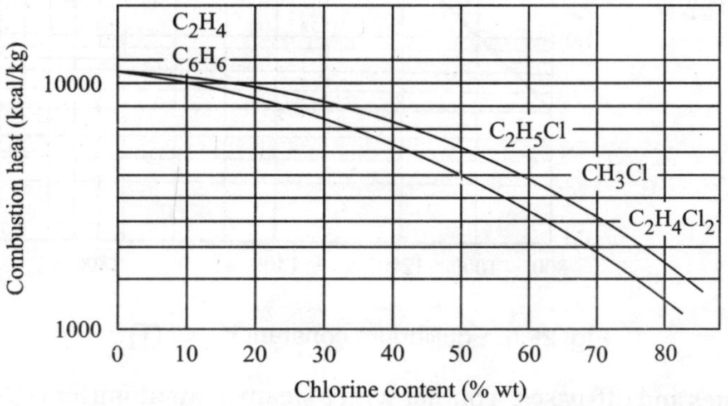

Fig. 25.4. Trend of calorific value versus chlorine content.

In the case of highly chlorinated hydrocarbons, the hydrogen content is not sufficient to convert chlorine into HCl. However, because the calorific value is also low, the supplementary fuel added to maintain the combustion temperature also provides sufficient hydrogen to make up the stoichiometric requirement. If necessary, water or diluted HCl solution is added to favour conversion.

Referring to Fig. 25.5, it may be noted from the given equilibrium conditions that a high reaction temperature, high water content and low oxygen content favour the formation of HCl and reduce the chlorine content in the gas mixture. This means that incineration should take place under almost stoichiometric conditions. In practice, such theoretical requirements must be adapted to the technical feasibility and the economic aspects of an incineration system and have to be optimised accordingly. The applicable maximum temperatures are limited by the resistance and the costs of the refractory. A high steam content in the combustion gases burdens the production of highly concentrated hydrochloric acid. A certain amount of excess oxygen should be allowed to ensure complete oxidation.

Equipment

Different types of combustion chamber systems are used for incineration of chlorinated hydrocarbons. Specific design will depend on the application.

$$K_p = \frac{(P_{HCl})^2 \, (P_{O_2})^{1/2}}{(P_{H_2O}) \, (P_{Cl_2})}$$

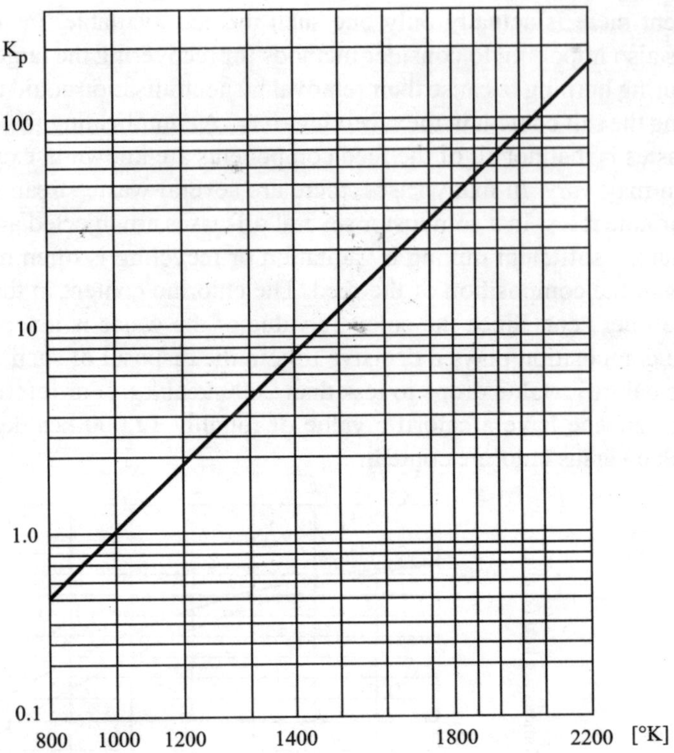

Fig. 25.5. Equilibrium constant K_p = f (T).

For both, liquid wastes and off-gases, a burner with a steam or air atomiser is used. Its design provides for very fine atomisation of the waste. The supplementary fuel is gas, which is introduced through a number of tips located concentric to the atomiser. In cases of chlorinated off-gas disposal, velocity and location of tips have to be carefully selected for good turbulence around the burner. Combustion air inlets must cope with atomised spraying cone and off-gas inlets for an appropriate mixing. Another purpose of the burner arrangement is to have a hot flame core through which chlorinated compounds flow for complete cracking before reaching the furnace operating temperature.

Before the flue gases are treated further they have to be cooled. This can be achieved through direct contact with water or hydrochloric acid whereby the temperature drops to the dew-point. Such quenching requires special care in material selection (at high cost and heavy maintenance) and dilutes the flue gas, reducing the partial pressure of the HCl to be recovered. Additional disadvantages are loss of water and waste of flue gas heat content.

At present the simplest solution for cooling the flue gas is to use a fire tube waste heat boiler which produces medium pressure saturated steam.

It is found from experience that by maintaining tube wall temperatures below 300°C and above the dew point of the flue gases, carbon steel can be used without any corrosion problem. However, a critical item is the tube sheet of the boiler which must be protected with suitable refractory inserts.

Major problems occur during start-up and shut-down when the system goes through the dew point range. The normal operation is to light off and fire with a clean fuel until steady state conditions have been reached and then fire the wastes. This same approach should be used during shutdown so that at no time will gases containing HCl be contained in the system. For the production of concentrated hydrochloric acid, the known methods of isothermal or adiabatic absorption are employed. As a rule, incinerator and quench system are followed by an isothermal absorber or cooler in which essentially the acid contained in flue gas is recovered and the temperature and equilibrium conditions of the off-gas are adjusted to provide, as far as possible, optimum conditions in the second absorber. Here, the hydrogen chloride produced in the plant is actually recovered.

The second absorption stage is either an isothermal or adiabatic absorber. The isothermal absorbers are usually tube bundle heat exchangers made in graphite with water-carrying steel shells. The adiabatic absorbers are packed towers. The maximum concentration that can be reached depends primarily on the steam/hydrogen chloride ratio in the incinerator off-gas and this in fact, depends on the chlorine content in the waste. In any case absorbers are designed to produce dilute hydrochloric acid at concentrations ranging from 10 to 28 per cent.

In order to recover HCl acid at higher concentrations, it is necessary to remove water from the solution by an extractive distillation process. Solvents that may be used for this purpose are sulphuric acid, calcium chloride or magnesium chloride. By this method acid concentrations upto 100 per cent be reached. Fig. 25.5 shows a liquid and off-gas wastes disposal unit where all HCl is recovered, concentrated to anhydrous and routed back to VCM production.

In order to meet the requirements of permissible HCl and Cl emissions under all operating conditions, the exhaust flue gas is again washed in a two stage scrubber before dis-charge into the atmosphere. In the first stage, which is operated with water, residual HCl is removed and in the second stage, caustic liquor is used to remove any free available chlorine.

Emission levels less than 10 mg/Nm^3 to HCl and 2 mg/Nm^3 of Cl_2 can be achieved in this way. A typical flow diagram is represented in Fig. 25.6.

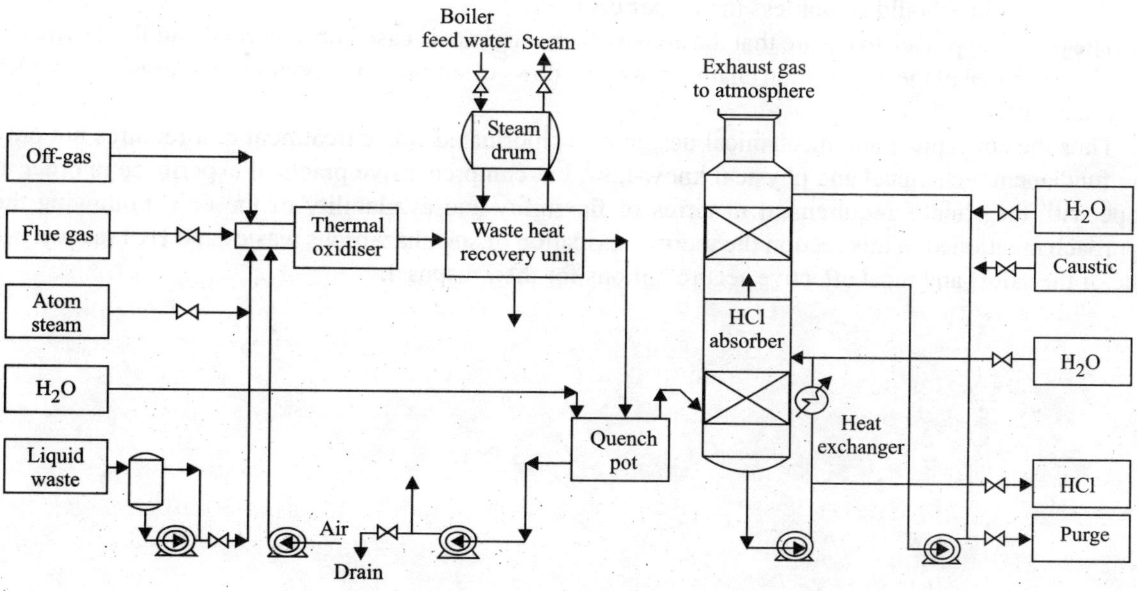

Fig. 25.6. Controlling of HCl and Cl_2 emission levels.

Dioxine Formation

Being known carcinogens, dioxines are a matter of particular environmental concern and the subject of strict regulations. Consequently, avoiding dioxine formation must be of major concern in waste treatment. Dioxine formation is related to the presence of some dioxine precursors in the streams to be incinerated, such as chlorinated aromatic hydrocarbons.

During the incineration of chlorinated hydrocarbons, several design parameters must be selected to avoid the formation of dioxines.

In the design of the thermal oxidiser and the relevant flue gas treatment unit, selection of certain process parameters will minimise dioxine formation within much more stringent limits than required by environmental regulations. These parameters are the following:

1. Combustion temperature.
2. Residence time in the combustion chamber.
3. Good mixing inside the incinerator.
4. O_2 content in flue gases.

The criteria to select the values for these parameters are:

1. Combustion temperature: Due to a high chlorine content in the streams, a high combustion temperature is necessary (normally above 1259°C).
2. Residence time: The residence time must be 2 seconds or more in the combustion chamber. This residence time coupled with a high temperature will allow a very high destruction efficiency (higher than 99.9999 per cent).
3. Mixing: To achieve thorough mixing in the combustion chamber, a high turbulence burner is required as well as distributor appropriately located for off-gases with respect to chamber geometry.
4. Oxygen content: To keep the desired oxygen content in the flue gas excess air must be employed; this value should be not less than 3 per cent.

The proper approach to ensure that the above criteria are met is based on a thermal and fluid dynamic simulation model of the combustion chamber, which allows to simulate, temperature and pressure profiles very accurately.

Thus the conceptual and mechanical design of a chlorinated waste treatment unit requires not only the fundamental chemical and physical know-how, but comprehensive practical experience in order to cope with the client's requirement in terms of flexibility and availability of the unit. Following the approach mentioned in this section the thermal oxidation of such hazardous wastes and HCl recovery is one of the safest and most effective secure options for their disposal.

26

Waste Minimisation in Textile Industry

INTRODUCTION

The Indian textile industry is engaged in different levels of production for various kinds of fibre. These include yarn preparation and processing, fabric preparation and processing and finished goods preparation. The finished goods preparation section includes ready-made garments, handicrafts, carpets, etc.

Cotton and synthetic fibres together account for about 97 per cent of the total yarn (produced and imported) in the textile industry. An outline of the manufacture process of textiles from fibre to finished stage, irrespective of the type of fibre used, can be broadly divided into three principle steps: (i) yarn manufacture; (ii) fabric preparation; and (iii) fabric processing.

The general principles of yarn production are much the same, irrespective of the type of fibre. The fibres are initially freed from impurities such as sand, dirt and stalk. Based on the requirements, they are then mixed, blended and drawn out in the form of slivers in which they are fairly parallel, but not twisted. This is followed by attenuation and further mixing by a combination of laps or slivers and still more attenuation. Processes are then employed to eliminate the very short fibres and to ensure to a large degree that the fibres in the form of slivers are all within certain extreme limits of length. The process is called carding. Thereafter, further parallelisation of the fibres is attained by combing and repeated drawing (or drafting). In drawing slivers will not hold unless a small degree of twist is introduced. The sliver is then known as a roving. The twist is given so that the resultant roving has sufficient strength to prevent breakage in its attenuation. Finally, the uniform fibre in the form of the roving is further drafted and twisted into yarn. The typical flowchart for yarn manufacture is given in Fig. 26.1.

Waste is generated mainly during the initial stage of cleaning the fibres and in the form of short fibres during the carding stage. Waste generated during the cleaning of cotton fibres is mainly in the form of twigs, leaves and earth. Raw wool fibre contains as much as 50 per cent impurities in the form of natural grease and suint (perspiration). These impurities are removed by scouring the fibre in soap solution containing alkali. About 25 per cent of raw silk consists of silk gum, which is removed by boiling the silk in a strong soap liquor.

FABRIC PREPARATION

Yarns manufactured from textile fibres by the method described above is used for the production of fabrics. The types of fabrics manufactured are: (i) woven fabrics; (ii) knitted fabrics; and (iii) non-woven fabrics.

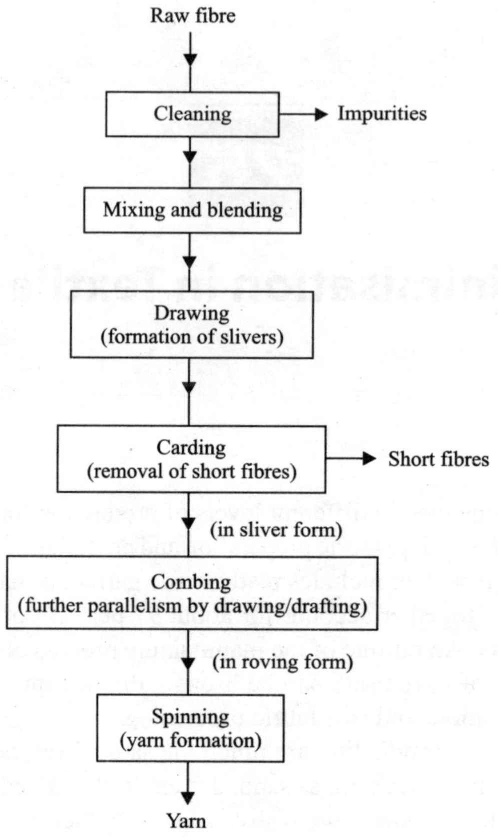

Fig. 26.1. Flow diagram of yarn manufacture.

Fabric Processing

Textile processing is done to improve the appearance and serviceability of the fabric in many ways. The main operation carried out in this section include:

Desizing

In this process, the sizing ingredients are removed from the grey fabric by dissolving them. This is necessary as the presence of sizing agents on the fabric hampers the penetration of other chemicals in the subsequent stages. Depending upon the sizing agent applied, about 10–20 per cent of the weight of the fabric is constituted by the sizing agent. The waste generated by the removal of these agents is organic and highly biodegradable in nature. This step is carried out mainly for cotton fabric.

Scouring

The scouring process is carried out to remove impurities such as wax, fatty acids, oils, etc., present in the fabric. Scouring is carried out in alkaline conditions under high pressure and temperature. Alkaline waste with high BOD and COD is generated at this stage.

Mercerising

Mercerising increases the tensile strength, lustre and dye affinity of the fabric. The operation is accomplished by impregnating the cotton fabric with cold sodium hydroxide solution. This causes swelling of the fibre, which results in an increase in the dye intake. When increase in tensile strength is the main criteria, mercerising is carried out in a tenter frame. Waste generated at this step is highly alkaline in nature. The fabric is taken for mercerising depending upon the finishing requirements. This is done only for cotton fabrics.

Bleaching

Bleaching is carried out to destroy the colouring matter and whiten the fabric. Different chemicals such as hypochlorites, hydrogen peroxides, etc., are used as bleaching agents. The process conditions during bleaching vary with the type of agent used. The waste-water from this section is alkaline in nature and contains chlorides and dissolved solids.

Dyeing

Dyeing is carried out to impart a solid all-over shade to the fabric. The main purpose of dyeing is to anchor dyestuff molecules to the textile fibres. Dye shades are obtained by treating the textile material with a solution containing dyestuff and other auxiliary chemicals. Dyeing waste-water is characterised by the presence of unexhausted dyestuff and other chemicals. The waste-water generally has high TDS, colour and BOD concentrations.

Printing

Printing produces coloured patterns on the fabric. It is carried out using print paste containing dyestuff, pigments and other auxiliary chemicals. Different types of printing, that is, block printing, screen printing, etc., are used to produce the coloured patterns on the textile materials. Large volumes of coloured waste-water with a high BOD are generated from the print section.

Finishing

When the textile material has been scoured, bleached, dyed and printed, it has to be finished. This stage includes the final operations necessary for making the textile presentable and attractive. The finishing operations include:

1. Drying: Drying removes the moisture from the fabric using the drying machine.
2. Providing dimensional stability: This is one of the most important finishing operations. The fabric, which is in a distorted condition, is brought to the required dimensions of width and length on the stenter.
3. Calendering: A kind of lustrous skin is formed on the fabric surface during calendering. The damp fabric is pressed hard against a hot, polished metal surface until it dries.
4. Softening: After calendering, the fabric becomes a little stiff. Breaking down this stiffness is called softening. The fabric is led through the softening machine so that it touches the studded rollers lightly and drags them around. In this way, the surface of the fabric is lightly disturbed making it much softer. Apart from the above mentioned finishing operations many other special properties depending upon requirements such as water properties, fire resistance, etc. are provided. These are done by passing the fabric through a trough of chemical (for providing special finishes) followed by drying/stentering.

Depending upon the type of fabric to be processed and the final product, any or all of the above processing operations can be carried out. Each of these operations is water-intensive and involves consumption of huge quantities of water and chemicals.

PRODUCTION PROCESS

This section covers the relevant aspects of the various unit operations carried out in this industry.

Textile processing mainly involves: (i) pre-treatment; (ii) dyeing; (iii) printing; and (iv) finishing.

Pre-treatment Process

The pre-treatment process is carried out to prepare the textile material for subsequent processing, which includes dyeing, printing and finishing. In the pre-treatment stage of synthetic fabric, the three-stage processing of desizing, alkaline extraction and bleaching gives the best results.

Desizing

Sizes are applied in a process to improve the strength and bending behaviour of fibres during fabric weaving. These include natural, synthetic as well as semi-synthetic sizes. The grey fabric in case of synthetic textile consists of soil and water-soluble synthetic sizes such as polyvinyl alcohol (PVA), carboxyl methyl cellulose (CMC) and polyacrylytes. The desizing process involves removal of these sizes. In addition to the sizes, a number of impurities are present in the fabric which can be partly extracted during the desizing process. In a simple desizing process, a cold wash which can be either static or running is given to remove the water-soluble impurities.

Alkaline extraction

This process is carried out for the efficient extraction of foreign substances which have been partially removed at the desizing stage. It is carried out either batchwise or continuously by impregnation/steaming process or by prolonged heat treatment at high temperature and pressure.

During extraction, swelling of fibres takes place which improves the dye uptake of the fabric in the subsequent stages. The oils, etc., get hydrolysed, the degree of saponification depending upon the temperature and time of reaction.

Bleaching

Alkaline extraction does not completely remove all the foreign materials from the fabric. Rather, they are 'chemically degraded' and have to be oxidatively and hydrolytically decomposed and eliminated in the subsequent bleaching stage. The whiteness of the textile materials is improved by an oxidative or reductive decomposition of these foreign materials.

Pre-treatment Auxiliaries

Desizing

As desizing for synthetic fabric can be done by a static/running cold wash, auxiliaries are not required at this stage.

Alkaline extraction

The various chemicals used for alkaline extraction include:

1. Wetting agents/detergents: They are used to promote rapid penetration of the liquor and so overcome natural resistance to wetting by alkaline liquor and emulsification of lipophilic impurities.
2. Alkali (sodium hydroxide): It is used for saponification of oil/grease.
3. Complexing agents: Chemicals such as ethylene diamine tetra acetic acid (EDTA) are used for complexation of hardness formers and heavy metal ions.
4. Dispersing agents: Chemicals such as polyacrylates and phosphoric are used for dispersion of insoluble degradation products, soil and lime soaps.
5. Reduction agents: Chemicals such as dithionite are used for reducing impurities.

Bleaching

1. Bleaching agents: Chlorine (Cl_2), sodium hypochlorite (NaOCl) and chloride of lime ($CaOCl_2$) are three of the oldest bleaching agents. However, hydrogen peroxide is used more widely now.
2. Bleaching stabilisers: They are used to inhibit the decomposition of the bleach active perhydroxy anion formed during the peroxide bleaching process and ensure a high oxidation potential over the whole bleaching time.
3. Surfactants: Surfactants used in bleaching should have emulsifying, dispersing and wetting properties which promote the removal of hydrophobic impurities and soil and assist in the transport of reaction products formed during bleaching.
4. Fluorescent whitening agents: If the bleaching treatment is unduly prolonged or is otherwise made too severe, it is likely that the fibres will be attacked so much as to be impoverished. FWAs which have an affinity for the fibres, to which they are applied usually in aqueous solution. In synthetic fabrics, they are often introduced into the fibres during their manufacture.

Environmental Aspects of Pre-treatment

Desizing

In the desizing process, 90 per cent of the sizing chemicals are disposed off in the waste stream, making it one of the industry's largest waste streams. They contribute large amounts of BOD and COD loads which can be as high as 6,00,000 ppm. Synthetic sizes which are not biodegradable can pass the treatment system and are linked to aquatic toxicity in receiving waters.

Alkaline extraction

As sequesterants they are able to release heavy metals from particles and sediments and transfer them into stable water-soluble compounds. The use of sequestering agents can be avoided by performing an extraction step with formic or hydrochloric acid before bleaching, eliminating hardness and heavy metal from the material.

Surfactants

Surfactants and related compounds such as detergents, emulsifiers and dispersants are used in the formulation of chemicals used for almost all textile processes.

Bleaching

Hypochlorite bleaching is detrimental for all fibres and finishings containing amino groups. It also results in adsorbable organic halides (AOX). The decomposition of hydrogen peroxide which takes places during bleaching reactive forms only water and oxygen.

Dyeing

Dyeing is employed to give an all-over shade to the fabric. It basically involves diffusion of the dye molecules into the textile fabric which imparts the required colour. In the dyeing process these dye particles quickly come into contact with the surface of each fibre, form a thin layer and pass slowly by solution from this into it.

Methods of dyeing

There are essentially two techniques available for dyeing the textile material. These are:

Batch technique

The liquor and the textile are placed together in a vessel and the required amount of dye is added.

Continuous technique

The dye is dissolved or dispersed in the liquor. A definite quantity of the liquor is locally applied to the textile.

Dyeing Process

Polyester (PES) fibre is quantitatively the most important synthetic fibre. Its inexpensive production from petrochemical raw material and excellent textile properties alone and in combination with other natural fibres make polyester fabrics universally acceptable.

Fixation rate

The fixation rate of dyeing varies with the type of dye and the dyeing fabric, thus determining the concentration of dye in the waste-water.

Dyeing rate

In dyeing polyester fabric, the dyeing rate can be increased by raising the dyeing temperature.

Dispersing agents

Vat and disperse dyes must be finely dispersed in the dye bath. The main purpose of adding dispersants is to break down large dye particles into small ones, thus facilitating their penetration into textile material and absorption thereby.

Levelling agents

Uniform dyeing, the basic requirement of colouring, is endangered by differences in temperature and concentration. Levelling agents provide uniform distribution of the dye in the textile so that dyeing is level with a similar shade and depth of colour. The three typical groups of levelling agents are:

1. Dyeing accelerants (carriers): They increase the rate of diffusion of the dye thereby improving migration of dye.
2. Ethoxylated products: They have a retarding effect on dyes in the absorption phase and are, therefore, used in rapid dyeing or shading at high temperature.
3. Special levelling agents: These produce equal rates of absorption of dye mixtures during the heating phase and increase dye migration in the high temperature phase.

Wetting agents

Wetting agents are not required in the process of dyeing synthetic fibres.

Complexing agents

Complexing agents are added to the dye bath to combine with multivalent cations, especially the calcium, magnesium and iron salts which are carried into the liquor by textile material. EDTA and the related chelating agent DTPA, NTA (nitrilotriacetate) and derived phosphoric acids are very effective complexing agents.

pH regulator

Dyeing with dispersed dyes is carried out at a pH 4.5, at which dispersed dyes are the most stable. The acid-releasing agents are salts of halocarboxylic acids which decompose on heating.

Dyeing accelerants

Dyeing accelerants (carriers) are used in the dyeing of synthetic fibres by the exhaustion process to give an increased rate of absorption of the dispersed dye onto the fibre, more rapid diffusion into the fibre and better dyeing yield.

Reducing agents

They are used in post-treatment for improving the colour fastness of the textile dyes and printed with disperse dyes by the reductive decomposition of dye adhering to the surface.

Stripping agents

While dyeing and printing, any defect may occur which must be rectified for economic reasons. Defects can results from non-uniform dyeing properties of the textile, process error or soiling during transportation of material. Before the material can be correctly dyed or printed, the original dye must normally be stripped. Stripping fabric dyed with dispersed dyes is very difficult and is possible only at high temperatures. These textiles can be brightened by using high concentration of non-ionic levelling agents and carriers, but some fabric damage can occur.

Printing

Printing is a process by which coloured patterns are produced on the fabric. Ornamental patterns consisting of one or several colours can be reproduced on fabric by local applications of sharply defined areas of dye preparation (printing paste) either to undyed or previously dyed cloth.

The overall printing process involves the following activities.

Pre-treatment

The appropriate pre-treatment of textile, prior to printing, is crucial to the success of the process. It is necessary that the fabric shape be stabilised. This may include pre-shrinking, removal of tension introduced during weaving, stabilisation of the weave structure, and straightening of warps and wefts in the thread direction.

Printing

The fabric is printed by the screen process on tables or with flat or rotary machines, as well as by the roller printing method. It can be glued either to a back gray or directly to the rubber blanket depending upon the fineness of the material and intricacy of the design. There are two types of printing:

1. Dye printing: Application of dyes.
2. Pigment printing: Application of coloured pigments.

In principle, the main distinction between dye printing and pigment printing is that pigments do not have any affinity for the fibre whereas the dyes have an affinity.

Drying

The drying operation is carried out to prevent smearing of the print as it passes over the guide rollers. Various drying methods are practised, including ones based on hot air or combustion gases from burners, direct contact with fabric (cylinder drying) and radiation.

Fixation

Fixation is the process which enables dyes to diffuse into the fibres. As synthetic fibres are strongly hydrophobic and considerably less swelling, a certain amount of effort is required for carrying out the fixation process.

Washing

Washing of a printed textile is intended to accomplish removal of thickeners, unfixed dyes and auxiliaries. Unfixed dyes may be either soluble or insoluble, but both cases present the damage of staining or bleeding. In the washing process, the fabric is thoroughly rinsed (warm or cold) and then treated with the alkaline reducing agent. At a batch temperature of 40–50°C, most of the thickener and unfixed dye is either dissolved or reductively decomposed. Because of low temperature, neither dye nor decomposition products are absorbed onto the fabric.

Print Auxiliaries

Textile printing auxiliaries are defined as monocolouring products used at all stages of the printing process from preparation to the final wash. As mentioned before, the essential auxiliaries for pigment printing are necessary to adhere the pigments permanently to the fibre. The auxiliaries remain on the fibre and give colour fastness. In case of dye printing, the auxiliaries are generally removed from the fibre in the final wash.

Auxiliaries for dye printing

The various auxiliaries used for dye printing include those used for preparation of print paste, after-treatment agents, and printing adhesives. These are:

1. Thickening agent.
2. Fixation (hydrotopic) auxiliaries.
3. Dispersing agent.
4. Defoamer.
5. Acids.
6. Oxidising agents.
7. Printing adhesives.
8. After-treatment agent.

Auxiliaries for pigment printing

The pigment printing recipe consists of a thickening system, a binder, fixing agents, hand modifiers and other auxiliaries which are used to ensure a problem-free production process and to meet functional properties of the finished pigment printed article.

Finishing Process

In the finishing process, the fabric is passed through a bath containing the finishing ingredients. This is followed by passing it through a pair of squeeze rolls to squeeze out as much treating solution as possible before passing the fabric into drying devices such as a tenter loop or a cage drier. The drier is often followed by a high temperature curing device (if curing is needed).

Wet pick-up may vary from 60 to 100 per cent depending upon the composition and construction of the fabric and the composition and pressure of squeeze rolls. Driven initially by attempts to conserve energy in the drying step, but later by economic and environmental considerations, low wet pick-up methods of finish applications have been proposed. These include expression methods in which the fabric is saturated with a relatively large amount of finishing solution after which the excess is removed by special squeeze rolls, a vacuum extractor or air or steam jets. The other method is topical, in which a limited amount of treating solution is applied to the fabric by devices such as engrave roll, spray, open and closed foam applicators.

With these systems, pick-up as low as 12 per cent has been reported on 100 per cent synthetic fabrics. In addition to savings due to reduced energy consumption for drying, low wet pick-up methods also provide savings in chemical costs, improved hand and less migration of finish. However, decreasing the wet pick-up increases the degree of control required for uniform pick-up, thus making the control mechanisms for low wet pick-up systems critical to their success. For this reason, the expression methods which are somewhat easier to control and tend to give more uniform treatment are more widely used commercially than the topical methods.

Finishing Auxiliaries

The various auxiliaries used in the finishing process are:

1. Cross-liking agents.
2. Catalysts.
3. Hand builders.
4. Softeners.

ENVIRONMENTAL ISSUES

The textile industry poses numerous problems to the environment which include impurities in waste-water and exhaust gases and toxic residues on the textile. As a result, it is confronted by continually growing number of environmental regulations and restrictions.

Waste Streams of Concern

Textile manufacturing produces aqueous air and solid waste streams that must be managed, recovered, treated and/or disposed of.

Water

Water is used extensively throughout textile processing operations and the textile finishing industry is one of the sectors with the highest specific water consumption. Water consumption figures vary widely among the unit processes.

For the same unit operations, water consumption can vary depending upon the type of equipment used. The water consumption figures for various types of fabric is given in Table 26.1.

Table 26.1. Water consumption (including treatment).

Textile	*Consumption; m^3 per tonne fibre material*
Cotton fabric	80–240
Cotton woven goods	70–180
Woolen fabric	100–250
Polyacrylic fabric	10–70

Of the total water consumed, 88.4 per cent is discharged as waste-water and the balance 11.6 per cent represents evaporation loss. Studies have revealed that fresh water and waste-water cost an average of 5.2 per cent of the turnover (only finishing work).

Besides water, fibres contain impurities which contribute to pollution in the industry's effluents. Most of the impurities present in the fibre, such as metals and hydrocarbons, are applied intentionally as a part of the spin finish to improve physical properties and workability of fibres. These finishes are typically removed before final processing and thus, contribute to pollution in the effluent.

The composition of waste-water depends to a large extent on the properties of the dyed substrates, dyes, textile auxiliaries and chemicals used which make the statement of average impossible. As a rule, textile waste-water has an alkaline pH, a higher temperature, high conductivity and a poor BOD:COD ratio (that is, lower biodegradability). Typical BOD:COD value lies between 1:25 to 1:5. The organic load of waste-water originates largely from pre-treatment of chemicals; BOD values of upto 210 kg/tonne have been obtained in case of alkaline boiling out of polyester.

The effluent includes aqueous discharges from fibres preparation, fabric preparation, dyeing and finishing operations. Of particular concern are dyes which are often sources of metals, salts and colour in the effluent sizes which have high BOD and COD levels and surfactants which are strongly linked to aquatic toxicity.

Air

Although the textile industry is a relatively minor source of HAP (hazardous air pollution) compared to other manufacturing industries, air emissions have been identified as the second greatest pollution problem (after aqueous effluents) for the industry. As many different types of commodity and speciality chemicals are used to manufacture textiles, characterisation and management of air emissions is a challenging responsibility.

Emissions consist of both point sources and fugitive emissions. Point sources include high temperature coating, drying and curing ovens which emit volatile organic compounds (VOCs), boilers which are a source of particulates, nitrous oxides and sulphur dioxide and bulk storage tanks for commodity and speciality chemicals. Fugitive air emissions results from spills, solvent-based cleaning, waste-water treatment plant operations and warehouses used to store finished fabric.

Solid

Solid waste is the largest waste stream produced (by volume) following aqueous effluent wastes generated. It includes fibre wastes (reworkable and non-reworkable), packaging waste, selvage, trimmings and sludge from waste-water treatment. The quantity of solid waste generated varies widely among the different industrial units depending upon the size and type of the textile operations, the nature of waste and the efficiency of the machinery used.

Table 26.2 provides a list of solid wastes generated during textile processing.

Table 26.2. Solid wastes produced in textile manufacture.

Ash	*Fabric waste*	*Plastic packaging*
Cardboard	Metal drums	Rags
Card waste	Paper (office, computer)	Sweeps
Carpet waste	Paper (packaging)	Waste treatment sludge

WASTE MINIMISATION

Waste minimisation (WM), in recent times, has emerged as an attractive proposition to tackle the environmental problems posed by rapid industrialisation and is being accepted worldwide. Besides reducing pollution, it also improves the process efficiency, thus reducing the cost of production. The textile processing sector in India is characterised by high consumption of resources like water, fuel, dyestuffs and chemicals, which, combined with low process efficiency, result in substantial wastage of resources. In this scenario, the concept of WM holds specific relevance to the industry. To adopt WM in the industry, it is important to get familiarised with the concept of WM.

Waste minimisation can be defined as: 'A new and creative way of thinking about products and processes that make them. It is achieved by the continuous application of strategies to minimise the generation of wastes and emissions'. The need for waste minimisation in the context of textile processing industry is elaborated below.

Chemical and Auxiliaries Conservation

The textile processing industry uses a wide variety of chemicals and auxiliaries in huge quantities. A typical synthetic textile processing industry consumes about 350 to 500 kg of chemicals per tonne of fabric processed. In the textile processing industry, unlike other manufacturing industries, only about 15 to 20 per cent of the chemicals used during processing appear in the product while the remaining find their way out to the environment in the form of waste generated from the industry. Taking into account the ever-increasing cost of these chemicals and the high pollution load caused by them, the industry cannot afford to continue losing them in the form of waste. Specific consumption of these chemicals and auxiliaries can be considerably reduced by adopting WM techniques. For instance, re-using spent dye liquor (polyester dyeing with disperse dyes), which has 80–95 per cent auxiliary chemicals added initially, remains unexhausted for the preparation of fresh dye liquor.

Water Conservation

The textile processing industry is a water-intensive industry. The ratio of water consumed in proportion to the amount of fabric processed is very high and is in the range of 15–20 m^3/1000 m of fabric.

In India, most of the textile processing units rely on ground water to meet their process demand. In many of the important textile processing centres water level has gone down from 15 m to around 300 m in the last 20 years due to excessive withdrawal of ground water. Apart from increasing water shortage, excessive withdrawal of water compared to the recharging rate has caused highly saline and hard water from other nearby water sources, such as the sea and rivers, to infiltrate the ground water aquifers. There is also an indiscriminate discharge of effluents into surface water resources and on land. Due to this, the water quality has deteriorated drastically and is unfit for use in the process in many regions. Units located in these regions have to procure water through tankers from other area, treat it and then use it in the process. In these regions, water has become a highly priced commodity. This has increased the share of water cost in the total processing cost from 0.5 per cent in earlier years to about 15 per cent as of now. As more regions are becoming water scarce, there is an urgent need to save water. Adopting WM techniques will help in conserving, recycling and re-using water and ultimately in reducing water consumption in the unit considerably.

Energy Conservation

In the textile processing industry, both thermal and electrical forms of energy are used and account for about 15 to 20 per cent of the fabric's total processing cost. In order to meet thermal and electrical energy demands, fuel is consumed in large quantities.

Thermal energy in the process house is generated from boilers and is used in stenters, dryers, thermic fluid heaters, etc. which normally operate at low efficiencies, resulting in proportionally high fuel consumption and level of emissions. Adoption of waste minimisation can improve the operational efficiency of the equipment resulting in both reduced energy cost and reduced emissions.

Pollution Control

Apart from large quantities of water and energy, the textile processing industry also uses a wide variety of chemicals such as acids, alkalis, organic solvents, dyes, and pigments, polymers, surfactants, etc. A considerable portion of these chemicals appear in the waste stream at various stages of processing. This waste, if discharged without treatment, can cause grave damage to the environment. Emission resulting from the burning of fuel result in air pollution. The treatment of waste-water from textile processing houses poses two major problems:

1. High volume of waste streams which require large and expensive treatment systems. For SMEs, this is of prime importance as they also face space shortage and non-availability of funds.
2. 'Difficult to treat' characteristics which make the treatment technically complex and tedious.

Hence, the first step in tackling these environmental problems lies in reducing the volume of waste generated. This should be followed by efforts to minimise wastage of chemicals and auxiliaries to reduce waste toxicity and treatment complexity. Adoption of waste minimisation aims at meeting both these requirements, facilitating the treatment of waste at a lower cost as well as ensuring compliance with regulatory norms.

Waste minimisation can be used as a tool for improving a company's public image by highlighting the steps it has taken to protect the environment. Once waste minimisation becomes an integral part of a company's activities, statements such as 'processed in a green environment' or 'green/eco-friendly product' can be used to increase the product's market share and acceptability.

WASTE MINIMISATION POTENTIAL

The large quantity of waste generated from the textile processing industry can be attributed to several reasons such as obsolete and inefficient technology, untrained manpower, poor operation and maintenance practices, lack of proper production planning, non-availability of comprehensive literature on waste minimisation, etc. Apart from these, there is a general belief amongst the process houses the 'addition of excess chemicals improves processing quality'. Also, very often, extra addition of a chemical at one stage, warrants the need for the addition of other chemicals, to counteract the previous chemical's effects in the next stage. This not only results in increased resource consumption but also in higher quantum of waste generated. 'Waste attracts more waste' holds good in such circumstances. They reduce the company's profitability and sometimes even result in substandard processing quality. Moreover, as mentioned before, a very small percentage of chemicals added during processing is consumed and the rest finds its way into waste streams. Hence, the potential for waste minimisation in this sector is very high. Waste minimisation, apart from improving process efficiency, reduces the cost of processing and curtails the pollution load.

WASTE MINIMISATION TECHNIQUES

Waste minimisation is best achieved by minimising the generation of waste at the source itself. After exhausting the at-source reducing potential, attempts should be made to utilise the waste generated within the unit. The techniques which are available in these two areas are listed below:

Source Reduction

Technology change

1. Use of jet dyeing machine instead of beam dyeing machine in which low material-to-liquor ratio can be maintained.
2. Use of direct natural gas firing system instead of thermic fluid heating system for drying operations.

Input material change

1. Replace acetic acid with formic acid or mineral acids.
2. Replace sodium hydrosulphite with thiourea dioxide.

Better process control

1. Proper adjustment of dampers in the stenters.
2. Maintaining proper temperature, pressure and time conditions in the ageing process.

Equipment modification

1. Provision of a scrubber on print machine for better washing of the print blanket.
2. Pressure switch for viewing glass light in jet dyeing machine.

Product modification

1. Avoid stiff finish (given by urea formaldehyde) in packing ribbons.

Good housekeeping

1. Provision of self-closing taps.
2. Use of a tray for collecting print paste spills while adding paste to squeeze.
3. Semi-automatic flat-bed screen printing.
4. Fill the print paste transfer buckets to 80 per cent capacity only, to avoid spillages during manual transfer.

Waste Utilisation

Recycle/Re-use

1. Re-use of steam condensate.
2. Recycle print blanket wash water.

Input material recovery

1. Recovery of ethyl acetate solvent from the solvent cleaning waste.
2. Provision of a doctor blade on print machine to recover adhering print paste.

By-product formation

1. Use of rags for the manufacture of mattresses.

WASTE MINIMISATION OPPORTUNITIES

The experience of waste minimisation studies in textile dyeing and printing industries has shown that there are numerous proven waste minimisation measures which can be implemented to tap the waste minimisation potential. Most importantly, these measures, besides increasing the profitability of the unit, enable tremendous levels of reduction in waste-water volume, pollution load and toxicity.

Here, 54 waste minimisation opportunities have been described. These can be implemented in other similar units with slight modifications. For the sake of clarity, these have been grouped as per the following waste minimisation techniques:

Source Reduction

1. Good housekeeping.
2. Process change/modification:-
   (a) Input material change/substitution.
   (b) Better process control.
   (c) Equipment modification.
   (d) Technology change.
   (e) Product change/modification.

Recycling

1. On-site recovery/Re-use.
2. Creation of useful by-product.

For all the measures, the technical requirements, economic viability and anticipated environmental benefits have been discussed. The economic viability is expressed in terms of investment (I), annual savings (S), annual operating cost if any (O), and payback period (P). Since the measures were implemented in various units with different capacities, the results have been discussed with respect to one unit, the details of which are given below:

The overall results have been presented in Table 26.3.

In some cases it was not possible to quantify savings (shown as NQ) due to the following reasons:

1. Wide variation in designs, shades, cloth quality, etc. making representative averaging difficult.
2. Difficulty in assessment of impact in the absence of extensive implementation.
3. Lack of measured data due to lack of instrumentation.
4. Intangible nature of benefits such as reduced maintenance, improved safety, etc.

Many of the measures listed above are common to almost all the small and medium scale units in India and can be implemented with slight modifications. There are a number of water conservation measures which at the present cost seem economically unviable but with increasing water shortage and rising prices, these may take the top priority.

Table 26.3. Results of adopting waste management.

Waste minimisation measure	*Requirements*	*Economics*	*Anticipated benefits*	*Remarks*
Generic and good housekeeping				
Regular repair of leaking water lines, valves and pumps and other equipments	Regular maintenance	S = NQ	Reduced water consumption Improved work environment	Shift gradually from breakdown to preventive maintenance practices
Avoiding preparation of excess print paste by estimating actual required quantity before hand based on print design	Nil	I = Nil S = NQ	Reduced left over paste (which many time has to be thrown if not used for long)	Can be done by print master At present in most of the units no such estimation is made
Filling of the print paste charging buckets to about 3/4th of its capacity so as to avoid spills	Instructions to the colour kitchen incharge	I = Nil S = Rs. 18,000/ann to Rs. 36,000/ann P = Immediate	Reduced spills and thus losses Reduced pollution load Reduced accidents because of improved shop floor condition	About 5–10 kg print paste spills occurs while transfer with full bucket
Provision of lift so as to reduce spillage due to manual handling	Lift Training of operator	I = Rs. 1,20,000 S = Rs. 1,00,000/ann O= Rs. 40,000/ann P = 2 years	Reduced print paste loss Reduced pollution load Improved work environment Reduced labour requirement Reduced accidents	Applicable if the print paste is to be transferred to different floors
Provision of overhead moving rail with stirrer in colour kitchen so as to avoid print paste spillages which is commonly encountered in situations when stirrer is stationary and filled drums are moved for paste preparation	Overhead rails and a stirrer	I = Rs. 15,000 S = Rs. 18,000/ann to Rs. 25,000/ann P = 1 year	Reduced spillage Ease of operation and reduced fatigue to workers Reduced pollution load Improved shop floor environment	About 5–7 kg print paste per day spills while shifting drums
Instead of transferring print paste manually by mugs to the charging bucket, the same can be filled under gravity flow by making a provision of a tap in the drum	Slight modification in drum	I = Rs. 4,000 for 20 drums S = Rs. 10,000/ann P = 6 months	Reduced print paste loss Reduced pollution load Reduced floor washing frequency	Diameter of the tapping to be decidede based on the flow of the print paste
Blower and water pump of the printing machines should be switched off when printing is not in operation (while switching over to next printing batch)	Proper supervision	I = Nil S = Rs. 66,825/ann	Reduced electricity consumption	Savings calculation basis: 6 HP load per machine connected for a minimum of 3 hrs/d when printing operation is not done

(Contd...)

Waste minimisation measure	*Requirements*	*Economics*	*Anticipated benefits*	*Remarks*
Proper steam distribution and net-working so as to avoid pressure and temperature variation	Pipes, header, steam traps, etc.	I = NQ S = NQ	Better quality and reduced off specification products Reduced steam consumption	In small scale units expansion is normally in adhoc manner resulting in temporary arrangements
Input material changes				
Acetic acid by catalyst DD quantity reduced by 50%	Nil	S = Rs. 14/kg of acetic acid used	Reduced COD load Reduced cost	
$Na_2S_2O_4$ (hydro) by thiourea dioxide	Nil	S = Rs. 26/kg of hydro used	Reduced COD load Reduced cost	
Kerosene and NI detergent by acrylic thickeners in pigment-based print paste preparation	Nil	S = Rs. 0.62/kg of paste prepared	Reduced toxic vapour emission Improved work environment Reduced fire hazards Reduced COD load in wash water	
Replace citric acid with citric W	Nil	S = Rs. 45/kg of citric acid used	Reduced COD and BOD load	
Non ionic detergent (Nonyl phenol ethoxylate based) by alpha olefin sulphonate based Ginasol 6836	Nil	S = Rs. 46/kg of NI detergent used	Avoidance of toxic metabolites biodegradation products	
Lyogen DFT by Hicoleveller BJD (levelling agent)	Nil	S = Rs. 30/kg of lyogen DFT used	Reduced pollution load	
Use of grey having less oil and other impurities	Nil	S = NQ	Reduced COD load	Supplier (weaver) to be properly selected and instructed
Acid black 52 metal complex dye by acid black 210 (for nylon)	Nil	S = Negligible	Reduced chromium levels in effluent	For nylon Local availability has to be ensured
EDTA based sequestering agent by HEDP based	Nil	S = Rs. 35/kg of EDTA base softener	Reduced cost Reduced pollution load	
Penta chlorophenol (preservative) based printing gum by dichloro-phenol based gum	Nil	S = Negligible	Reduced toxicity Reduced fire hazards No threat to microbes growth in ETP	
Phenol by DEG (di-ethylene glycol)	Nil	S = Rs. 10/kg of phenol used	Avoidance of toxic compound Reduced health hazards Reduced treatment cost Ease to meet regulatory compliance (phenol)	

Oxalic acid partially by HCl	Nil	S = Rs. 12/kg of oxalic acid used	Reduced COD load Reduced cost	Savings calculation done for 'substituting 1 kg oxalic acid by 0.3 kg oxalic acid and 1.0 kg 30 per cent HCl'
CCl_4 based stain remover by detergent	Nil	S = Negligible	Reduced toxicity Elimination of O_3 depleter	
Better process control				
Optimisation of boiler damper and pulley settings to reduce energy losses	Combustion analyser	I = Rs. 10,000 S = Rs. 42,000 P = 3 months	Reduced fuel consumption	Excess O_2 reduced from 4% to 2%
Before carrying out neutralisation pH should be checked and accordingly neutralisation chemical to be added	pH metre/pH strips	I = Rs. 1,000 S = NQ	Reduced chemical consumption and pollution load	Normally in most of the units chemical addition is based on experience and is often in excess of the required quantity
Use of high pressure jet nozzles for screens washing	High pressure nozzles	I = Rs. 1,500 S = Rs. 1,05,000/ann P = 1 month	Reduced wash water quantity by 33% (from 150 m^3/day to 100 m^3/day)	Each nozzle costs Rs. 100
Introduction of static cold wash instead of running wash before scouring	3 dip tanks	I = Rs. 15,000 S = Rs. 96,500/ann P = 2 months	Reduced hydro requirement by 50% in scouring	Hydro consumption was reduced by 1.0 gpl.
Avoidance of drying unit operation between print washing and stenter finish	Proper production scheduling Storage tanks	I = Rs. 50,000 Drying cost = Rs. 0.03/m S = Rs. 52,000/ann P = 1 year	Fuel saving Reduced air pollution	
Water meter installation to control material to liquor ratio and for further data generation on water consumption	Water meter	I = Rs. 1,5000/per water meter S = NQ	Better control over process Raw material consumption may reduce	
Neutralisation (before whitening) batch can be used for whitening	Nil	I = Nil S = Rs. 22,600/ann P = Immediate	405 m^3/ann saving in water consumption Saving of oxalic acid @ 100 kg/ann which was used earlier for whitening operation Energy saving of Rs. 75/batch for heating dyeing bath	Water @ Rs. 7/m^3 $(COOH)_2$ @ Rs. 29/kg

(Contd...)

Waste minimisation measure	*Requirements*	*Economics*	*Anticipated benefits*	*Remarks*
Reduction in HCl concentration in print washings by 30%	Nil	I = Nil S = Rs. 1,600/ann P = Immediate	Reduced HCl consumption by 1.8 kg/day Reduced pollution load	Calculation for the reduced from 1.7 gpl (33% HCl) to 1.2 gpl
Closing excess length of vacuum slit (before stenter) opening by an adhesive tap	Adhesive tape	I = Nil S = NQ P = Immediate	Electricity saving Better moisture removal and Reduced fuel consumption	Proper supervision is required
Improvement and close watch on boiler operation by on line flue gas analysis	2 analysers	I = Rs. 6,80,000 S = Rs. 45,000/ann P = 2 years	Reduced fuel combustion Reduced air pollution	
Equipment modification				
Provision of a vacuum slit before stenter so as to reduce moisture content	Vacuum slit, a vacuum pump and piping	I = Rs. 70,000 S = Rs. 4,70,000/ann P = 2 months	Reduced fuel consumption Increased production capacity	In almost all the small scale units stenters are the bottle neck in the capacity utilisation
Use of a press switch for view glass light in jet dyeing machine	Press switch	I = Rs. 30 S = Rs. 3,500/ann P = Immediate	Reduced electricity consumption by one unit per day per machine	
Increase number of chamber in stenter from 3 to 5 and increase the fabric speed	Stenter chambers and additional accessories Additional space	I = Rs. 4,000,000 S = Rs. 6,30,000/ann P = 9 months	Increased production capacity by 20% Reduced specific fuel consumption	Profit margin 30 paise/m Space is often a constraint in SMEs
Reduction in print paste loss through edge carryout on printing machine by covering the excess width of screen by adhesive tape	Adhesive tape	I = Negligible S = Rs. 90,000/ann P = Immediate	Reduced print paste loss Reduced washing water requirement Reduced pollution load	About 1 to 2% of the total paste applied per day can be saved Calculations for the savings of about 25 kg paste per day
Technology change				
Replace 6 no. of electric resistance heaters by 2 no. of infra-red-heaters for print blanket heating on printing machines	IR heaters	I = Rs. 24,000 S = Rs. 6,70,000/ann P = < 1 month	Reduced electricity consumption of 8 unit per hour Better work environment	IR heater @ Rs. 4,000 per heater
Use of direct natural gas firing system in stenter instead of thermic fluid heating system	Combustion chamber Gas burners	I = Rs. 8,10,000 S = Rs. 12,50,000/ann P = 8 months	Avoided heat losses due to double transfer of heat Reduced fuel consumption Reduced maintenance	Resale price of thermopack system (Rs. 8,50,000) has been taken into consideration while calculating net investment

Automated print washing instead of manual	Motor Washing tanks Accessories like gears, shafts, rollers	I = Rs. 1,20,000 O= Rs. 45,000/ann S = Rs. 1,20,000/ann P = 8 months	Reduced workforce by 10 persons Reduced water consumption by 20% Reduced time Reduced chemicals consumption	For calculations, savings due to reduced manpower (@ Rs. 1,000/ month) has been considered
Pneumatic operation of screen printing instead of manual (colour dispensing still manual)	Compressors, solenoid valves and electrical panel, piping, etc.	I = Rs. 13,20,000 O= Rs. 1,90,000/ann S = Rs. 11,75,000/ann P = 16 months	Reduced manpower from 12 to 7 per machine per shift Reduced print paste spills by about 0.1%	For calculation savings due to reduced manpower in two shifts has been taken into consideration
Automatic colour kitchen Standardise colour and gum stock preparation Computer controlled system for delivering the exact quantities of colour and gum for a given recipe	Pneumatic print paste transfer arrangement Ultrasonic mixer for preparing homogenous mixtures Computer controlled system for delivering exact quantities of gum and colours	I = Rs. 38,00,000 O= Rs. 50,000/ann S = Rs. 5,35,000/ann P = 8 years	Reduced manpower by 5 persons in each shift Reduced print paste handling loss by about 30% (60 kg/d) Reduced pollution load Improved work environment Reduced off-specification processing by about 0.5% Product quality improved	Appears economically infeasible but with increasing treatment cost, manpower shortage and increasing demand for better quality products it may become necessary in the days to come
Installation of computers for colour matching	Computer colour matching system	I = Rs. 10,00,000 O= Rs. 20,000/ann S = Rs. 6,00,000 P = 20 months	Reduced colour consumption by about 3 to 5% Reduced defective processing by about 0.5%	Calculations done for about 3% reduction in colour consumption
Recycle/Re-use/Recovery				
Recycle of treated waste-water to non-critical areas	Effluent treatment plants	I = Rs. 30,00,000 O= Rs. 3,00,000/ann S = Rs. 8,00,000/ann P = 8 years	Reduced fresh water use by about 400 m^3/d	The water can be used for blanket washing, screens, squeeze washing, floor washing, etc.
Collection of steam condensate from jet dyeing machines and zero-zero finish machine and re-using it as boiler feed water	Condensate conveyance system Storage tank Pump	I = Rs. 70,000 S = Rs. 1,30,000/ann P = 7 months	Heat value recovery Reduced water consumption Reduced soft water requirement Reduced hydraulic load on ETP	Savings have been calculated for about 2000 m^3/ann of condensate recovery Condensate price (including heat value) = Rs. 65/m^3

(Contd...)

Waste minimisation measure	*Requirements*	*Economics*	*Anticipated benefits*	*Remarks*
Re-use of spent scouring liquor twice	Storage tank Filter Valves Piping	I = Rs. 70,000 S = Rs. 55,000/ann P = 15 months	Heat recovery worth Rs. 13.50 per recovered batch Reduced alkali requirement by 20% Reduced pollution load	Check out oil content, it may re-deposit if present in excess quantity Assuming recovery of about 70% alkali (0.4 kg) per batch
Dye bath re-use	3 Storage tanks 9 Valves Piping	I = Rs. 75,000 O= Rs. 7,000/ann S = Rs. 80,000/ann P = 12 months	Heat recovery worth Rs. 13.5 per recovered batch Reduced auxiliary chemical consumption Reduced pollution load	Number of time of re-use is fixed depending upon level of oil content dye fabric and also on type of shade
Reuse of neutralisation (after scouring and washing) bath of dyeing batches	Storage container Pipe, valves Pump	I = Rs. 25,000 S = Rs. 6,000/ann P = 4 years	Reduced acid consumption Reduced water consumption Reduced pollution load	Calculations basis: reusing the liquor for a day per machine
Recycling blanket wash water	Centralised collection and a pump (1HP) Filter	I = Rs. 4,500 O= Rs. 5,000/ann S = Rs. 1,00,000/ann P = 17 days	Reduced water consumption from 60 m^3/d to 10 m^3/d per machine	Calculations are for one printing machine The water reduction considered as combined effect of WM options
Print paste recovery from screens and squeeze (after each lot of printing) by a simple wiper	Wiper	I = Rs. 1,000 for 50 wipers S = Rs. 3,50,000/ann	Print paste recovery Reduced pollution load Reduced water requirement for washing	Saving calculation done for recovery of about 100 kg paste per day from printing dept
Print paste recovery from the blanket carryover by a doctor's knife. Use this collection for dark shades preparation	Doctor's knife and collection bucket	I = Rs. 1,000 S = Rs. 1,44,000/ann	Print paste recovery Reduced pollution load Reduced water requirement for washing	@ Rs. 12/kg, 12 T/month collection from 7 machines
Recovery of ethyl acetate solvent from the print blanket solvent wash waste by distillation	Steam distillation chamber Steam	I = Rs. 40,000 O= Rs. 30,000/ann S = Rs. 3,12,000/ann P = 2 months	Recovery of ethyl acetate (70%) Reduced pollution load	
Recycle of non-contact cooling water from zero-zero finishing machine	Collection tank, pump, piping	I = Rs. 20,000 O= Rs. 10,000/ann S = Rs. 1,15,000/ann P = 3 months	Water consumption reduced by about 55 m^3/d	

By-product				
Use of fabric cuttings for mattress manufacturing	Nil	I = Nil S = Rs. 6,000/ann	Avoided waste disposal problem	The cuttings can be sold to mattress manufacturers
Use of solid residue from distillation of print blanket solvent cleaning waste as support fuel in boiler (9,000 kcal/kg) (200 kg/month)	Nil	I = Nil S = Rs. 10,000/ann problem	Reduced fuel consumption Avoided waste disposal	If the existing combustion chambers do not have facility for charging solid waste, required modifications has to be done
Product change				
Avoiding a particular stiff paper finish (given by formaldehyde resins) in packing ribbons	Nil	I = Nil S = Rs. 15 per kg of resin used P = Immediate	Avoided use of a known carcinogenic Avoided free formaldehyde emissions, thus improved work environment and reduced health hazards	

Note: NQ = Not quantifiable; I = investment; S = savings; P = payback period; O = operation cost.

This list is not the end of all the waste minimisation measures. Since wide variations exist in this industrial sector there could be many more such generic as well as many industry-specific measures which need to be identified and developed using a methodical step-by-step approach.

BARRIERS TO WASTE MINIMISATION AND ENABLING MEASURES

Despite the tremendous scope, waste minimisation is yet to gain a strong foothold in this sector. The industry continues to consume large quantities of water, fuel and chemicals coupled with low process efficiency. As a result, a major portion of raw materials consumed goes down the drain, posing a grave environmental hazards. Several reasons, which are characteristic features of the textile industry, have been identified which are proving to be a major hindrance in the adoption of waste minimisation measures. It is essential that these bottlenecks be precisely noted down and measures be evolved which would assist in overcoming them. This would facilitate effective implementation of the waste minimisation programme. The barriers faced by the Indian textile processing industry and the corresponding enabling measures are: (i) attitudinal barriers; (ii) organisational barriers; (iii) trade barriers; (iv) technical barriers; (v) economic barriers; and (vi) other barriers.

Attitudinal Barriers, Catalysts and Enabling Measures

Barriers

1. Lack of awareness on industrial pollution.
2. Ignorance of environmental issues.
3. Emphasis on maximising production rather than productivity.
4. Complacent attitude towards existing process/production conditions.
5. Hesitant in taking up risks involved.
6. Low participation of workers in WM programme.
7. 'I am doing the best' belief.

Enabling measures

1. Increase awareness on WM/environmental issues.
2. Involve workers in decision-making.
3. Acknowledge worker's efforts.
4. Formulate incentive schemes for workers.
5. Encourage experimentation for WM options.
6. Review WM measures on regular basis using simple indicators.
7. Increase interaction among similar kind of industries.

Organisational Barriers, Catalysts and Enabling Measures

Barriers

1. One-man show; middle (supervisory) level missing.
2. Loose management structure.
3. Production on ad-hoc basis.
4. Labour-intensive: workers employed on contract basis.
5. Inappropriate profit sharing.
6. Improper documentation of inventory and production data.

Enabling measures

1. Delegation of authority to sub-ordinates.
2. Induction of technically sound persons at supervisory levels.
3. Proper production planning.
4. Right wage for the right person.
5. Recruitment of regular skilled workforce on the payroll.

Trade Barriers, Catalysts and Enabling Measures

Barriers

1. Production on job-order basis.
2. Poor quality of input raw material.
3. Wide range of shades involved in dyeing/printing.
4. Industry mainly catering to local markets.

Enabling measures

1. Setting up of integrated plants covering weaving, dyeing and printing.
2. Ensuring good quality of raw materials from supplier.
3. Standardisation of fabric shades.
4. Promote marketing in the international market.

Technical Barriers, Catalysts and Enabling Measures

Barriers

1. Lack of:-
   (a) Proper guidance on WM.
   (b) Technically sound professionals.
   (c) Skilled workforce.
   (d) Laboratory analysis facility.
   (e) Adequate in-plant waste usage opportunities not available.
2. Erratic power supply.
3. Relevant technical literature on eco-friendly chemicals not readily available.
4. Fabric quality and shade: highly sensitive processing parameters.
5. Highly water-intensive process steps.
6. Non-applicability of technology developed abroad in Indian conditions.
7. Difficulties in production re-scheduling for implementation.

Enabling measures

1. Conducting of training and awareness workshops on WM.
2. Setting up of laboratory with basic analysis facility and pilot scale study facilities; assistance can be extended by local organisations.
3. Provision of regular power supply/captive power generation.
4. Promoting circulation of relevant technical literature.
5. Development of indigenous WM measures.
6. Encouraging waste exchange among industrial units.

Economic Barriers, Catalysts and Measures

Barriers

1. Non-availability of adequate funds.
2. Low financial returns on certain WM measures.
3. Availability of cheap unskilled labour, making automation less attractive.
4. Changing excise duties.

Enabling measures

1. Providing soft loans.
2. Planned investment.
3. Incentive schemes for industries going in for WM.
4. Training of workforce for specific jobs.
5. Formulating long-term industrial policy.

Other Barriers, Catalysts and Enabling Measures

Barriers

1. Abundant water supply, making water conservation less attractive financially.
2. Lack of availability of space.
3. Lack of regulations on environmental management systems and solid waste.

Enabling measures

1. Imposition of water cess on industries to restrict water usage.
2. Modernisation of existing plants.
3. Formulation of area-wise voluntary WM groups.
4. Evolution of load-based emission standards for textile industry.

IDENTIFY THE CAUSES OF WASTE GENERATION

The causes of waste generation are given as under:

Cause Analysis

Process step/unit	*Causes (operation)*
Scouring and weight reduction	Impurities addition to grey fabric by the grey suppliers
	High liquor ratio
	Use of high pollution load exerting hydrocarbons
	Use of excess quantity acid
	Use of high pollution load exerting $(COOH)_2$
	Poor grey washing practices
	Poor alkali exhaustion in saponification having about 30–40% re-usable alkali content
Dyeing	Discharge spent dye liquor having re-usable auxiliaries intact
	Use of chemicals generating high pollution load

(Contd...)

Process step/unit	*Causes (operation)*
Printing	Excess quantity of paste preparation than required
	Once matching is over, print paste adhered to the screens and squeezes is not collected back but washed off
	Inefficient washing system
	Printing on portions of blanket not covered by fabric
	Use of high COD load exerting print gum and containing phosphates
	Unwanted chemicals addition
Heat loss from electric resistance heater for blanket drying	Large distance between blanket and heater
	Poor heater selection
Stenter drying	Inefficient mode of heat transfer that is, natural gas-thermic fluid—hot air instead of natural gas—hot air
Ageing	Waste of residual heat in the pressurised ageing vessel
Print wash	Improper ageing due to sub-optimal pressure and temperature in the ageing vessel
Shrink control	No system for steam condensate and cooling water collection and re-use

GENERATING WM OPPORTUNITIES

Based on the detailed cause analysis, number of waste minimisation opportunities developed along with technique used and the anticipated benefits are listed below:

Waste Minimisation Opportunities

Waste minimisation measure	*Waste minimisation technique*	*Anticipated benefits*
Fabric purification		
Cold wash running		
Dip fabric in fresh water in the trolley before loading on the jet	BPC	Water soluble impurities dissolved and removed before being loaded on machine
		Reduced pollution load, hence, reduced chemical consumption in subsequent stages
		Hydro uses in the scouring stage can be avoided
Make running wash static	BPC	Reduced water consumption
Spent scouring liquor		
Reduce quantity of caustic by 20%	IMC	Reduced chemical consumption
Reduce quantity of detergent	IMC	Reduced ETP load
Replace sodium hydrosulphite by Diasyn HF		Reduced pollution load
		Reduced chemical cost
		Reduced chemical consumption
Recycle scouring liquor after adding make-up chemical	RC	Reduced chemical consumption
		Reduced pollution load

(Contd...)

Waste minimisation measure	*Waste minimisation technique*	*Anticipated benefits*
Replace non-ionic detergent by Ginasol 6836	IMC	Reduced COD load on ETP Reduced chemicals cost
Increase customer awareness on oil content in grey fabric	IMC	Improved quality of grey fabric resulting in reduced chemical consumption and pollution load
Provide 5 minute dragout time after every process discharge	BPC	Reduced carryover of chemicals and impurities to the next stage
Steam condensate from all process steps		
Insulation of heat exchanger	EM	Reduced heat loss
Coolant water from all process steps		
Re-using cooling water in the subsequent hot wash	RU	Reduced water consumption Reduced fuel consumption
Avoid cooling	BPC	Reduced process time and water consumption
Spent acid liquor		
Replace oxalic acid with inorganic acid	IMC	Reduced costs Reduced COD load on ETP
Reduce quantity of acid	IMC	Reduced pollution load Reduced costs
Eliminate acid wash	BPC	Reduced water and chemicals consumption Reduced process time Reduced pollution load
Use spent acid wash as dye bath (only for whitening)	BPT	Reduced acid consumption in the dye bath Reduced water consumption and pollution load
Avoid use of sodium nitrite	IMC	Reduced chemical cost
Avoid cold wash	BPC	Reduced water consumption and pollution load
Dyeing		
Spent dye liquor		
Replace acetic acid with formic acid/catalyst D	IMC	Reduced COD load on ETP
Eliminate carcinogenic acid and direct dyes	PC	Reduced toxicity level in the waste generated
Replace Lyogen DFT with HICO leveller BJD	IMC	Lower COD/BOD Lower chemical cost
Replace metal complex dye with lower chrome content dye	PC	Reduced heavy metal concentration in waste generated
Provide pressure switch for view glass lights	EM	Reduced power consumption
Re-use spent dye liquor for dyeing by addition of dyestuff and reduced quantities of auxiliaries. Use dyeing matching pressure for pumping to insulated overhead tank	RU	Reduced pollution load
Use of soft water for preparation of dye bath	IMC	Higher dye exhaustion level rates resulting in lower dye consumption and reduced pollution load

(Contd...)

Waste minimisation measure	*Waste minimisation technique*	*Anticipated benefits*
Insulation of jet dyeing machine	EM	Reduced heat loss resulting in reduced fuel consumption
Eliminate/substitute citric acid in stenter with inorganic acid (hydrochloric acid)	IMC	Reduced chemical cost Reduced COD load on ETP
Replace EDTA sequestering agent with HEDP	IMC	Reduced chemical cost Reduced chemical consumption Reduced COD load on ETP
Operation at a lower M:L ratio from 1:3 to 1:22	BPC	Reduced energy, water and chemical consumption
Reduce quantity of defoamer	BPC	Reduced chemical consumption and pollution load
Heat loss from stenter		
Use of direct gas firing for heating in stenter/ drier instead of heating it by thermic fluid heater (for both stenter and drier)	TC	Reduced fuel consumption
Provision of vacuum slit for reducing moisture content of fabric	EM	Reduced fuel consumption
Increase the number of chambers from three to five	EM	Increase the speed of movement of fabric resulting in increased production
Reduce air inflow to senter by using lower size pulley for the fans along with damper adjustment	EM	Reduction in electrical and thermal energy consumption
Recycling the hot air from the cooler (at the end of stenter) into the stenter's combustion chamber for use as combustion air	RC	Reduced heat loss resulting in lower fuel consumption
Printing blanket wash		
Recover ethyl acetate from blanket wash	RC	Reduced consumption of ethyl acetate Reduced pollution load
Substitute ethyl acetate for blanket wash with fatty acids (toluene and zyolene)	IMC	Reduced evaporation loss due to increase in boiling temperature and this facilitates better recovery
Recycle blanket wash water	RC	Reduced water consumption
Provide doctor blade on blanket	RU	Recovery of print paste Reduced water consumption for washing of blanket
Restrict printing to area on blanket covered by fabric	BPC	Reduced consumption of print paste Reduced water consumption for blanket wash
Print paste preparation		
Reduce/eliminate ammonium sulphate in the print paste	IMC	Reduced chemical consumption and pollution load
Replace citric acid in the print paste with citric-W	IMC	Reduced pollution load
Replace print gum containing pentachlorophenol by dichlorophenol	IMC	Reduced toxicity levels
Replace kerosene in print paste by acrylic thickener	IMC	Reduced chemical cost and consumption Reduced pollution load

(Contd...)

Waste minimisation measure	*Waste minimisation technique*	*Anticipated benefits*
Quantity of print paste to be prepared as per design coverage and matching length	BPC	Reduced wastage of print paste resulting in reduced print paste consumption and pollution load
Spillages		
Attach stirrers on an overhead moving rail to reduce drum movement	EM	Reduced spillages of print paste
Switch over from manual to pneumatic operation for screen printing	TC	Savings in labour Uniform printing can be done
Fill the paste transfer buckets only up to 80% of its capacity	BPC	Reduced print paste spillages
Provision of lift for movement of print paste between the different floors	EM	Reduced print paste spillages Convenience to the workers
Tiling of kitchen floor and keeping it clean	GH	Easy movement of drums Reduced spillages
Provision of valve arrangement in the drum for taking out paste	GH	Convenient to the workers Reduced wastage of print paste
Provision of drag out tray to the worker to collect dripping paste	GH	Reduced print paste wastage Helps to keep the shop floor clean
Screen, squeeze, container wash		
Scrape off the print paste properly from screen before washing	RC	Recovery of print paste Reduced water consumption for screen washing
Pre-soaking of screens before washing	BPC	Reduction in water consumption in screen washing
Provision of high pressure jet nozzles to wash screens	EM	More efficient screen washing Reduced water consumption for screen washing
Switch over to 2 IR heaters from 4 electric resistance heaters used per print machine and reduce distance between the heater and the blanket	EM	Reduced heat loss resulting in reduced power consumption
Steam loss from ageing		
Partial re-use of steam in the next ageing vessel	RU	Reduced heat loss resulting in reduced heat consump tion
Re-use of steam for heating boiler feed water	RU	Reduced heat loss resulting in reduced heat consumption
Print washings		
Optimum pressure and temperature conditions to be maintained in the ageing vessel. Higher pressure, temperature and increased ageing time for darker shades	BPC	Better level of dispersion of dye into the fabric Reduced print paste contents in the washings
Counter-current washings to be given	RU	Reduced water consumption Reduced hydro consumption Reduced pollution load
Providing squeezing arrangement between two washing stages (automisation of print washing)	BPC	Reduced carryover of impurities and chemicals from one stage to the other

(Contd...)

TURNING WASTE INTO FIBRE

It may soon be possible to produce a low cost, high-value, high-strength fibre from a biodegradable and renewable waste product for air filtration, water filtration and agricultural nanotechnology. The achievement is the result of using the recently perfected technique of electrospinning to spin nanofibres from cellulose.

'Cellulose is the most abundant renewable resource polymer on earth. It forms the structure of all plants,' says *Margaret frey*, an assistant professor of textiles and apparel at Cornell. They have developed some new solvents for cellulose, which can produce fibres using the technique known as electrospinning.

The technique of electrospinning cellulose on the nanoscale involves dissolving cellulose in a solvent, squeezing the liquid polymer solution through a tiny pinhole and applying a high voltage to the pinhole. The technique relies on electrical rather than mechanical forces to form fibres. Thus, special properties are required of polymer solutions for electrospinning, including the ability to carry electrical charges.

The charge pulls the polymer solution through the air into a tiny fibre, which is collected on an electrical ground. The fibre produced is less than 100 nanometers in diameter.

The new technique is now possible because of a new group of solvents that can dissolve cellulose. Whenever cotton is converted to fabric and garments, fibre (cellulose) is lost to scrap or waste. At present it is largely discarded or used for low value products, such as cotton balls, yarns and cotton batting.

Producing a high-performance material from reclaimed cellulose material will increase motivation to recycle these materials at all phases of textile production and remove them from the waste stream. It has been found that electrospinning produces nonwoven mats of nanofibres, which could provide nanoscale pores for industrial filters.

Producing very-small diameter fibres from cellulose could have myriad applications that would exploit the surface area of nonwoven mats of nanofibres and the possibility of controlling molecular orientation and crystalline structures of nanoscale fibres. Possible applications might include air filtration, protective clothing, agricultural nanotechnology and biodegradable nanocomposites.

Another application is using the electrospun cellulose mats to absorb fertilisers, pesticides and other materials. These materials would then release the materials at a desired time and location, allowing targeted application.

CASE STUDY—TEXTILE INDUSTRY (RECOVERY OF CAUSTIC SODA)

In a reputed textile mill located in Mumbai, producing over 2 lakhs kgs of cloth per month, the raw cloth is processed in caustic soda during the process of mercerisation. In this process, the cloth is first processed in concentrated caustic soda and later washed with fresh water to minimise the carry over of caustic soda along with the cloth. The weak caustic solution generated out of this washing was normally having about 2 to 3 per cent of caustic and used to be drained off into the effluent treatment plant (ETP) for further treatment and discharge into the sewerage.

Thermax energy performance services (TEPS) conducted detailed study of the existing process and made various modifications in the process along with addition of control system such that the concentration of weak liquor flowing out from the machine is maintained at around 6 per cent. This weak liquor is collected in an underground tank and is further passed through a triple effect evaporator for recovering the caustic at 25 per cent concentration. This recovered caustic is centrifuged for removal of sizing chemicals and is directly used in the process in place of fresh caustic. The steam supplied to the evaporator is recovered as hot condensate to be used as boiler feed water. As this condensate is alkaline, usage of this as boiler feed water results in reduction of dosing in the boiler feed water. Additional quantity of hot water is generated in the evaporator system, which is used in the process. Systems have been laid to distribute this hot process water throughout the plant. The modifications made on the merceriser machine has also resulted in reduction of fresh water consumption resulting in additional savings. A full-fledged data acquisition system (DAS) is installed to on-line monitor and record the operation of the plant.

The financing arrangement to implement this project was made by thermax energy performance services (TEPS) and the same is being paid back fully from the savings generated out of operation of this project. The balance saving, over and above the debt re-payment, is being equally shared between the customer and TEPS. Operation summary is given in Table 26.4.

Table 26.4. Operation summary of various parameters.

Parameters	*Unit*	*Quantity*	
		Design (Note–1)	*Actual (Note–2)*
Weak liquor concentration at inlet	%	6.0	5.4
Weak liquor flow rate	kg/hr	2778	2108
Strong liquor concentration at outlet	%	25.0	26.3
Strong liquor flow rate	kg/hr	667.0	422.5
Caustic recovered	kg/hr	166.8	111.1
Steam consumed	kg/hr	750.0	604.1
Hot water generated	kg/hr and °C	12920 and 58.0	5809 and 58.0
Alkaline condensate generated	kg/hr and °C	2861 and 60.0	2290 and 58.4

Note: (1) The capacity of the caustic recovery plant (CRP) was selected based on the peak capacity of the textile plant; and (2) the current textile plant capacity is well below the peak capacity and hence the utilisation of CRP is not 100 per cent.

The economic summary of units is given in Table 26.5.

Table 26.5. Highlights of the economic summary of the unit.

Parameters	*Units*	*Actual after implementation*
Total project expenditure	Rs. lakhs	120.0
Net monthly savings (average over last 6 months)	Rs. lakhs	5.80
Simple pay-back period	Years	1.72

Original Situation

The subject plant has got two units in Mumbai. The process houses at both the units use caustic soda for mercerising cotton and polyester cotton blended fabric. The process of mercerising consist of treating cloth with high concentration caustic soda of concentration varying between 15 to 25 per cent for about 1 to 3 minutes. Stretching the cloth after impregnation or preventing its shrinkage during impregnation imparts lustre to the cloth. The cloth is further washed in the recuperator followed by washing with fresh water in washing sections. To minimise the loss of caustic, it is advisable to have true counter-current flow between cloth and water. The mercerisers in both the plants did not have true counter-current flow pattern between cloth and water. The water from the primary washing stage was collected in dilute caustic soda tanks and part of this was used for diluting fresh concentrated caustic soda to be used for mercerisation but the bulk of it was drained to the ETP.

Modifications Implemented

Based on detailed study of the operation of merceriser system the following modifications have been made in the plant:

1. In order to minimise the cost of caustic soda recovery it is desirable to have an optimum concentration of 60 per cent for the weak liquor discharged from the mercerisers. To achieve this objective, the water spray system in the mercerisers was modified and on-line concentration analysers and controllers were installed. This system controls the actual quantity of water pumped into the recuperator, which in turn maintains the caustic concentration in weak liquor. Along with this, the washing system was modified to bring in true counter-current washing in the mercerisers.
2. The weak liquor from the mercerisers is collected in a new weak liquor storage tank.
3. To recover the caustic from the weak liquor, a triple effect evaporation system was installed. In this evaporator, the caustic was recovered at 25 per cent concentration.
4. A centrifuge was installed to remove sizing chemicals from the recovered caustic.
5. System for collection of alkaline condensate and feeding the same to boiler was installed.
6. System for collection and distribution of hot water generated during evaporation was installed.
7. A full-fledged data acquisition system (DAS) was installed to on-line monitor and record the operation of the plant. Software for regular generation of MIS on plant operation was also installed.

Types of Saving

The implementation of the above project has resulted in savings in operation cost in the following areas.

1. The purchase of fresh caustic is reduced due to recovery:
2. Hot water generated during the evaporation process reduces the consumption of steam during process heating.
3. Alkaline condensate generated during evaporation process is used as boiler feed water resulting in reduced chemical dosing.
4. The modification implemented on mercerisers has resulted in reduction of fresh water consumption in washing. Consequent to this, the steam consumption in mercerisers for heating the water is reduced.
5. Additional revenue is generated by recovering caustic from weak liquor purchased from outside.

Execution of Project

The project was executed within four months from the date of approval of the report. After commissioning of the project, TEPS is regularly monitoring the operation of the plant to make sure that the maximum possible saving is generated out of the system. The results of operation of the system are in line with the projection made in the Feasibility Report and re-payment of loan is being done from the savings generated. The balance of savings is being shared between TEPS and the customer.

Caustic Recovery Plant CRP is Different from Other Conventional CRPs

This CRP is Different from Other Conventional CRPs as per the details given below:

1. In case of conventional caustic recovery plants (CRP), normally the objective is only to recover the caustic. In this CRP, the objective was to recover caustic at the cheapest cost. To achieve this objective, the following additional features are built in:-
   (a) Modification of mercerisers to discharge weak liquor at regulated concentration, the additional benefit being reduction in fresh water consumption.
   (b) System for 100 per cent utilisation of alkaline condensate generated is provided.
   (c) System for 100 per cent utilisation of hot water generated during process is also provided. Provision of these two systems make the plant more than 90 per cent efficient in terms of energy consumption.
   (d) Accurate flow and concentration measurement instrumentation along with on-line data acquisition system (DAS) is provided for proper plant operation.
2. Project engineered and executed under turnkey agreement by TEPS.
3. Finance for implementation of this project is arranged by TEPS and the debt re-payment is 100 per cent from savings generated from operation of this plant.
4. Measurement and verification services for the plant provided by TEPS.
5. Performance guarantee is provided by TEPS.

References

Alabester, J.S., *Treatment of Agro-Industrial Waste,* Butterworths, London.

Allen, S.E., Grimshaw, H.M., Parkinson, J.A. and Christopher, *Chemical Analysis of Ecological Materials,* Blackwell Scientific Publications, Oxford.

Arceivala, K.J., *Waste-water Treatment and Disposal,* Marcel Dekker Inc., New York.

Baker, J.M., *Ecological Effects of Oil Pollution,* Applied Science Publishers, London.

Besselievre, E.B. and Schwartz, M., *The Treatment of Industrial Wastes,* McGraw-Hill, Tokyo.

Bordne, R.B. and Winter, J., *Microbiological Methods for Monitoring the Environment-Water Cincinnati,* Ohio, USA.

Brown, R.L., *Pesticides in Clinical Practice, Charles C. Thomas,* Spring field, New York.

Bryan, G.W., *Effects of Pollutants on Aquatic Organisms,* Cambridge University Press, Cambridge.

Bryan, G.W., *Pollution Prevention,* Academic Press, London.

Budyko, M.I., *Food Processing Waste,* Progress Publishers, Moscow.

Cattabeni, S., Cavallaro, A. and Galli, G., *Dioxin-Toxicological and Chemical Aspects,* S.P. Medical and Scientific Books, New York.

Connell, D.W. and Miller, G.J., *Chemistry and Ecotoxicology of Pollution,* John Wiley &Sons, New Y ork.

Coolingwood, R.W., *Biological Indicators of Water Quality,* John Wiley &Sons, New York.

Curds, C.R. and Hawkes, H.A. (eds.), *Ecological Aspects of Used Water Treatment, The Organisms and their Ecology,* Academic Press, London.

Curtis, J.T., *Pollution Prevention,* University of Wisconsin Press, Madison.

Dix, H.M., *Environmental Pollution,* John Wiley &Sons, New York.

Dugan, P.R., *Biochemical Ecology of Water Pollution,* Plenum Publishing Corporation, London.

Goldberg, E.D., *Treatment of Tanner Waste,* Gordon and Breach, Science Publishers, New York.

Hesese, P.R., *A Textbook of Soil Chemical Analysis,* Prentice-Hall, London.

Holmes, J.W. and Talsma, T., *Land and Stream Salinity,* Elsevier Scientific Publishing Co., Amsterdam.

Jackson, M.L., *Soil Chemical Analysis,* Prentice-Hall, London.

James, A. and Evison, L., *Biological Indicators of Water Quality,* John Wiley &Sons, New York.

Ledbetter, J.O., *Air Pollution: Analysis,* Marcel Dekker, New York.

Lehr, J.H., Tyler, E.G, Wayne, A.P. and Jack, D., *Treatment of Dairy Wastes,* McGraw-Hill, New York.

Lenihan, J. and Fletcher W.W.(eds.), *Utilisation of Waste from Pulp and Paper Mill,* Blackie, Glasgow and London.

Liptak, B.G., *Environmental Engineering Handbook, Vol. 2. Air Pollution,* Chilton Book Co., USA.

Liptak, Bel., G., *Biotreatability of Industrial Dye Wastes,* Chilton Book Company, Radnor, Pennsylvania.

Lowman, F.G., *Recent Developments for Effluent Water Treatment,* National Academy of Sciences, Washington DC.

Mason, C.F., *Biology of Freshwater Pollution,* Longman Group Ltd., London.

McCaull, J. and Crossland, J., *Water Pollution,* Harcourt Brace Jovanovich, New York.

Michael, P., *Ecological Methods of Field and Laboratory Investigations,* Tata McGraw-Hill, New Delhi.

NEERI, *Air Quality Monitoring—A Course Manual,* NEERI, Nagpur.

Nyer, E.K., *Groundwater Treatment Technology, 2nd ed.,* Van Nostrand Reinhold, New York.

Odum, E.P., *Fundamentals of Ecology,* W.B. Saunders and Co., New York.

Odum, E.P., *Management of Solid Waste,* Saunders, Philadelphia.

Painter, D.E., *Air Pollution Technology,* Reston Publishing Co., Reston, Virginia.

Phillips, D.J.H., *Quantitative Aquatic Biological Indicators*, Applied Science Publishers, London.

Rao, K.L., *India's Water Wealth,* Orient Longman, New Delhi.

Raw, J.G. and Wooten, D.C., *Environment Impact Analysis Handbook*, McGraw Hill, New York.

Reid, G.K., *Ecology of Inland Waters and Estuaries,* Reinhold Publishing Corporation, New York.

Rudd, R.L. and Herman, S.G., *Environmental Toxicology of Pesticides*, Academy Press, New York.

Rudd, R.L., *Pesticides and the Living Landscape,* University of Wisconsin Press, Madison, Wisconsin.

Sawyer, C.N. and McCarty P.L., *Chemistry for Environmental Engineering*, McGraw-Hill, Tokyo.

Sax, N.I., *Industrial Pollution,* Van Nostrand Reinhold, New York.

Schrowebel, J., *Chemistry of Air Pollution,* Pergamon Press, New York.

Snell, I.D. and Snell, C.T., *Handbook of Industrial Pollution,* D.Van Nostrand, New York.

Southwood, T.R.E., *Ecological Methods,* with Special Reference to the Study of Insect Population, Methuen, London.

Stumm, W. and Morgan, J.J., *Aquatic Chemistry,* John Wiley & Sons, New York.

Stumm, W. and Stumm-Zollineger, E., *Water Pollution Microbiology,* Wiley Interscience, New York.

Teal, J.M., *Effects of Petroleum Hydrocarbons in Marine Organisms,* Pergamon Press, New York.

Tebbutt, T.H.Y., *Principles of Water Quality Control*, Pergamon Press, Oxford, London.

Vinogradov, A.P., *Geochemistry of Chemical Elements in Soils,* Consultants Bureau, New York.

Vollenweider, R.A., *A Manual of Methods for Measuring Primary Production in Aquatic Environment,* IBP Handbook No. 12 (ed. II), Blackwell Scientific Publications, New York.

WHO, *Health Hazards of the Human Environment,* World Health Organization, Geneva.

Index